Marine Mammals in a Changing World, 2nd Edition

Marine Mammals in a Changing World, 2nd Edition

Guest Editors

Alexander Werth
Matthew S. Savoca

MDPI

Basel • Beijing • Wuhan • Barcelona • Belgrade • Novi Sad • Cluj • Manchester

Guest Editors
Alexander Werth
Department of Biology
Hampden-Sydney College
Hampden-Sydney
USA

Matthew S. Savoca
Department of Oceans
Hopkins Marine Station,
Stanford University
Pacific Grove
USA

Editorial Office
MDPI AG
Grosspeteranlage 5
4052 Basel, Switzerland

This is a reprint of the Special Issue, published open access by the journal *Oceans* (ISSN 2673-1924), freely accessible at: https://www.mdpi.com/journal/oceans/special_issues/89DASAG907.

For citation purposes, cite each article independently as indicated on the article page online and as indicated below:

Lastname, A.A.; Lastname, B.B. Article Title. *Journal Name* **Year**, *Volume Number*, Page Range.

ISBN 978-3-7258-7542-9 (Hbk)
ISBN 978-3-7258-7543-6 (PDF)
https://doi.org/10.3390/books978-3-7258-7543-6

Contents

About the Editors

Alexander Werth

Alexander Werth is Patterson Professor of Biology at Hampden–Sydney College in Virginia, USA. He earned his degrees from Duke University and Harvard University and worked as a postdoctoral research fellow with the Department of Wildlife Management in Utquiagvik, Alaska, USA. His research focuses on the functional morphology and biomechanics of feeding in marine mammals, especially suction feeding in toothed whales and dolphins, and filter feeding in baleen whales. He also publishes on the evolution of complexity, evolution pedagogy, and the philosophy of science.

Matthew S. Savoca

Matthew S. Savoca is a Research Associate at the Hopkins Marine Station of Stanford University and a Principal Investigator at the California Marine Sanctuary Foundation of Monterey, California. He earned a Bachelor's degree in Applied Ecology from Cornell University and a PhD in Ecology from the University of California at Davis. His research focuses on marine ecology, and especially on how pollutants (including plastics and other contaminants) alter the biology of marine organisms, particularly mammals and birds. He has participated in many collaborative research projects around the globe and published many articles on science and science education.

Preface

Marine mammals (cetaceans: whales, dolphins, and porpoises; pinnipeds: seals, sea lions, and walruses; sirenians: manatees and dugongs; sea otters and polar bears) are fascinating animals whose bodies, behaviors, and ecology reflect the myriad of ways in which their ancestors left the terrestrial habitats of typical mammals to revert to the marine habitat. As these marine mammals evolved and continue to evolve, every aspect of their biology has changed, especially their physical morphology, anatomy, and biomechanics, as well as their behavioral, physiological, and biomechanical processes, including immunological and other crucial responses to environmental stressors. However, their marine environment continues to change in rapid and unexpected ways due to anthropogenic (human) impact. This includes ever-increasing exposure to plastics, tourists, and climate change, among many other potential environmental factors. The twelve published Special Issue articles in this Reprint collection all involve studies on how various marine mammals face environmental challenges and how they continue to adapt to them. All of the research projects described in these twelve papers demonstrate how advances in marine mammal science enable researchers to better understand marine ecosystems. The articles by Ellier et al. and Anderson et al. on harbor porpoise populations in the Salish Sea of the Eastern Pacific, and by Ulfsson et al. about Swedish harbor porpoises, clearly document how the philopatric allegiances of these small marine mammals are changing as a result of human-induced alterations in the porpoises' habitats due to changes in temperature, food availability, and other human impacts such as vessel traffic and noise—all major factors influencing porpoise biogeography. The article by Girón-Castaño et al. on how humpback whales are being affected by marine tourism, and the article by Aulich et al. on how fin whale migrations around Australia are also changing, further document human impact on cetaceans. Likewise, the article by MacDougall and Robinson shows how prey availability is affecting the biology and biogeography of minke whales around Scotland, just as the biogeographic range of Irrawaddy dolphins in Malaysia reflect human impact (demonstrated by Ali and Rajamani). Saville et al. show how the gray seal's biology and habitat is being altered by storms and other climate disturbances. In terms of pollutants, Hart et al. showed how bottlenose dolphins are commonly ingesting plastics and plasticizing chemicals, while Werth et al. demonstrated that mysticetes (baleen whales) are highly susceptible to plastic ingestion due to their effective keratinous oral filter. D'Agnese et al. show how fur seal gut microbiomes relate to salmon aquaculture in Tasmania—another reflection of humans impacting marine mammals. Finally, Almunia et al. document a new kind of killer whale behavior, demonstrating how plastic this species' behavioral repertoire can be, and how we continue to learn more about how their behavior changes. Together, these articles highlight varied marine mammal species, regions, and research perspectives, but the common thread uniting these articles is change; this began as evolution over time, but is increasingly in terms of short-term accommodation to new threats to the rapidly changing marine habitats where cetaceans, pinnipeds, and other marine mammals are found. Like these marine habitats, marine mammals will continue to change and evolve over time—so long as their populations continue to survive in our ever-changing world.

Alexander Werth and Matthew S. Savoca
Guest Editors

oceans

MDPI

Article

Interactions Between Tourism Vessels and Humpback Whales in the Gulf of Tribugá, Colombia

Laura Valentina Girón-Castaño [1,2], Ann Carole Vallejo [2,*], Isabel C. Avila [1,3] and Alan Giraldo [1]

1 Grupo de Investigación en Ecología Animal, Department of Biology, Universidad del Valle, Cali 760032, Colombia; laura.giron@correounivalle.edu.co (L.V.G.-C.); isabelc.avila@gmail.com (I.C.A.); alan.giraldo@correounivalle.edu.co (A.G.)
2 R&E Ocean Community Conservation, Oakville, ON L6M 5H5, Canada
3 Institute for Terrestrial and Aquatic Wildlife Research (ITAW), University of Veterinary Medicine Hannover, Foundation, 25761 Büsum, Germany
* Correspondence: carole.vallejo@oceancommunityconservation.org

Abstract: Whale-watching is a growing ecotourism activity in Colombia that offers economic benefits but may pose behavioral risks to humpback whales (*Megaptera novaeangliae*) if not properly managed. Although preventive recommendations exist to promote sustainable practices, their effectiveness remains largely unquantified in the Colombian North Pacific. This study evaluates adherence to whale-watching regulations and describes humpback whale behavioral responses to vessel presence in the Gulf of Tribugá, a key breeding and calving area. Data were collected from tourism vessels during July and September 2023, documenting 236 whales across 99 groups—71% of which included calves. The predominant whale response to vessels was neutral (74%), while evasive behaviors occurred in 22% of encounters, particularly among mother–calf pairs. Surface-active behavior was infrequent (22%) and most observed in calf–escort groups, serving as a supplementary indicator of behavioral state. Compliance with whale-watching recommendations varied: although most encounters involved a single vessel (57%) and averaged 16 min in duration, only 14% of vessels maintained the recommended minimum distance of 100 m. These findings highlight persistent gaps in regulatory adherence and underscore the need for strengthened enforcement and adaptive management to ensure sustainable whale-watching practices in Colombia's North Pacific.

Keywords: aerial displays; *Megaptera novaeangliae*; recommendations; Tropical Eastern Pacific; whale-watching

1. Introduction

Humpback whales (*Megaptera novaeangliae* Borowski, 1791) are charismatic marine mammals that inhabit oceans worldwide. These cetaceans exhibit a conspicuous range of surface-active or aerial behaviors, including breaching, flippering, and lobtailing, observed in both their feeding and breeding grounds [1]. This acrobatic repertoire is believed to serve multiple functions, including the removal of external parasites, courtship, communication, and social dynamics [2,3]. Furthermore, these behaviors are believed to be energetically costly [4]. Their frequent occurrence on breeding grounds—where adults typically fast and would be expected to conserve energy—suggests that such displays may play a significant role in the species' communication repertoire [5].

In Colombia, humpback whales from Stock G are found along the Pacific coast between June and November [6]; however, some individuals have been observed as early as May, with others lingering through December [7]. These months provide favorable conditions for mating, breeding, and calf-rearing during the austral winter [8]. Due to their coastal habits [9] and proximity to areas of high human activity, these whales face multiple threats [1] and are considered the marine mammals inhabiting the largest risk area globally [10]. Although the species is classified as Least Concern worldwide [11], its conservation status in Colombia is Vulnerable [12]. Humpback whales in the region face numerous threats, including unregulated whale-watching practices, which exacerbate conservation challenges [10,13].

Whale-watching is a rapidly expanding industry worldwide [1,14], providing economic benefits and fostering cultural values within local communities [15,16]. It is considered a cornerstone of the economy along the Colombian Pacific coast [17]. Additionally, whale-watching activities have contributed to scientific knowledge on humpback whales [18], making the species one of the most extensively documented cetaceans [13,19]. In the Colombian Pacific Ocean, whale-watching began in 1994 in Bahía Málaga [20] and has since expanded to multiple locations, including Isla Gorgona, Gulfs of Tribugá and Cupica, and Tumaco Bay [21]. Since 2001, preventive guidelines for sustainable and responsible whale-watching have been developed, including The Responsible Whale-Watching Guide for Aquatic Mammals in Colombia [17,21]. These recommendations advise approaching whales from behind and to the side, maintaining a minimum distance of 100 m, limiting encounters to a maximum duration of 30 min, ensuring that no more than three vessels engage with a group of whales at any given time, and keeping engines on throughout interactions [17].

Despite the numerous benefits derived from whale-watching, concerns remain regarding the negative effects of inadequate regulation and implementation of management practices on the behavior and health of targeted animals in the short, medium, and long term [18,22]. Multiple studies have examined behavioral changes in whales exposed to whale-watching vessels. Some of these studies have utilized private or research vessels as observation platforms [23–30], while others have employed commercial tourism vessels as platforms of opportunity. Although the latter approach is limited to detecting impacts occurring near the vessel [31], it nonetheless contributes to the sustainable development of the practice [16]. In Colombia, some studies have assessed these impacts [32–36], with most research focusing on Bahía Málaga.

Similarly, adherence to whale-watching recommendations has been researched [21,34,35]. Still, uncertainty persists regarding the behavior of humpback whales and the extent to which whale-watching guidelines are followed. This gap is particularly evident in areas such as the Gulf of Tribugá, located in Nuquí, Chocó, in the Colombian North Pacific, where whale-watching tourism has expanded significantly over the past decade [19]. In this region, the municipality of Nuquí plays a pivotal role in tourism logistics and expedition planning, serving as a departure point for vessels traveling north or south within the Gulf of Tribugá.

The need to evaluate the ecological and regulatory impacts of whale-watching activities in Colombia has been formally recognized since 2007 [37], and reaffirmed as a strategic priority in the National Action Plan for the Conservation of Colombia's Aquatic Mammals 2022–2035 [13]. In response, this study investigates the degree of compliance with whale-watching regulations in the Gulf of Tribugá by analyzing vessel behavior and its influence on humpback whale responses. Specifically, we test three hypotheses: (1) that evasive responses increase with decreasing vessel distance, (2) that mother–calf pairs exhibit heightened sensitivity compared to other group types, and (3) that vessel operators frequently deviate from recommended practices, particularly in terms of approach distance, encounter duration, and vessel number per sighting. Surface-active behaviors are documented as supplementary indicators of behavioral state but are not central to the compliance analysis.

Given the ecological, economic, and cultural significance of humpback whales in Colombia, the findings aim to support evidence-based conservation strategies and inform regulatory improvements for sustainable whale-watching in the region.

2. Materials and Methods

2.1. Study Area

The Gulf of Tribugá is part of the North Pacific Chocoan Coastal Environmental Unit (UAC-PNCh) and was designated as a marine protected area in 2015 under the Regional Integrated Management District of the Gulf of Tribugá-Cabo Corrientes. It has also been recognized as a Hope Spot and is included within the Tribugá-Cupica-Baudó Biosphere Reserve. As the largest gulf in Colombia [38], it spans 103,100 ha, extending from Utría National Natural Park in the north (6°11′ N, 77°09′ W) to Cabo Corrientes in the south (5°53′ N, 77°28′ W).

The area forms part of the Chocó Biogeographic Region, one of the most biodiverse in the world. It is characterized by high relative humidity (80–95%), frequent cloud cover, and an annual precipitation of approximately 10,000 mm, influenced by the Serranía del Baudó and the Intertropical Convergence Zone [39]. The mean annual air temperature is 28 °C, while the mean annual sea temperature is 20.5 °C, and oceanographic variables are periodically modulated by the El Niño-Southern Oscillation (ENSO). The width of the continental shelf ranges from 1 to 3 km and the turbidity in the study site is high, due to the confluence of multiple rivers [40].

The Gulf of Tribugá forms part of the humpback whale migratory corridor, with Utría National Natural Park and its southernmost boundary considered essential nursery areas for the species [39]. As stated previously, this population belongs to humpback whale Breeding Stock G, which feeds in the Antarctic Peninsula and along the coast of Chile, and breeds in southwestern Central America and northwestern South America [41]. The population abundance is estimated at roughly 11,785 individuals, with an average annual growth rate of 5% between 2006 and 2018 [41].

2.2. Data Collection

Data were collected between July and September 2023 as part of the "Responsible Whale and Dolphin Watching Research Project" and the "Community Science Program" of the R&E Ocean Community Conservation Foundation (https://www.facebook.com/reoceancom/ accessed on 19 February 2025). Observations were conducted by a single researcher aboard commercial whale-watching vessels operating in the study area. Vessels ranged from 7 to 9 m in length and were powered by one or two outboard engines rated between 40 and 90 horsepower. Variability in vessel type and engine power was not controlled for; therefore, interpretations of whale behavioral responses should be made with caution.

Sampling employed focal follows of humpback whale (*Megaptera novaeangliae*) groups, with continuous surface behavior recording throughout each encounter. Each focal follow was conducted by trained observers positioned on board the research vessels, who systematically recorded whale behavior, group composition, and vessel interactions using standardized protocols. Group composition was determined based on spatial proximity and behavioral synchrony among individuals [3,38]. Whales were categorized as adults (A) or calves (C). The adult nearest to a calf was presumed to be its mother (M), while additional accompanying adults were classified as escorts (E). Based on these criteria, four group types were defined: A (solitary adult), 2A (two or more adults), M+C (mother–calf pair), and M+C+≥1E (mother–calf pair with one or more escorts). For each encounter, the number of individuals and group type were recorded. Definitions of all variables are provided in Table S1.

Behavioral responses to vessel presence were documented when a whale-watching vessel approached a group within 100 m, initiating the focal follow. Observations concluded once the vessel exceeded 100 m from the group. Due to limited underwater visibility, only surface-active behaviors were recorded. Initial reactions to vessel approach were classified as *neutral*, *evasive*, or *approach*, based on changes in movement direction and spatial orientation between whales and vessels during the first five minutes post-encounter (modified from [3]).

In this study, evasive responses were defined strictly as changes in spatial positioning relative to the vessel. Behaviors such as prolonged dives, increased surfacing intervals, or long-distance underwater movement were not categorized as evasive, but rather as neutral. This classification was informed by previous studies employing consensus-based behavioral coding [3] and reflects the absence of baseline data on dive patterns prior to vessel encounters, which could otherwise introduce interpretive bias.

Additional surface-active behaviors were recorded throughout each observation period. Adult dive duration (in seconds) was calculated as the mean of up to three consecutive dives, defined as the interval between surfacing, within the first 15 min of each encounter. The frequency of five aerial displays—chin breach, tail breach, spin breach, lobtail, and flipper slap—was documented using standardized ethograms (adjusted from [42]). The identity of the performing individual (adult or calf) was noted. In groups with multiple adults, a single focal whale was selected for behavioral recording to maintain consistency.

Compliance with whale-watching regulations established by the Ministry of Environment and Sustainable Development (MADS) [17] was assessed using the following criteria: (1) approach maneuver, categorized as *direct* (frontal or rear) or *indirect* (lateral or angled); (2) initial distance between vessel and whale group at encounter onset (0–50 m, 50–100 m, $\geq$100 m); (3) number of vessels simultaneously present during the encounter (including the research vessel); (4) duration of the encounter (in minutes); and (5) engine status during the interaction (on or off). The duration of encounters involving other non-research vessels was not recorded. Vessel operators were not informed of the specific data being collected; instead, their engagement with researchers was framed as part of an environmental education and cetacean research initiative, without disclosing detailed study objectives.

2.3. Data Processing and Statistical Analysis

Data were categorized into five subsamples to facilitate analysis: (C-I) total number of encounters, (C-II) number of sampled days, (C-III) encountered group classes, (C-IV) whales' initial reactions at the beginning of encounters, and (C-V) mean dive duration (Table S2). Figures and tables indicate the sample size (n) and the corresponding data category (C-I to C-V) used in each analysis. Descriptive statistics were applied to characterize humpback whale behavior during whale-watching encounters. Values are reported as mean ($\overline{x}$) and standard deviation (SD), or as median (M) and interquartile range (IQR), where applicable ($n \geq 4$).

Behavioral variables were analyzed across four group classes: A (solitary adult), $\geq$2A (two or more adults), M+C (mother–calf pair), and M+C + $\geq$1E (mother–calf pair with one or more escorts), as group composition is considered an explanatory factor for behavioral variation [3]. The behavior rate was defined as the number of surface-active behaviors per minute per individual within each group class. Additionally, the relative frequency of encountered group types was standardized by the monthly number of encounters to account for temporal sampling variation.

To assess independence between behavioral responses and explanatory variables, pairwise comparisons were conducted. Behavioral responses included initial reaction, frequency of surface-active behaviors, and mean dive duration. These were tested against whale-watching compliance indicators—approach maneuver, encounter distance, engine status, and number of vessels present—as well as group class (A, ≥2A, M+C, and M+C+≥1E) (Table S3).

Assumptions of normality and homogeneity of variance were evaluated using the Shapiro–Wilk and Levene tests, respectively, with a significance level of $\alpha = 0.05$. As the data did not meet parametric assumptions, non-parametric tests were applied. Chi-square (χ^2) tests were used to examine independence among categorical variables, with test statistics compared against critical ${\chi^2}_1$ values to determine significance. Spearman's rank correlation coefficient was employed to quantify associations between pairs of quantitative variables. Mann–Whitney U tests were used to compare medians of binomial variables, while Kruskal–Wallis tests assessed differences in medians across multiple groups. Post hoc analyses were performed using Dunn's multiple comparison test with Bonferroni correction and Holm adjustment. All statistical analyses were conducted in RStudio© (version 2023.6.0.421), utilizing the car, nlme, dplyr, and PMCMRplus packages.

3. Results

Observations were conducted between 11 July and 26 September 2023 along the Gulf of Tribugá, spanning from Utría National Park in the north (06°00′36.7″ N, 077°21′08.3″ W) to La Roñosa in the south (05°35′6.22″ N, 077°30′50.76″ W) (Figure 1). Over the 43-day study period, 39.5% of encounters occurred north of Nuquí, while 60.5% occurred to the south. A total of 107 humpback whale sightings were recorded, with an effective sampling effort of 29 h and 43 min (excluding search time). Monthly effort was distributed as follows: July (16 encounters, 5.18 h), August (50 encounters, 10.62 h), and September (41 encounters, 13.92 h).

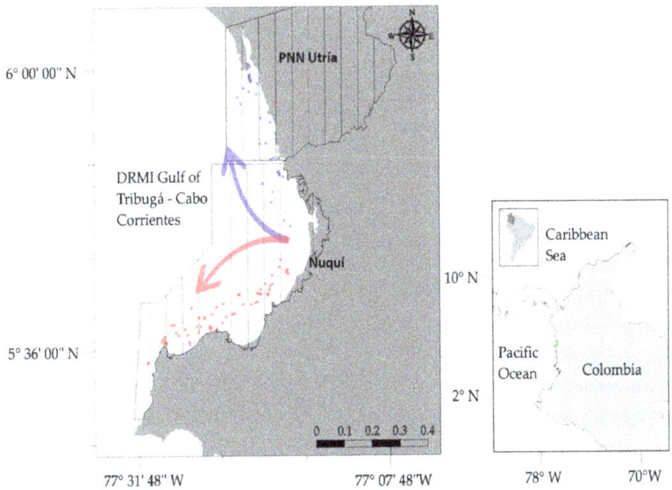

Figure 1. Study area (**left**) and its geographical location (in green) in South America and Colombia (**right**). Location of the geographical coordinates, indicated by crosses, where encounters with humpback whales occurred. Arrows indicate the routes northwards (blue) or southwards (red) within the Gulf of Tribugá, taken by the commercial whale-watching vessels that sailed from Nuquí. DRMI = Regional Integrated Management District.

3.1. Whale Groups

A total of 236 humpback whales were observed, comprising 71 calves (C) and 165 adults, categorized as mothers (M), escorts (E), or solitary adults (A). The mean number of whales per encounter was 2.4 ± 1.2. Ninety-nine distinct groups were recorded, with calf-containing groups being most prevalent (Table 1). In July, M+C pairs and ≥2A groups were encountered at equal rates (1.16 groups/month). A and ≥2A groups were the least frequent across all months, while M+C and M+C+≥1E groups were most common in August.

Table 1. Classes of humpback whale groups encountered during the 2023 breeding season in the Gulf of Tribugá. A = groups of one adult whale, ≥2A = groups of two or more adult whales, M+C = mother–calf pairs and M+C+≥1E = M+C pairs accompanied by one or more escorts. The category of data is C-III (see Table S2).

Whale Group Class	A	≥2A	M+C	M+C+≥1E
Group Class (n = 99)				
	9.1%	20.2%	51.5%	19.2%
Group Classes Encountered Per Hour Per Month (n = 99)				
July	0	1.16	1.16	0.77
August	0.6	0.80	2.30	0.80
September	0.23	0.46	1.68	0.54

3.2. Whale Behavior

Most whale groups (73.6%) exhibited a neutral response to vessel approach, while 22.2% responded evasively and 4.2% approached the vessel (Table 2). Initial reaction was independent of group class (χ^2_6 = 3.25, p = 0.78) but significantly associated with initial encounter distance (χ^2_4 = 10.39, $p < 0.05$). Evasive responses were more frequent at distances of 0–50 m (χ^2_1 = 2.37, $p > 0.05$), though not statistically significant.

Table 2. Whale behavior during whale-watching encounters. A = groups of only one adult whale, ≥2A = groups formed by two or more adult whales, M+C = mother–calf pairs and M+C+≥1E = M+C groups accompanied by one or more escorts. $\overline{x}$ ± SD = mean ± standard deviation. The categories of data were C-I, IV and V (Table S2).

Whale Behavior Descriptors	Summary Statistics
Initial Reaction (N = 72; C-IV)	
Evasive	22.2%
Neutral	73.6%
Approach	4.2%
Dive Duration in Seconds of Adults ($\overline{x}$ ± SD; n = 58; C-V)	
All groups	104.54 ± 121.26
A	167.7 ± 204.71
≥2A	65.67 ± 41.79
M+C	109.03 ± 130.28
M+C+≥1E	125.14 ± 113.04
Rate of Surface-Active Behavior in each Month (n = 107; C-I)	
July	0
August	4.34
September	6.21

Mean dive duration did not differ significantly among group classes (Kruskal–Wallis: H_3 = 1.28, p = 0.73). Across all adult whales, mean dive time was 104.54 ± 121.26 s (Table 2). Although not a primary focus, resting behavior was opportunistically recorded in three encounters (2.8%), exclusively among M+C pairs.

Surface-active behavior occurred in 21.5% of encounters, with no aerial displays observed in July. These behaviors emerged in August and peaked in September. Lobtailing was the most frequent display, followed by spin breaching and chin breaching. Chin and tail breaching were observed only in groups containing calves. Calves exhibited the highest frequency of surface-active behaviors, particularly when accompanied by escorts (E) (Figure 2). Among adult-only groups, solitary adults (A) displayed the highest behavior rates.

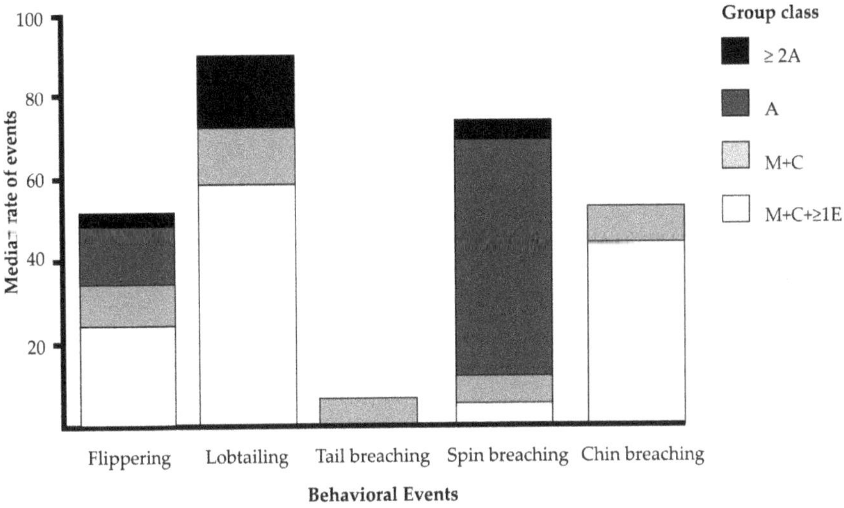

Figure 2. Median rate of the five behaviors by group class. A = groups of only one adult whale (n = 2), ≥2A = groups formed by two or more adult whales (n = 4), M+C = mother–calf pairs (n = 20) and M+C+≥1E = M+C pairs accompanied by one or more escorts (n = 13).

3.3. *Vessel Behavior*

Approach maneuvers were indirect in 90.3% of encounters and direct in 9.7%. Vessels maintained a distance of 50–100 m in 50% of encounters, while 36.2% occurred within 50 m and 13.8% at ≥100 m. A single vessel was present in 57% of encounters, two vessels in 16.8%, three in 14%, and more than three in 12.2%. Median encounter duration was 15 min (IQR = 16 min; $\bar{x} \pm$ SD = 16.98 ± 13.28 min). Half of the encounters lasted 8–24 min, with 25% exceeding 24 min. Engines remained on in approximately 75% of encounters and were turned off at least once in the remainder (Tables 3 and 4).

Group class was significantly associated with several vessel–whale interaction metrics. Initial distance and approach maneuver varied by group class ($\chi^2_3 = 7.71$, $p > 0.05$; $\chi^2_6 = 18.19$, $p < 0.01$). Vessels maintained distances ≥ 100 m more frequently from A groups ($\chi^2_1 = 13.29$, $p < 0.05$), while ≥2A groups were more often approached directly ($\chi^2_1 = 3.73$, $p < 0.05$). Encounter duration and vessel count also varied by group class (Kruskal–Wallis: $H_3 = 10.6$, $p < 0.01$; $H_3 = 3.24$, $p < 0.05$). Encounters with M+C+≥1E groups lasted longer (M = 22.5 min; IQR = 14.75–27 min) than those with A groups (M = 7 min; IQR = 4.75–10.5 min; $p = 0.009$; Figure 3A). The highest number of vessels was also recorded for M+C+≥1E groups (M = 3; IQR = 1–3.25), differing significantly from A groups (M = 1; IQR = 1–1; $p < 0.05$) and M+C pairs (M = 1; IQR = 1–2; $p < 0.03$; Figure 3B). The only encounter involving nine vessels occurred with an M+C+≥1E group.

Table 3. Recommendations for approaching maneuver and encounter duration for whale-watching by the Responsible Whale-watching Guide of Aquatic Mammals in Colombia (MADS, 2017), with respective rates of adherence by tour operators that sailed from Nuquí obtained in this study.

Recommendations for Responsible Whale-Watching	Adherence Rate
Approach as slowly and progressively as possible, never make fast or sudden approaches	Not assessed in the present study
Approach from the rear at a 45° angle, using the posterior half of the animal as a reference. Never approach completely from the front or rear	90.29%
Never approach closer than 100 m to whales	13.83%
Vessels in the waiting zone (200 m from the targeted whale group) should keep a path parallel to the animals and should be positioned on only one side of the group	Not assessed in the present study
Leave an escape route for the animals	Not assessed in the present study
When staying 100–200 m from whales, maintain a constant trajectory and speeds should not exceed 5 kn or 9 km/h	Not assessed in the present study
If one individual or group approaches the vessel, keep the engine in neutral	Not assessed in the present study
Keep the engine on during whale-watching	78.00%
If the need to stop during an approaching maneuver arises, reduce speed gradually	Not assessed in the present study
If there is only one vessel present during an encounter, restrict the encounter duration to 30 min	$\overline{x}$ = 17.14 min Minimum = 0.17 min Maximum = 91 min
If there is more than one vessel, restrict the encounter duration to 15 min	$\overline{x}$ = 16.79 min Minimum = 1 min Maximum = 48 min
Restrict the number of vessels present during an encounter to no more than three for one individual or group	87.85%
Whale-watching should be performed by vessels in the order of their arrival	Not assessed in the present study

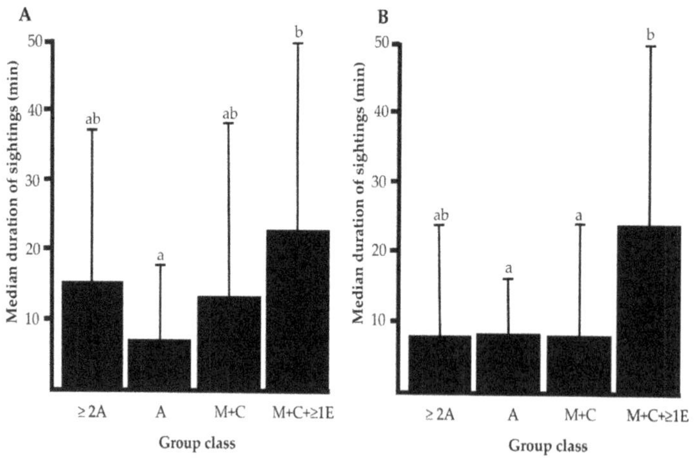

Figure 3. (**A**) Median (+q3) duration of encounters (n = 102) and (**B**) median of the number of vessels present (n = 104) by group class. Different letters above the error bars (a,b) indicate statistically significant differences in the median rate of a given behavior between group classes. Bars that share at least one letter (e.g., 'ab' and 'a') are not significantly different. Bars with different letters and no overlap (e.g., 'a' vs. 'b') differ significantly.

Table 4. Characterization of whale-watching vessel behavior with whales in the Gulf of Tribugá. $\bar{x} \pm$ SD = mean ± standard deviation. All data are from C-I (see Table S2).

Vessel Behavior Descriptors	Summary Statistics
Approaching Maneuver (n = 103)	
Direct	9.71%
Indirect	90.29%
Distance at the Beginning of the Encounters (n = 94)	
0–50 m	36.17%
50–100 m	50%
≥100 m	13.83%
Number of Vessels Present During Encounters (n = 107)	
One	57%
Two	16.8%
Three	14%
More than three vessels	
Four	7.5%
Five	3.7%
Nine	1%
Duration of Encounters (min; n = 105; C-I)	
Minimum	1
Maximum	91
Median	15
Interquartile range	16
Mode	15
$\bar{x} \pm$ SD	16.98 ± 13.28
Engine's Status (n = 100)	
On (throughout encounter)	78%
Off (at least once during encounter)	22%

Encounters involving surface-active behaviors lasted, on average, 4.44 min longer than those without such displays. However, the nine-vessel encounter did not involve surface-active behavior. These differences were not statistically significant (Mann–Whitney U: U = 1131, n = 105, p = 0.15; U = 969, n = 107, p = 0.98; Table 5).

Table 5. Characterization of vessel behavior, according to the occurrence of surface-active behaviors. $\bar{x} \pm$ SD = mean ± standard deviation. The category for data is C-I (see Table S2).

	Encounters with Surface-Active behaviors	Encounters Without Surface-Active Behaviors
Duration of Encounters (min; n = 105)		
$\bar{x} \pm$ SD	19.94 ± 11.87	15.5 ± 13.72
Number of Vessels Present During Encounters (n = 107)		
$\bar{x} \pm$ SD	2.01 ± 1.74	1.82 ± 1.10
Maximum	9	5

4. Discussion

The high proportion of whale-watching encounters involving mother–calf (M+C) pairs (70.7%) suggests that the Gulf of Tribugá may serve as an important nursery area for humpback whales. This rate is comparable to that reported in Bahía Málaga (86%), a recognized breeding ground in the Colombian Pacific [34]. In contrast, Bahía Solano, Chocó, exhibits a lower proportion of M+C pairs (20%) and calf-containing groups (49.6%), likely reflecting its role as a migratory corridor for Stock G [43]. The increasing number of M+C pairs observed over time in this study aligns with the seasonal progression of the breeding period.

All four group classes exhibited predominantly neutral responses to vessel approaches. However, the absence of overt behavioral changes does not preclude disturbance, as physiological stress may precede visible reactions [44]. A tendency toward evasive behavior was noted when vessels approached within 0–50 m, differing from previous findings in the region where evasive responses were more frequent at 50–100 m [35]. These discrepancies may stem from methodological differences in defining "neutral" responses (see Methods). Additionally, sampling from whale-watching vessels may bias observations toward individuals tolerant of vessel proximity, potentially underestimating true rates of avoidance.

Mean dive durations for three group classes were shorter than those reported in the absence of vessels in Bahía Málaga [34], consistent with patterns observed in other breeding areas [22]. Although this study did not directly compare dive times between vessel-present and vessel-absent conditions, the findings suggest that whale-watching activity may influence short-term dive behavior in the Gulf of Tribugá, warranting further investigation.

The observation of resting M+C pairs during whale-watching encounters raises concern and requires further investigation. Adult humpback whales are particularly vulnerable during the breeding season, relying on stored energy reserves from polar feeding grounds [45]. Previous studies have shown that vessel presence reduces resting time in breeding areas [34,46–48]. Reduced rest may compromise maternal energy budgets and calf development, limiting nursing opportunities and increasing energetic costs associated with heightened activity [44,49,50]. These disruptions could ultimately affect reproductive success.

Surface-active behaviors were recorded in only 25% of encounters, with the highest frequency occurring in September, coinciding with the peak of the breeding season. Calves exhibited the highest rates of surface-active behavior, consistent with other studies in the Colombian Pacific. M+C+$\geq$1E groups also displayed elevated frequencies of these behaviors and were encountered more often and by more vessels. This pattern aligns with previous reports from the southern extent of Stock G's breeding range [51] and earlier studies in the Gulf of Tribugá [35].

Contrary to prior findings [35], flippering was not exclusive to adult-only groups in this study. While behavior rates were not explicitly analyzed in relation to vessel presence, previous research in the Colombian Pacific has yielded mixed results. Some studies suggest that surface-active behavior decreases in response to vessel disturbance [48], potentially reflecting transitions to evasive states such as traveling [46]. Others report increased activity, possibly as a communicative response to elevated noise levels [35,45].

The interpretation of surface-active behavior is context-dependent: newborn calves and juveniles are likely engaged in the acquisition, reinforcement, and refinement of motor and social skills critical for their development [52], while in adults, such behaviors may relate to courtship, group dynamics, or acoustic signaling [2,5]. Further research is needed to identify which natural behaviors are altered by tourism vessel presence and to assess the influence of environmental and social factors on behavioral variability [5]. Investigating physiological indicators and core behavioral states will be essential for understanding sub-lethal impacts.

Regarding compliance, tour operators in Nuquí adhered to four of five recommended whale-watching guidelines [17]. Compared to Bahía Málaga, where up to 14 vessels may observe a single group [19] and 51.6% of encounters involve 2–5 vessels [34], the Gulf of Tribugá exhibited lower vessel density, with a single vessel present in 57% of encounters and a maximum of nine vessels per group. Encounter durations were also shorter (16.98 $\pm$ 13.28 min), aligning with regulatory recommendations and remaining below durations reported in Bahía Málaga (29 $\pm$ 12.6 min in 2008 [34]; 42.8 $\pm$ 6.8 min in 2019 [18]). Even though observation durations in this study remained within the recommended thresholds, future studies need to account for cumulative exposure time, as prolonged or repeated interactions may lead to long-term impacts on whales.

Although the presence of a researcher may have influenced operator behavior, the imperfect adherence to guidelines suggests that operators acted as they typically would. This supports the validity of the compliance data. However, only a minority of vessels maintained the recommended ≥100 m distance from whales, mirroring patterns observed in Bahía Málaga [34]. The reasons for non-compliance with this specific guideline fall outside the scope of this study, but it is notable that some operators have failed to adhere even after receiving warnings. Prior research indicates that valuing whale-watching guidelines does not guarantee adherence [35]. Strengthening operator understanding of the ecological rationale behind these recommendations is essential for promoting sustainable whale-watching practices.

These findings underscore the need for governmental authorities to incorporate behavioral sensitivity thresholds into marine spatial planning and whale-watching regulations, particularly by enforcing temporal and spatial restrictions during breeding and nursing periods to reduce vessel-induced disturbances.

5. Conclusions

In summary, the findings of this study support all initial hypotheses. First, operator compliance with whale-watching guidelines was partial. Although most operators adhered to four out of five recommendations, maintaining the minimum 100 m distance was the least respected—mirroring patterns reported in other Colombian breeding grounds. Second, vessel proximity was associated with behavioral changes in humpback whales in the Gulf of Tribugá, particularly among mother–calf pairs. Alterations in dive duration and surface-active behaviors were observed, even in the absence of overt avoidance responses, suggesting subtle yet ecologically significant effects.

The study also demonstrated the utility of commercial tourism vessels as platforms of opportunity for behavioral monitoring, offering a cost-effective and scalable alternative to dedicated research vessels. Key management gaps were identified, prompting recommendations for proactive strategies, including the implementation of standardized monitoring tools, enhanced operator education, and targeted outreach programs to promote sustainable whale-watching practices in the region. Although most operators demonstrated general awareness of whale-watching guidelines, the consistent failure to maintain an appropriate distance highlights a gap between knowledge and practice. Notably, the presence of researchers onboard did not significantly alter operator behavior, suggesting that observed compliance levels reflect standard operating norms rather than observer-induced bias. Nevertheless, the results of this study contribute to public awareness and provide a scientific basis for refining whale-watching regulations to minimize disturbance. They also underscore the value of integrating commercial vessels into long-term monitoring programs, particularly in data-limited regions.

Furthermore, our findings reinforce the ecological importance of the Gulf of Tribugá as a potential nursery area for humpback whales, as evidenced by the high proportion of mother–calf pairs observed during whale-watching encounters. Behavioral observations indicate that vessel proximity may influence whale behavior, particularly among resting dyads, which are energetically constrained during the breeding season. The targeting of these vulnerable groups by tourism activities raises concerns about potential impacts on maternal investment and calf development.

To strengthen conservation outcomes, self-monitoring initiatives promoted by environmental and maritime authorities should be reinforced through the development of standardized evaluation frameworks. Given the vulnerability of breeding humpback whales and the increasing popularity of whale-watching in the Gulf of Tribugá, there is an urgent need to enhance regulatory enforcement, expand monitoring capacity, and invest in education and outreach. These measures are essential to ensure that whale-watching activities remain ecologically sustainable, safeguarding both humpback whale populations and the socioeconomic well-being of coastal communities. Proactive management is critical to prevent the intensification and unsustainable expansion of whale-watching operations—an issue already documented in other Colombian breeding areas.

Supplementary Materials: The following supporting information can be downloaded at: https://www.mdpi.com/article/10.3390/oceans6040067/s1, Table S1: Definition of the variables used to characterise humpback whale behaviour during encounters and adherence to preventive whale-watching recommendations in the Gulf of Tribugá; Table S2: Categories of data used for descriptive analysis; Table S3: Results of statistical tests (Chi-square (X^2), Spearman's rank correlation coefficient (S), Kruskal–Wallis (H) and Mann–Whitney U (U), ANOVA of the generalised linear model is included (GLM)). Values of test statistics, the sample size (n), degrees of freedom (df) and p-values (P) are given. Asterisks (*) denote statistical significance ($\alpha = 0.05$).

Author Contributions: Conceptualization, L.V.G.-C., A.C.V., I.C.A. and A.G.; methodology, L.V.G.-C. and A.C.V.; software, L.V.G.-C.; validation, A.C.V., I.C.A. and A.G.; formal analysis, L.V.G.-C., A.C.V., I.C.A. and A.G.; investigation, L.V.G.-C. and A.C.V.; resources, L.V.G.-C., A.C.V. and A.G.; data curation, L.V.G.-C. and A.C.V.; writing—original draft preparation, L.V.G.-C. and A.G.; writing—review and editing, A.C.V., I.C.A. and A.G.; visualization, L.V.G.-C., A.C.V. and A.G.; supervision, A.C.V., I.C.A. and A.G.; projects administration, A.C.V. and A.G. funding acquisition, A.C.V. and A.G. All authors have read and agreed to the published version of the manuscript.

Funding: This research received funding from R&E Ocean Community Conservation Foundation, Grupo de Investigación en Ecología Animal—Universidad del Valle, and the Corporation Center of Excellence in Marine Sciences—CeMarin. An APC waiver was granted by the journal's editorial board.

Data Availability Statement: The original contributions presented in this study are included in the article/Supplementary Material. Further inquiries can be directed to the corresponding author.

Acknowledgments: We extend our sincere gratitude to Oscar E. Murillo-García and Wilmar Alexander Torres L. for their invaluable support in statistical analyses. We also wish to thank Don Santiago, Duvian, Don Varón, Don José Manuel, Jose, and Don Danubio, whale-watching tour operators in Nuquí, for their valuable collaboration in conducting field records. Additionally, we express our appreciation to the Los Riscales Community Council (Nuquí) for supporting this research in the Gulf of Tribugá. We also would like to thank the anonymous reviewers for their insightful and constructive comments, which have significantly enhanced the quality and rigor of this article.

Conflicts of Interest: The authors declare no conflicts of interest.

References

1. Fleming, A.; Jackson, J. *Global Review of Humpback Whales (Megaptera novaeangliae)*; NOAA Technical Memorandum NMFS-SWFSC-474; NOAA: La Jolla, CA, USA, 2011. Available online: https://repository.library.noaa.gov/view/noaa/4489 (accessed on 25 June 2023).
2. Meynecke, J.-O.; de Biej, J.; Barraqueta, J.-L.M.; Seyboth, E.; Dey, S.P.; Lee, S.B.; Samanta, S.; Vichi, M.; Findlay, K.; Roychoudhury, A.; et al. The Role of Environmental Drivers in Humpback Whale Distribution, Movement and Behavior: A Review. *Front. Mar. Sci.* **2021**, *8*, 720774. [CrossRef]
3. Fabietti, Y.; Spadaro, C.; Tigani, A.; Giglio, G.; Barbuto, G.; Romano, V.; Fedele, G.; Leonetti, F.L.; Venanzi, E.; Barba, C.; et al. Surface Behaviours of Humpback Whale *Megaptera novaeangliae* at Nosy Be (Madagascar). *Biology* **2024**, *13*, 996. [CrossRef]
4. Whitehead, H. Why Whales Leap. *Sci. Am.* **1985**, *252*, 84–93. [CrossRef]

5. Kavanagh, A.S.; Owen, K.; Williamson, M.J.; Blomberg, S.P.; Noad, M.J.; Goldizen, A.W.; Kniest, E.; Cato, D.H.; Dunlop, R.A. Evidence for the Functions of Surface-Active Behaviors in Humpback Whales (*Megaptera novaeangliae*). *Mar. Mammal Sci.* **2017**, *33*, 313–334. [CrossRef]
6. Modest, M.; Irvine, L.; Andrews-Goff, V.; Gough, W.; Johnston, D.; Nowacek, D.; Pallin, L.; Read, A.; Moore, R.T.; Friedlaender, A. First description of migratory behavior of humpback whales from an Antarctic feeding ground to a tropical calving ground. *Anim. Biotelemetry* **2021**, *9*, 42. [CrossRef]
7. Avila, I.C.; Dormann, C.F.; García, C.; Payán, L.F.; Zorrilla, M.X. Humpback whales extend their stay in a breeding ground in the Tropical Eastern Pacific. *ICES J. Mar. Sci.* **2020**, *77*, 109–118. [CrossRef]
8. Capella, J.; Gibbons, J.; Flórez-González, L.; Llano, M.; Valladares, C.; Sabaj, V.; Vilina, Y.A. Ciclo migratorio de ballenas jorobadas individualizadas del estrecho de Magallanes: Indicios sobre la duración de la migración y filopatría en los destinos. *Rev. Chil. Hist. Nat.* **2008**, *81*, 547–560. [CrossRef]
9. Pelayo-González, L.; Oviedo, L.; Márquez-Artavia, A.; Herra-Miranda, D.; Pacheco-Polanco, J.D.; Bessesen, B.; Guzmán, H.M. Habitat use of south-eastern Pacific humpback whales (*Megaptera novaeangliae*; Borowski 1781) in Golfo Dulce, Costa Rica. *Mar. Freshw. Res.* **2022**, *73*, 1465–1474. [CrossRef]
10. Avila, I.C.; Giraldo, A. Risk areas for marine mammals in Colombia. *Rev. Biol. Trop.* **2022**, *70*, 96–113. [CrossRef]
11. Braulik, G.T.; Taylor, B.L.; Minton, G.; Notarbartolo di Sciara, G.; Collins, T.; Rojas-Bracho, L.; Crespo, E.A.; Ponnampalam, L.S.; Double, M.C.; Reeves, R.R. Red-list status and extinction risk of the world's whales, dolphins, and porpoises. *Conserv. Biol.* **2023**, *37*, e14090. [CrossRef] [PubMed]
12. Ministerio de Ambiente y Desarrollo Sostenible. *Lista de Especies Silvestres Amenazadas de la Diversidad Biológica Continental y Marino-Costera de Colombia—Resolución 0126 de 2024 Expedida por el Ministerio de Ambiente y Desarrollo Sostenible, v1.0*; Dataset/Checklist; Ministerio de Ambiente y Desarrollo Sostenible—MADS: Bogotá, Colombia, 2024. [CrossRef]
13. Trujillo, F.; Caicedo, D.; Diazgranados, M.C.; Avila, I.C. *Plan de Acción Nacional para la Conservación de los Mamíferos Acuáticos de Colombia 2022–2035*; Avello-Castiblanco, G.C., González-Delgadillo, A.M., Quintero-Gil, J.A., Eds.; Ministerio de Ambiente y Desarrollo Sostenible: Bogotá, Colombia, 2022; ISBN 978-958-5551-82-4.
14. Pacheco, A.S.; Sepúlveda, M.; Corkeron, P. Whale watching impacts: Science, human dimensions and management. *Front. Mar. Sci.* **2021**, *8*, 737352. [CrossRef]
15. Suárez-Rojas, C.; León, C.J. A scientometrics review of fifty years of whale watching tourism research. *Mar. Policy* **2025**, *176*, 106658. [CrossRef]
16. Suárez-Rojas, C.; Hernández, M.M.G.; León, C.J. Sustainability in whale watching: A literature review and future research directions based on regenerative tourism. *Tour. Manag. Perspect.* **2023**, *47*, 101120. [CrossRef]
17. Ministerio de Ambiente y Desarrollo Sostenible [MADS]. *Guía de Avistamiento Responsable de Mamíferos Acuáticos en Colombia*; Ministerio de Ambiente y Desarrollo Sostenible: Bogotá, Colombia, 2017.
18. Gray, C.; Schuler, A.R.; Parsons, E.C.M. Recent advances in whale watching research: 2019–2020. *Tour. Mar. Environ.* **2022**, *17*, 113–120. [CrossRef]
19. Avila, I.C.; Ortega, L.F.; Pretel, C.; Mayor, G. A decade of whale watching in an important tourist destination in the Pacific coast of Colombia: Challenges for proper management. *Lat. Am. J. Aquat. Mamm.* **2021**, *16*, 23–32. [CrossRef]
20. Soto-Cortés, L.V.; Luna-Acosta, A.; Maya, D.L. Whale watching Management: Assessment of Sustainable Governance in Uramba Bahía Málaga National Natural Park, Valle del Cauca. *Front. Mar. Sci.* **2021**, *8*, 575866. [CrossRef]
21. Arias-Gaviria, D.; Muñoz-Vargas, C.A.; Avila, I.C. *Guía de Avistamiento Ballenas Jorobadas en Colombia*; Ministerio de Industria, Comercio y Turismo, Viceministerio de Turismo, Dirección de Calidad y Desarrollo Sostenible del Turismo: Bogotá, Colombia, 2011.
22. Currie, J.J.; McCordic, J.A.; Olson, G.L.; Machernis, A.F.; Stack, S.H. The impact of vessels on humpback whale behavior: The benefit of added whale watching guidelines. *Front. Mar. Sci.* **2021**, *8*, 601433. [CrossRef]
23. Eiras, C.N.; Costa-Silva, S.; Melo, T.H.; Verissimo, L.; Marcondes, M.C. Unusual behaviour of humpback whale (*Megaptera novaeangliae*) mothers and calves. *Aquat. Mamm.* **2021**, *47*, 330–336. [CrossRef]
24. Zeh, J.M.; Adcock, D.L.; Perez-Marrufo, V.; Cusano, D.A.; Robbins, J.; Tackaberry, J.E.; Jensen, F.H.; Weinrich, M.; Friedaender, A.S.; Wiley, D.N.; et al. Acoustic behavior of humpback whale calves on the feeding ground: Comparisons across age and implications for vocal development. *PLoS ONE* **2024**, *19*, e0303741. [CrossRef]
25. Cárdenas, S.; Gabela-Flores, M.V.; Amrein, A.; Surrey, K.; Gerber, L.R.; Guzmán, H.M. Tourist Knowledge, Pro-Conservation Intentions, and Tourist Concern for the Impacts of Whale watching in Las Perlas Archipelago, Panama. *Front. Mar. Sci.* **2021**, *8*, 627348. [CrossRef]
26. Liu, J.; Wang, S.; Law, R.; Chen, J.; Yang, L. Growing threat of marine tourism activities to whales: Evidence from vessel trajectory big data mining. *J. Sustain. Tour.* **2025**, 1–20. [CrossRef]
27. Affatati, A.; Scaini, C.; Scaini, A. The role of operators in sustainable whale watching tourism: Proposing a continuous training framework. *PLoS ONE* **2024**, *19*, e0296241. [CrossRef]

28. Santos-Carvallo, M.; Barilari, F.; Pérez-Alvarez, M.J.; Gutiérrez, L.; Pavez, G.; Araya, H.; Anguita, C.; Cerda, C.; Sepúlveda, M. Impacts of Whale watching on the Short-Term Behavior of Fin Whales (*Balaenoptera physalus*) in a Marine Protected Area in the Southeastern Pacific. *Front. Mar. Sci.* **2021**, *8*, 623954. [CrossRef]
29. Henderson, A.F.; Hindell, M.A.; Wotherspoon, S.; Biuw, M.; Lea, M.A.; Kelly, N.; Lowther, A.D. Assessing the viability of estimating baleen whale abundance from tourist vessels. *Front. Mar. Sci.* **2023**, *10*, 1048869. [CrossRef]
30. Melo, M.A. Patrón de Efectos a Corto Plazo en el Comportamiento de la Ballena Jorobada (*Megaptera novaeangliae*) Causado por las Embarcaciones Turísticas en Zonas de Alimentación y Reproducción. Bachelor's Thesis, Pontificia Universidad Javeriana, Bogotá, Colombia, 2021.
31. Bejder, L.; Samuels, A. Evaluating the Effects of Nature-Based Tourism on Cetaceans. In *Marine Mammals: Fisheries, Tourism and Management Issues*; Gales, N., Hindell, M., Kirkwood, R., Eds.; CSIRO Publishing: Collingwood, VIC, Australia, 2003; pp. 229–257.
32. Herrera, J.C.; Flórez-González, L.; Avila, I.C.; Falk, P.; Capella, J.; Tobón, I.C. Efecto de las embarcaciones de turismo en el comportamiento de grupos con cría de ballena jorobada (*Megaptera novaeangliae*), en Bahía Málaga, Colombia. In *Memorias del Taller de Trabajo Sobre el Impacto de las Actividades Antropogénicas en Mamíferos Marinos en el Pacífico Sudeste*; CPPS/PNUMA: Bogotá, Colombia, 2007.
33. Correa, L.M. Efecto a Corto Plazo de las Embarcaciones Turísticas y en Tránsito Sobre el Comportamiento en Superficie de *Megaptera novaeangliae* Durante la Temporada de Reproducción 2008, en Bahía Málaga y Alrededores (Valle del Cauca-Colombia). Bachelor's Thesis, Pontificia Universidad Javeriana, Bogotá, Colombia, 2009.
34. Avila, I.C.; Correa, L.M.; Parsons, E.C.M. Whale watching Activity in Bahía Málaga, on the Pacific coast of Colombia, and its effect on humpback whale (*Megaptera novaeangliae*) behavior. *Tour. Mar. Environ.* **2015**, *11*, 19–32. [CrossRef]
35. Zapetis, M.E.; Samuelson, M.M.; Botero Acosta, N.; Kuczaj, S.A. Evaluation of a developing ecotourism industry: Whale watching in the Gulf of Tribugá, Colombia. *Int. J. Comp. Psychol.* **2017**, *30*, 1–16. [CrossRef]
36. Rey-Baquero, M.P.; Huertas-Amaya, L.V.; Seger, K.D.; Botero-Acosta, N.; Luna-Acosta, A.; Perazio, C.E.; Boyle, J.K.; Rosenthal, S.; Vallejo, A.C. Understanding Effects of Whale watching Vessel Noise on Humpback Whale Song in the North Pacific Coast of Colombia With Propagation Models of Masking and Acoustic Data Observations. *Front. Mar. Sci.* **2021**, *8*, 623724. [CrossRef]
37. Flórez-González, L.; Ávila, I.C.; Capella, J.; Falk, P.; Félix, F.; Gibbons, J.; Guzmán, H.; Haase, B.; Herrera, J.; Peña, V.; et al. *Estrategia para la Conservación de la Ballena Jorobada del Pacífico Sudeste. Lineamientos de un Plan de Acción Regional e Iniciativas Nacionales*; Fundación Yubarta: Cali, Colombia, 2007.
38. Versiani, L.L.; Azevedo, C.S. Surface activity of Humpback whales *Megaptera novaeangliae* (Cetacea, Mysticeti) on the northern coast of Bahia, Brazil. *Rev. Bras. Zoociênc.* **2020**, *21*, 1–12. [CrossRef]
39. Díaz-Galeano, J.M.; Galeano, J. El entorno biogeofísico. In *La Pesca Artesanal en el Norte del Pacífico Colombiano: Un Horizonte Ambivalente*; Díaz, J.M., Guillot, L., Velandia, M.C., Eds.; Fundación MarViva: Bogotá, Colombia, 2016; pp. 15–28.
40. Velandia, M.; Durán, D. Marine Spatial Planning (PEM) in the Colombian North Pacific. *Rev. Costas* **2021**, *2*, 497–510. [CrossRef]
41. Félix, F.; Acevedo, J.; Aguayo-Lobo, A.; Avila, I.C.; Botero-Acosta, N.; Calderón, A.; Cáceres, B.; Capella, J.; Carnero, R.; Castro, C.; et al. *Humpback Whale Breeding Stock G: Updated Population Estimate Based on Photo-ID Matches Between Breeding and Feeding Areas*; IWC Paper SC/68C/ASI/02; International Whaling Commission (IWC): Cambridge, UK, 2021.
42. McCulloch, S.; Meynecke, J.O.; Franklin, T.; Franklin, W.; Chauvenet, A.L.M. Humpback whale (*Megaptera novaeangliae*) behaviour determines habitat use in two Australian bays. *Mar. Freshw. Res.* **2021**, *72*, 1251–1267. [CrossRef]
43. Rocha-Gómez, O. Behavioral Description on Surface and Characterization of Humpback Whales Population (*Megaptera novaeangliae*) during the Breeding Season 2006 in Colombian Pacific Coast. *Rev. Biodivers. Neotrop.* **2011**, *1*, 105–115. [CrossRef]
44. Villagra, D.; García-Cegarra, A.; Gallardo, D.I.; Pacheco, A.S. Energetic effects of whale watching boats on humpback whales on a breeding ground. *Front. Mar. Sci.* **2021**, *7*, 600508. [CrossRef]
45. Kettemer, L.E.; Rikardsen, A.H.; Biuw, M.; Broms, F.; Mul, E.; Blanchet, M.A. Round-trip migration and energy budget of a breeding female humpback whale in the Northeast Atlantic. *PLoS ONE* **2022**, *17*, e0268355. [CrossRef]
46. Amrein, A.M.; Guzman, H.M.; Surrey, K.C.; Polidoro, B.; Gerber, L.R. Impacts of whale watching on the behavior of humpback whales (*Megaptera novaeangliae*) in the coast of Panama. *Front. Mar. Sci.* **2020**, *7*, 601277. [CrossRef]
47. Chou, E.; Kershaw, F.; Maxwell, S.M.; Collins, T.; Strindberg, S.; Rosenbaum, H.C. Distribution of breeding humpback whale habitats and overlap with cumulative anthropogenic impacts in the Eastern Tropical Atlantic. *Divers. Distrib.* **2025**, *26*, 549–564. [CrossRef]
48. Stack, S.H.; Sprogis, K.R.; Olson, G.L.; Sullivan, F.A.; Machernis, A.F.; Currie, J.J. The behavioural impacts of commercial swimming with whale tours on humpback whales (*Megaptera novaeangliae*) in Hervey Bay, Australia. *Front. Mar. Sci.* **2021**, *8*, 696136. [CrossRef]
49. van Aswegen, M.; Szabo, A.; Currie, J.J.; Stack, S.H.; Evans, L.; Straley, J.; Neilson, J.; Gabriele, C.; Steel, D.; Cates, K.; et al. Maternal investment, body condition and calf growth in humpback whales. *J. Physiol.* **2025**, *603*, 551–578. [CrossRef]
50. Hoarau, L.; Dalleau, M.; Delaspre, S.; Barra, T.; Landes, A.E. Assessing and mitigating humpback whale (*Megaptera novaeangliae*) disturbance of whale watching activities in Reunion Island. *Tour. Mar. Environ.* **2020**, *15*, 173–189. [CrossRef]

51. Pacheco, A.S.; Silva, S.; Alcorta, B.; Balducci, N.; Guidino, C.; Llapapasca, M.A.; Sanchez-Salazar, F. Aerial behavior of humpback whales *Megaptera novaeangliae* at the southern limit of the southeast Pacific breeding area. *Rev. Biol. Mar. Oceanogr.* **2013**, *48*, 185–191. [CrossRef]
52. Ratsimbazafindranahaka, M.N.; Huetz, C.; Saloma, A.; Andrianarimisa, A.; Charrier, I.; Adam, O. Time budget and behavioral synchrony of humpback whale mother-calf pairs on a breeding ground in the southwestern Indian Ocean. *Mar. Mammal Sci.* **2024**, *40*, e13129. [CrossRef]

oceans

MDPI

Article

Fin Whale Acoustic Presence Increases by 3 d/y in the Migratory Corridor off Cape Leeuwin, Western Australia—An Indicator of Population Growth?

Meghan G. Aulich [1], Robert D. McCauley [1], Brian S. Miller [2] and Christine Erbe [1,*]

[1] Centre for Marine Science and Technology, Curtin University, Bentley, WA 6102, Australia; meghan.aulich@hotmail.com (M.G.A.); r.mccauley@cmst.curtin.edu.au (R.D.M.)
[2] Australian Antarctic Division, 203 Channel Highway, Kingston, TAS 7050, Australia; brian.miller@aad.gov.au
* Correspondence: c.erbe@curtin.edu.au

Abstract: The population of southern fin whales (*Balaenoptera physalus quoyi*) was severely depleted by 19th and 20th century whaling. Its conservation status remains 'vulnerable', as recovery has been slow. Over 19 years of underwater acoustic recordings from the Comprehensive Nuclear-Test-Ban Treaty Organization (CTBTO)'s hydrophones off Cape Leeuwin, Western Australia, were analyzed to monitor fin whales' annual migration from their Southern Ocean feeding grounds (where they spend the austral summer) to their tropical breeding grounds (where they spend the austral winter) and back. Northward migrants arrived ~2 d/y earlier (2002–2020). The number of hours with fin whale acoustic presence increased by ~49 h/y and the number of days with fin whale acoustic presence by ~3 d/y. Thus, by the end of the 19-year recording period, fin whales were acoustically present on 74 more days than at the beginning of recording. While changes in habitat function, climate, and ambient noise may affect migratory behavior, the most likely explanation is a post-whaling increase in the number of animals of this Southern Hemisphere subspecies.

Keywords: fin whale; *Balaenoptera physalus*; Cape Leeuwin; passive acoustic monitoring

1. Introduction

Most baleen whale species are migratory, occupying high-latitude, polar waters during the summer months before dispersing to low-latitude, temperate/tropical waters during the winter months [1]. The purpose of this migration is to take advantage of the seasonally highly productive feeding grounds in polar waters and the warmer and more sheltered calving grounds in sub-equatorial waters, where energy requirements and risk of predation are lower [1,2]. The fin whale (*Balaenoptera physalus*) is no exception to this [3].

Passive acoustic monitoring (PAM) has enabled the extensive identification of the species' distribution and migration patterns around the world—based on its stereotypical vocalizations [4–8]. The most widely identified call type of the fin whale, referred to as the 20 Hz pulse, is ideal for PAM, as the pulses are strong (source levels up to 190 dB re 1 μPa m) and of low frequency (18–42 Hz) [9–14]—ideal for long-range propagation. The pulses are short (~1 s) and produced in highly stereotyped rhythmic sequences with inter-pulse intervals (IPs) of 7–26 s. These repetitive sequences (songs) are believed to only be produced by males as a breeding display, while single, irregular 20 Hz pulses may be produced by all demographic cohorts in association with social behaviors [15,16].

A recent PAM study has identified the long-term seasonal distribution and migratory pathways of the Southern Hemisphere subspecies of fin whale (*B. p. quoyi*) from their summer grounds in Eastern Antarctic waters to their winter grounds in temperate-to-tropical Australian waters [7]. Fin whales are seasonally present in Eastern Antarctica at the Southern Kerguelen Plateau from February to June (late summer and autumn) when they begin their migration to Australia [7]. Once they reach the southwestern tip of Australia at Cape Leeuwin, they continue to follow the Western Australian coastline through the Perth Canyon and into temperate-to-tropical waters [7,17].

Globally, studies have revealed long-term shifts in the presence and timing of migratory cetaceans: Earlier arrival times have been reported for humpback (*Megaptera novaeangliae*) and blue (*Balaenoptera musculus)* whales to low-latitude regions [18–20], and later departure times have been reported for bowhead (*Balaena mysticetus*) and beluga (*Delphinapterus leucas*) whales from high-latitude regions [21,22]. Such shifts in migration timing have also been reported for the fin whale in the North Atlantic, with the animals arriving earlier in the Gulf of St. Lawrence by ~1 d/y and departing earlier by ~0.4 d/y from 1984 to 2010 [23]. Whether other populations of fin whale shift their migration timing is unknown.

The fin whale is listed as vulnerable on the International Union for Conservation of Nature (IUCN) Red List [24] after severe population decline during the industrial whaling era, with ~700,000 animals caught in the Southern Hemisphere [25]. There have been few studies of fin whales in the Southern Hemisphere [26] and most have focused on Antarctic waters [27], particularly in the Atlantic and Pacific sectors [28–33]. Only a handful of contemporary studies of fin whales have been conducted in the temperate Southern Indian Ocean, with some being opportunistic or covering small geographic and temporal scales [34–37]. There are even fewer contemporary reports of fin whales in Australian waters [7,17].

In this study, we aimed to quantify the migratory acoustic presence of fin whales at Cape Leeuwin, which presents the first and last contact point with the Australian continent on their migration from and to Antarctica. We were specifically interested in any long-term trends. We analyzed 19 years of PAM data and present statistical measures of arrival and departure times. We discuss potential drivers of fin whale migration variability. Understanding shifts in fin whale migration may inform conservation management at a national (Australian) and international level.

2. Materials and Methods

Passive acoustic data were collected from the Comprehensive Nuclear-Test-Ban Treaty Organization (CTBTO) acoustic monitoring station (HA01) southwest of Cape Leeuwin, Western Australia (Figure 1). CTBTO has installed low-frequency hydrophones (sampling frequency: 250 Hz) in the SOFAR channel in every ocean to conduct long-range acoustic monitoring [38]. The hydrophones are cabled to shore and offer a continuous recording capability, making them suitable not just for their intended purpose of detecting nuclear explosions, but also for civil and scientific purposes such as the environmental monitoring of ice and ocean temperature [39] and the detection of whales [40,41]. Continuous recordings from Cape Leeuwin were obtained for the period of 1 January 2002–31 December 2020. For data analysis, recordings were split into 1 h samples.

In order to detect fin whale 20 Hz pulses in these recordings, a combined automatic and manual detection and verification process was implemented (see details in [7,17]). Briefly, the automatic detector was a spectrogram cross-correlator using a 20 Hz pulse template from Western Australia. Its performance had previously been assessed at a different site (0.27% false positive files and 1.62% false negative files for fin whale 20 Hz presence/absence per file [7]). All files with automated detections were manually validated by checking that the detections were pulses downsweeping from ~40 Hz to 18 Hz with a 1 s duration and that they occurred in repetitive patterns. False positives were manually removed and false negatives added as appropriate. Finally, a process of 'bracketing' was carried out, whereby all audio files without automatic detections within ±3 h of a file with a verified detection were manually inspected for missed pulses, and pulses were added as appropriate. This process was iterated until all files surrounding a file with validated pulses were found to not contain fin whale pulses. The date and time of each fin whale pulse were noted.

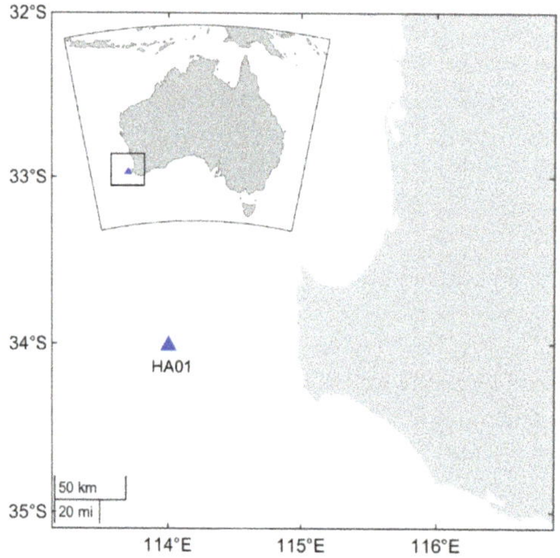

Figure 1. Location of the Comprehensive Nuclear-Test-Ban Treaty Organization (CTBTO) HA01 hydroacoustic monitoring station off Cape Leeuwin, Western Australia.

From the time series of verified fin whale 20 Hz pulse detections, we computed hourly presence/absence, the cumulative number of acoustic presence hours (i.e., pulse-hours) per day and per year, and the total number of days with fin whale pulses (i.e., pulse-days). To test for trends in fin whale acoustic presence at Cape Leeuwin, the first and last detection dates (of the fin whale season) and the total pulse-hours and pulse-days were each regressed on year. Simple linear regressions were performed using R statistical software version 4.2.2 [42].

3. Results

The fin whale 20 Hz pulse was detected across all acoustic recording years at Cape Leeuwin from 2002 to 2020 (Figure 2), with a total of 1233 days and 9938 h of acoustic presence. The detector performance included 0.81% false positive files and 4.51% false negative files before manual validation.

Across the 19 years of acoustic recordings at Cape Leeuwin, fin whale 20 Hz pulses were regularly detected from April to October (i.e., from austral autumn to mid-spring). Peaks in hourly acoustic presence were generally consistent across years, occurring in July and August (Figure 3). Acoustic detections did not split into two separate northern and southern migrations.

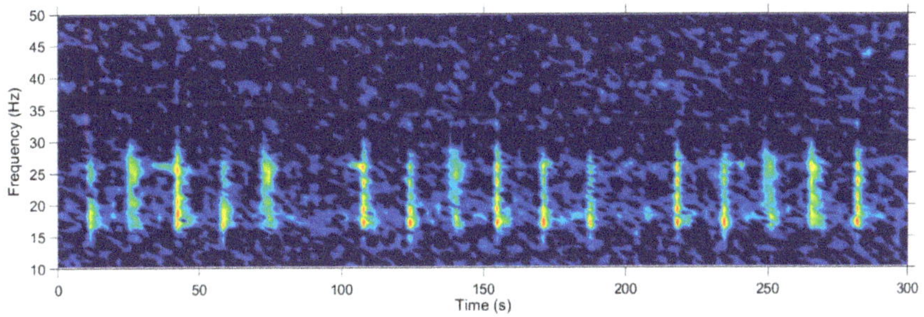

Figure 2. Spectrogram example of a fin whale 20 Hz pulse sequence. Recording from Cape Leeuwin (18 July 2012, 05:00). Spectrograms were calculated in 256-point Hann windows, 1 Hz and 1 s resolution, 50% overlap; sampling frequency: 250 Hz. Colors represent acoustic power (low: blue; high: red).

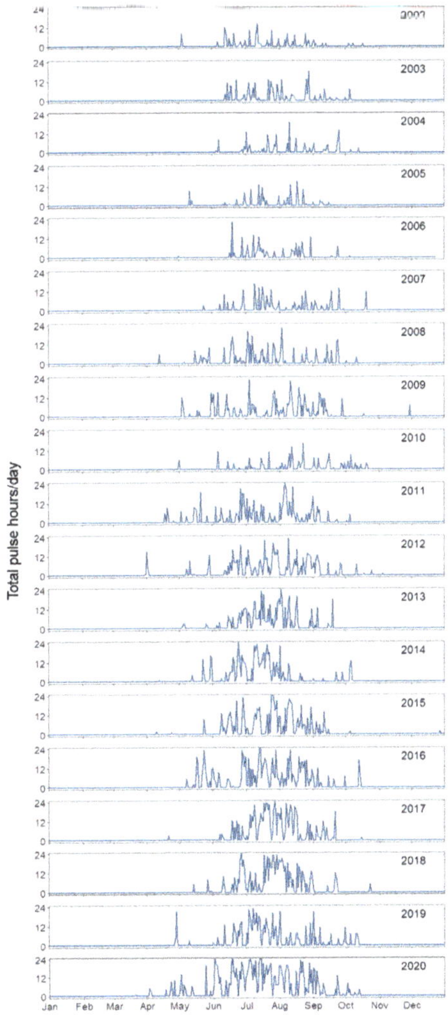

Figure 3. Line graphs of the total number of hours per day with fin whale 20 Hz acoustic presence (pulse-hours per day) across all years.

The first detection of fin whale acoustic presence shifted significantly over the study period, with earlier first detection by 2.14 d/y (Figure 4A). However, from 2013 to 2019, this trend waned, with a greater variation in the first detection date during this period (Figure 4A). The exception to this was in 2020, which observed the earliest first detection date of fin whale acoustic presence across all years (20th of March) (Figures 3 and 4A). The latest first detection date occurred on the 13th of June 2003 (Figures 3 and 4A). The last detection date of fin whale acoustic presence did not significantly shift across the study period (Figure 4B); acoustic presence commonly ceased in mid-October. An anomaly of last detection occurred in the year 2015, with fin whale acoustic presence detected much later in the year on the 27th of December (Figures 3 and 4B).

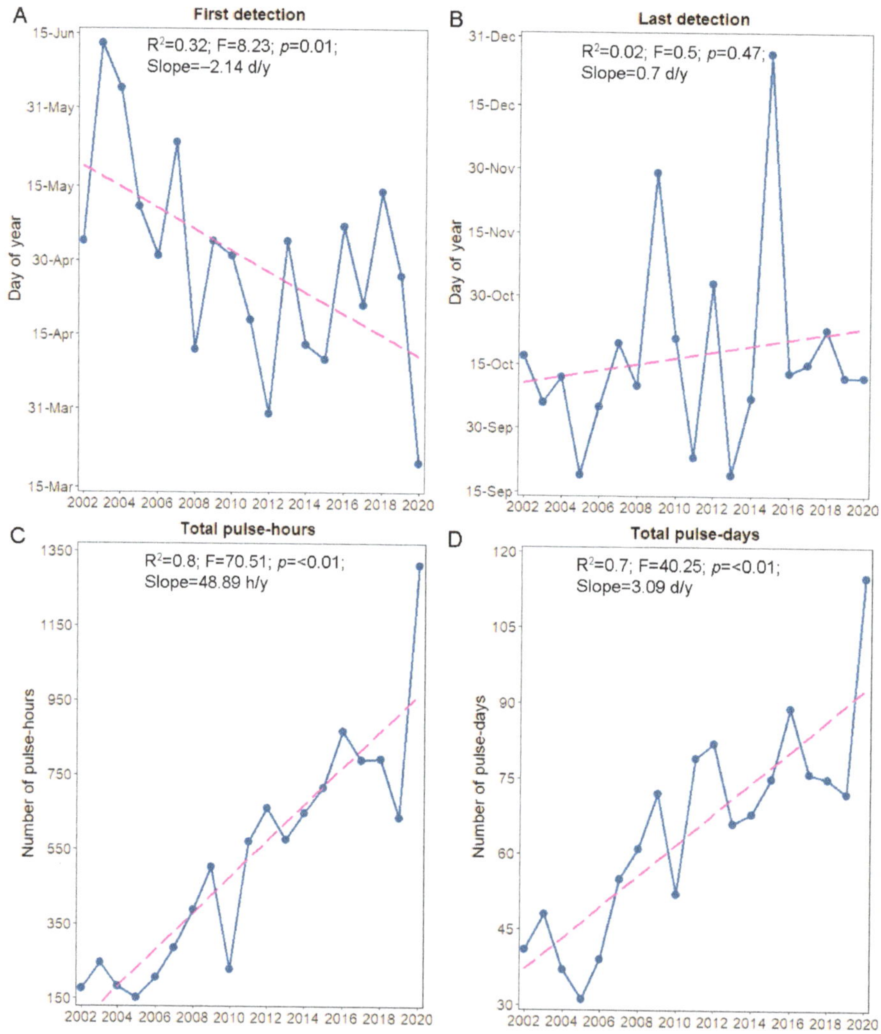

Figure 4. Scatterplots of fin whale (**A**) first detection, (**B**) last detection, (**C**) total pulse-hours per year, and (**D**) total pulse-days per year at Cape Leeuwin between 2002 and 2020. Regression statistics (R^2, F, *p*-value, and slope) are noted within each plot.

The total number of pulse-hours and pulse-days significantly increased from 2002 to 2020, with 48.89 more pulse-hours/y and 3.09 more pulse-days/y (Figure 4C,D). Both pulse-hours and pulse-days were low in 2002 with a total of 176 h and 41 d, respectively, before declining to a minimum of 153 pulse-hours and 31 pulse-days in 2005. Fin whale acoustic presence substantially increased in 2020, with a maximum of 1312 and 115 pulse-hours and pulse-days, respectively (Figure 4C,D).

4. Discussion

In this study, we identified 19 years (2002–2020) of fin whale seasonal acoustic presence at Cape Leeuwin, Western Australia, from austral autumn to mid-spring. No observations of year-round fin whale 20 Hz calls were made across the 19 years, providing no evidence of resident, non-migratory animals, in contrast to reports of resident populations in the Northern Hemisphere [6,43]. This long-term pattern reinforces Cape Leeuwin as a key location along the fin whale migratory pathway from Eastern Antarctic to Australian waters. Interestingly, PAM did not identify two separate migrations (northern vs. southern) at Cape Leeuwin. Some animals might have already traveled south, while others were still traveling north. Such an acoustic blurring of northern and southern migrations was observed for humpback whales in the Perth Canyon over a 16 y period, where a brief notch in the presence-hours per day was only seen in some years [18]. Alternatively, most adult males might cease their singing during the southern migration away from the breeding grounds to the feeding grounds. This hypothesis is supported by the rare acoustic detections in November and December (2009, 2015; Figure 3). Acoustically reduced, if not absent, southern migrations have also been reported for pygmy blue whales in the Perth Canyon [44]. Finally, singing males might travel farther from shore on a more direct great-circle route to their southern feeding grounds, with only mothers and calves hugging the coast for shelter. PAM further offshore or biotelemetry research might shed light on this.

At Cape Leeuwin, fin whale acoustic presence increased such that by the end of the 19-year recording period, fin whales were acoustically present on 74 more days than at the beginning (from 41 to 115 pulse-days). This increase in pulse-hours and pulse-days across the two decades may have been indicative of an increased number of individual fin whales in this region of the Southern Hemisphere. A recent study at the Western Antarctic Peninsula identified the occurrence of high numbers of fin whales (7909) at their historic feeding grounds at Elephant Island in 2018 [45] in comparison to density estimations in the Antarctic Peninsula in 2000 (4672 whales)—likely a sign of a recovering population after intense whaling. The fin whale population abundance, circumpolar south of 60°, was estimated to be 5445 whales between 1991 and 1998 [27]. However, no further abundance estimates are available for fin whales in Eastern Antarctic and Australian waters. Other species in Antarctic waters that suffered a similar population decline during the commercial whaling era are also reported to have increased in their abundance and population size. The western South Atlantic (WSA) humpback whale population recovered to 93% of their pre-industrial whaling era size [46]. The Antarctic blue whale population occupying South Georgia waters has increased in sightings and "D" calls [47]. If the increase in pulse-days is due to an increase in fin whale numbers, this may be the first preliminary indication of a recovering population of fin whales in Eastern Antarctic and Australian waters. Future research may consider the expansion of PAM over additional years and sites, visual surveys, photo-ID and mark-recapture, and genetic sampling to confirm fin whale population trends [48–51].

Alternatively, the trend of increasing acoustic presence per year at Cape Leeuwin may reflect an alteration in the animals' acoustic behavior. Fin whale vocalization rates are suggested to vary with acoustic behaviors such as reproductive displays, social displays, or feeding activities [9,16,52]. The reproductive acoustic displays of the fin whale can last up to 32 h [9], while social displays are irregular and inconsistent [16]. The observed increase in pulse detection rates at Svalbard Islands was likely due to this acoustic behavioral change, with the animals switching from irregular 20 Hz pulses to reproductive 20 Hz songs [53]. A similar shift in 20 Hz production rates at Cape Leeuwin may result in the animals vocalizing for longer time periods, resulting in greater acoustic presence hours and days per year.

Moreover, changes in the detection range of the acoustic system may affect pulse counts at Cape Leeuwin. The acoustic detection range could change with ambient noise, metocean and thus sound propagation conditions, fin whale swim depth and distance from shore, and call peak frequency. Some studies have shown a gradual decrease in low-frequency ambient noise at Cape Leeuwin, even after the removal of blue and fin whale calls [39,54,55]. The ambient noise in an area includes biotic (e.g., whale choruses), abiotic (e.g., wind), and anthropogenic noise (e.g., vessel traffic) and can affect the signal-to-noise ratio of a 20 Hz pulse, thereby affecting the detectability of the pulse [44]. The increasing temperature and acidification of the ocean might improve sound transmission but would affect signals and noises similarly [56]. If whales gradually migrated closer to shore or at new depths from which calls might better propagate to the Cape Leeuwin hydroacoustic station, then detectability would increase [40,57]. Finally, long-term shifts in call peak frequency [58] may lead to gradual, albeit small, changes in detection ranges.

The shift in fin whale first acoustic presence at Cape Leeuwin by ~2 d/y may be due to the animals varying their migration timing out of Antarctic waters. Recent studies have identified environmental variables which affect fin whale acoustic presence in regions around the world, such as sea surface temperature and chlorophyll-a concentration [59,60]. A long-term seasonal shift in fin whale first and last sightings at their feeding grounds in the Gulf of St. Lawrence, Canada was attributed to declining sea ice coverage and rising sea surface temperature in this region due to climate change [23]. Moreover, climate change is expected to impact Antarctic krill, reducing prey availability on fin whale feeding grounds [61,62], which might trigger earlier fin whale departure. The investigation of environmental variables that may drive fin whale migration out of Antarctic waters could shed light on the shift in the first acoustic presence of the fin whale along their migratory route at Cape Leeuwin.

The substantial increase in fin whale acoustic presence at Cape Leeuwin in 2020 may reflect an indirect effect of the COVID-19 pandemic. During the height of the pandemic in 2020, lockdown restrictions resulted in a reduction in anthropogenic activity such as commercial fishing, shipping, recreational boating, and tourism activities in Western Australia [63]. Throughout the lockdown period in 2020, there was a global increase in sightings of marine mammal species, which were likely related to changes in the behavior and distribution of the animals [64]. These changes were thought to be the result of a reduction in the physical disturbance and displacement of the animals or a reduction in underwater noise due to the decrease in vessel traffic [64]. As fin whales have been reported to be affected by vessel traffic and the resulting ambient noise [65,66], it is possible that the increase in the acoustic presence of the animals at Cape Leeuwin in 2020 was due to these environmental effects of the COVID-19 lockdowns. Similarly linked to the COVID-19 pandemic, the detection range of the acoustic systems at Cape Leeuwin may have been positively affected by a decrease in ambient noise due to reduced vessel traffic. Further analysis of the ambient noise environment and shipping activity combined with additional, consecutive acoustic recording years at this site may help to identify why there was a greater acoustic presence in 2020 in comparison to not only the prior but also the later years.

It is important to note that the first and last detections of fin whale vocalizations may not be a reliable representation of when fin whales arrive and depart Australian waters on their north- and southward migrations, respectively. The observations reported here only represent vocalizing fin whales and do not confirm the absence of the whales, as there may be animals that are present but silent. Visual survey, biotelemetry, and tagging with acoustic recorders may allow the determination of vocalization rates and quantify the uncertainty of physical presence derived from PAM only.

5. Conclusions

Our investigation of 19 years of fin whale acoustic presence at Cape Leeuwin, Western Australia, from 2002 to 2020 revealed a seasonal pattern of presence from austral autumn to mid-spring. The long-term consistency of presence at this southwesternmost point of the Australian continent reinforces this site as a key migratory point for the species as it heads for Australia from Antarctica, and vice versa, warranting further research to ultimately inform conservation policies in migratory corridors. Moreover, the observed increase in fin whale 20 Hz pulse-hours and pulse-days across the 19 years provides evidence of an increasing population of this vulnerable subspecies in the Southern Hemisphere, which remains to be confirmed with complementary survey methods to fully assess recovery under the IUCN.

Author Contributions: Conceptualization, M.G.A., R.D.M., B.S.M. and C.E.; methodology, M.G.A., R.D.M. and C.E.; software, M.G.A. and R.D.M.; validation, M.G.A.; formal analysis, M.G.A.; data curation, R.D.M. and C.E.; writing—original draft preparation, M.G.A.; writing—review and editing, M.G.A., R.D.M., B.S.M. and C.E.; supervision, R.D.M., B.S.M. and C.E. All authors have read and agreed to the published version of the manuscript.

Funding: This research received no external funding.

Data Availability Statement: Restrictions apply to the availability of these data. Data were obtained from the Comprehensive Nuclear-Test-Ban Treaty Organization (CTBTO) and are available https://www.ctbto.org/resources/for-researchers-experts/vdec (accessed on 19 June 2025) with the permission of the Comprehensive Nuclear-Test-Ban Treaty Organization (CTBTO).

Acknowledgments: We thank the Commission for the Comprehensive Nuclear-Test-Ban Treaty Organization (CTBTO) for the acoustic data from 2012 to 2020 and Geoscience Australia for the data from 2002 to 2011.

Conflicts of Interest: The authors declare no conflicts of interest. The material in this article is based on Chapter 2.3 in Meghan Aulich's dissertation "The Acoustic Ecology of the Fin Whale in Eastern Antarctic and Australian Water", Ph.D. degree, Curtin University, Perth, Australia, 2023.

Abbreviations

The following abbreviation is used in this manuscript:

PAM	Passive acoustic monitoring

References

1. Lockyer, C.; Brown, S. The migration of whales. In *Animal Migration*; Cambridge University Press Binghamton: New York, NY, USA, 1981; Volume 13, pp. 105–137.
2. Corkeron, P.J.; Connor, R.C. Why Do Baleen Whales Migrate? *Mar. Mammal Sci.* **1999**, *15*, 1228–1245. [CrossRef]
3. Mizroch, S.A.; Rice, D.W.; Breiwick, J.M. The fin whale, *Balaenoptera physalus*. *Mar. Fish. Rev.* **1984**, *46*, 20–24.
4. Simon, M.; Stafford, K.M.; Beedholm, K.; Lee, C.M.; Madsen, P.T. Singing behavior of fin whales in the Davis Strait with implications for mating, migration and foraging. *J. Acoust. Soc. Am.* **2010**, *128*, 3200–3210. [CrossRef]
5. Tsujii, K.; Otsuki, M.; Akamatsu, T.; Matsuo, I.; Amakasu, K.; Kitamura, M.; Kikuchi, T.; Miyashita, K.; Mitani, Y. The migration of fin whales into the southern Chukchi Sea as monitored with passive acoustics. *ICES J. Mar. Sci.* **2016**, *73*, 2085–2092. [CrossRef]
6. Širović, A.; Williams, L.N.; Kerosky, S.M.; Wiggins, S.M.; Hildebrand, J.A. Temporal separation of two fin whale call types across the eastern North Pacific. *Mar. Biol.* **2013**, *160*, 47–57. [CrossRef]

7. Aulich, M.G.; McCauley, R.D.; Miller, B.S.; Samaran, F.; Giorli, G.; Saunders, B.J.; Erbe, C. Seasonal distribution of the fin whale (*Balaenoptera physalus*) in Antarctic and Australian waters based on passive acoustics. *Front. Mar. Sci.* **2022**, *9*, 864153. [CrossRef]
8. Davis, G.E.; Baumgartner, M.F.; Corkeron, P.J.; Bell, J.; Berchok, C.; Bonnell, J.M.; Bort Thornton, J.; Brault, S.; Buchanan, G.A.; Cholewiak, D.M.; et al. Exploring movement patterns and changing distributions of baleen whales in the western North Atlantic using a decade of passive acoustic data. *Glob. Change Biol.* **2020**, *26*, 4812–4840. [CrossRef]
9. Watkins, W.A.; Tyack, P.; Moore, K.E.; Bird, J.E. The 20-Hz signals of finback whales (*Balaenoptera physalus*). *J. Acoust. Soc. Am.* **1987**, *82*, 1901–1912. [CrossRef] [PubMed]
10. Thompson, P.O.; Findley, L.T.; Vidal, O. 20-Hz pulses and other vocalizations of fin whales, *Balaenoptera physalus*, in the Gulf of California, Mexico. *J. Acoust. Soc. Am.* **1992**, *92*, 3051–3057. [CrossRef] [PubMed]
11. Garcia, H.A.; Zhu, C.; Schinault, M.E.; Kaplan, A.I.; Handegard, N.O.; Godø, O.R.; Ahonen, H.; Makris, N.C.; Wang, D.; Huang, W.; et al. Temporal–spatial, spectral, and source level distributions of fin whale vocalizations in the Norwegian Sea observed with a coherent hydrophone array. *ICES J. Mar. Sci.* **2019**, *76*, 268–283. [CrossRef]
12. Miksis-Olds, J.L.; Harris, D.V.; Heaney, K.D. Comparison of estimated 20-Hz pulse fin whale source levels from the tropical Pacific and Eastern North Atlantic Oceans to other recorded populations. *J. Acoust. Soc. Am.* **2019**, *146*, 2373–2384. [CrossRef] [PubMed]
13. Širović, A.; Hildebrand, J.A.; Wiggins, S.M. Blue and fin whale call source levels and propagation range in the Southern Ocean. *J. Acoust. Soc. Am.* **2007**, *122*, 1208–1215. [CrossRef] [PubMed]
14. Weirathmueller, M.J.; Wilcock, W.S.D.; Soule, D.C. Source levels of fin whale 20 Hz pulses measured in the Northeast Pacific Ocean. *J. Acoust. Soc. Am.* **2013**, *133*, 741–749. [CrossRef] [PubMed]
15. Croll, D.A.; Clark, C.W.; Acevedo, A.; Tershy, B.; Flores, S.; Gedamke, J.; Urban, J. Only male fin whales sing loud songs. *Nature* **2002**, *417*, 809. [CrossRef]
16. McDonald, M.A.; Hildebrand, J.A.; Webb, S.C. Blue and fin whales observed on a seafloor array in the Northeast Pacific. *J. Acoust. Soc. Am.* **1995**, *98*, 712–721. [CrossRef]
17. Aulich, M.G.; McCauley, R.D.; Saunders, B.J.; Parsons, M.J.G. Fin whale (*Balaenoptera physalus*) migration in Australian waters using passive acoustic monitoring. *Sci. Rep.* **2019**, *9*, 8840. [CrossRef]
18. Gosby, C.; Erbe, C.; Harvey, E.S.; Figueroa Landero, M.M.; McCauley, R.D. Vocalizing humpback whales (*Megaptera novaeangliae*) migrating from Antarctic feeding grounds arrive earlier and earlier in the Perth Canyon, Western Australia. *Front. Mar. Sci.* **2022**, *9*, 1086763. [CrossRef]
19. Szesciorka, A.R.; Ballance, L.T.; Širović, A.; Rice, A.; Ohman, M.D.; Hildebrand, J.A.; Franks, P.J.S. Timing is everything: Drivers of interannual variability in blue whale migration. *Sci. Rep.* **2020**, *10*, 7710. [CrossRef]
20. Avila, I.C.; Dormann, C.F.; García, C.; Payán, L.F.; Zorrilla, M.X. Humpback whales extend their stay in a breeding ground in the Tropical Eastern Pacific. *ICES J. Mar. Sci.* **2019**, *77*, 109–118. [CrossRef]
21. Laidre, K.L.; Heide-Jørgensen, M.P. Spring partitioning of Disko Bay, West Greenland, by Arctic and Subarctic baleen whales. *ICES J. Mar. Sci.* **2012**, *69*, 1226–1233. [CrossRef]
22. Hauser, D.D.W.; Laidre, K.L.; Stafford, K.M.; Stern, H.L.; Suydam, R.S.; Richard, P.R. Decadal shifts in autumn migration timing by Pacific Arctic beluga whales are related to delayed annual sea ice formation. *Glob. Change Biol.* **2017**, *23*, 2206–2217. [CrossRef] [PubMed]
23. Ramp, C.; Delarue, J.; Palsbøll, P.J.; Sears, R.; Hammond, P.S. Adapting to a Warmer Ocean—Seasonal Shift of Baleen Whale Movements over Three Decades. *PLoS ONE* **2015**, *10*, e0121374. [CrossRef] [PubMed]
24. Cooke, J.G. Balaenoptera physalus. The IUCN Red List of Threatened Species 2018: e.T2478A50349982. 2018. Available online: https://doi.org/10.2305/IUCN.UK.2018-2.RLTS.T2478A50349982.en (accessed on 13 July 2020).
25. Rocha, R.C.; Clapham, P.J.; Ivashchenko, Y.V. Emptying the oceans: A summary of industrial whaling catches in the 20th century. *Mar. Fish. Rev.* **2014**, *76*, 37–48. [CrossRef]
26. Edwards, E.F.; Hall, C.; Moore, T.J.; Sheredy, C.; Redfern, J.V. Global distribution of fin whales *Balaenoptera physalus* in the post-whaling era (1980–2012). *Mammal Rev.* **2015**, *45*, 197–214. [CrossRef]
27. Branch, T.; Butterworth, D. Estimates of abundance south of 60° S for cetacean species sighted frequently on the 1978/79 to 1997/98 IWC/IDCR-SOWER sighting surveys. *J. Cetacean Res. Manag.* **2001**, *3*, 251–270. [CrossRef]
28. Buchan, S.J.; Gutierrez, L.; Balcazar-Cabrera, N.; Stafford, K.M. Seasonal occurrence of fin whale song off Juan Fernandez, Chile. *Endanger. Species Res.* **2019**, *39*, 135–145. [CrossRef]
29. Burkhardt, E.; Van Opzeeland, I.; Cisewski, B.; Mattmüller, R.; Meister, M.; Schall, E.; Spiesecke, S.; Thomisch, K.; Zwicker, S.; Boebel, O. Seasonal and diel cycles of fin whale acoustic occurrence near Elephant Island, Antarctica. *R. Soc. Open Sci.* **2021**, *8*, 201142. [CrossRef]
30. Herr, H.; Hickmott, L.; Viquerat, S.; Panigada, S. First evidence for fin whale migration into the Pacific from Antarctic feeding grounds at Elephant Island. *R. Soc. Open Sci.* **2022**, *9*, 220721. [CrossRef]
31. Toro, F.; Vilina, Y.A.; Capella, J.J.; Gibbons, J. Novel coastal feeding area for eastern South Pacific fin whales (*Balaenoptera physalus*) in mid-latitude humboldt current waters off Chile. *Aquat. Mamm.* **2016**, *42*, 47–55. [CrossRef]

32. Viquerat, S.; Waluda, C.M.; Kennedy, A.S.; Jackson, J.A.; Hevia, M.; Carroll, E.L.; Buss, D.L.; Burkhardt, E.; Thain, S.; Smith, P.; et al. Identifying seasonal distribution patterns of fin whales across the Scotia Sea and the Antarctic Peninsula region using a novel approach combining habitat suitability models and ensemble learning methods. *Front. Mar. Sci.* **2022**, *9*, 1040512. [CrossRef]
33. Wood, M.; Širović, A. Characterization of fin whale song off the Western Antarctic Peninsula. *PLoS ONE* **2022**, *17*, e0264214. [CrossRef] [PubMed]
34. Dréo, R.; Bouffaut, L.; Leroy, E.; Barruol, G.; Samaran, F. Baleen whale distribution and seasonal occurrence revealed by an ocean bottom seismometer network in the Western Indian Ocean. *Deep Sea Res. Part II Top. Stud. Oceanogr.* **2019**, *161*, 132–144. [CrossRef]
35. Miller, B.S.; Masere, C.; Milnes, M.; Cleeland, J.; Lamb, T.; Maschette, D.; Welsford, D. Heard off Heard: Acoustic detections of sperm whales (*Physeter macrocephalus*) and other cetaceans off Heard Island. *JASA Express Lett.* **2024**, *4*, 061201. [CrossRef]
36. Sprogis, K.R.; Sutton, A.L.; Jenner, M.-N.; McCauley, R.D.; Jenner, K.C.S. Occurrence of cetaceans and seabirds along the Indian Ocean 110°E meridian from temperate to tropical waters. *Deep-Sea Res. Part II Top. Stud. Oceanogr.* **2022**, *205*, 105184. [CrossRef]
37. Todd, V.L.G.; Williamson, L.D. Cetacean distribution in relation to oceanographic features at the Kerguelen Plateau. *Polar Biol.* **2022**, *45*, 113–126. [CrossRef]
38. Lawrence, M.W. Acoustic Monitoring of the Global Ocean for the CTBT. In Proceedings of the Acoustics 2004, Gold Coast, Australia, 3–5 November 2004.
39. Park, J.; Haralabus, G.; Zampolli, M.; Metz, D. Low frequency ambient noise dynamics and trends in the Indian Ocean, Cape Leeuwin, Australia. *J. Acoust. Soc. Am.* **2023**, *153*, 2312. [CrossRef]
40. Gavrilov, A.; McCauley, R.; Salgado-Kent, C.; Tripovich, J.; Burton, C. Vocal characteristics of pygmy blue whales and their change over time. *J. Acoust. Soc. Am.* **2011**, *130*, 3651–3660. [CrossRef]
41. Leroy, E.C.; Royer, J.-Y.; Alling, A.; Maslen, B.; Rogers, T.L. Multiple pygmy blue whale acoustic populations in the Indian Ocean: Whale song identifies a possible new population. *Sci. Rep.* **2021**, *11*, 8762. [CrossRef]
42. R Core Team. R: A Language and Environment for Statistical Computing. 2022. Available online: https://www.R-project.org/ (accessed on 1 November 2022).
43. Morano, J.L.; Salisbury, D.P.; Rice, A.N.; Conklin, K.L.; Falk, K.L.; Clark, C.W. Seasonal and geographical patterns of fin whale song in the western North Atlantic Ocean. *J. Acoust. Soc. Am.* **2012**, *132*, 1207–1212. [CrossRef]
44. Erbe, C.; Verma, A.; McCauley, R.; Gavrilov, A.; Parnum, I. The marine soundscape of the Perth Canyon. *Prog. Oceanogr.* **2015**, *137*, 38–51. [CrossRef]
45. Herr, H.; Viquerat, S.; Devas, F.; Lees, A.; Wells, L.; Gregory, B.; Giffords, T.; Beecham, D.; Meyer, B. Return of large fin whale feeding aggregations to historical whaling grounds in the Southern Ocean. *Sci. Rep.* **2022**, *12*, 9458. [CrossRef] [PubMed]
46. Zerbini, A.N.; Adams, G.; Best, J.; Clapham, P.J.; Jackson, J.A.; Punt, A.E. Assessing the recovery of an Antarctic predator from historical exploitation. *R. Soc. Open Sci.* **2019**, *6*, 190368. [CrossRef]
47. Calderan, S.V.; Black, A.; Branch, T.A.; Collins, M.A.; Kelly, N.; Leaper, R.; Lurcock, S.; Miller, B.S.; Moore, M.; Olson, P.A. South Georgia blue whales five decades after the end of whaling. *Endanger. Species Res.* **2020**, *43*, 359–373. [CrossRef]
48. Tardy, C.; Ody, D.; Gimenez, O.; Planes, S. Abundance of fin whales (*Balaenoptera physalus*) in the north-western Mediterranean Sea, using photo-identification and microsatellite genotyping. *Mar. Ecol.* **2023**, *44*, e12737. [CrossRef]
49. Degollada, E.; Amigó, N.; O'Callaghan, S.A.; Varola, M.; Ruggero, K.; Tort, B. A novel technique for photo-identification of the fin whale, *Balaenoptera physalus*, as determined by drone aerial images. *Drones* **2023**, *7*, 220. [CrossRef]
50. Viquerat, S.; Herr, H. Mid-summer abundance estimates of fin whales Balaenoptera physalus around the South Orkney Islands and Elephant Island. *Endanger. Species Res.* **2017**, *32*, 515–524. [CrossRef]
51. Víkingsson, G.A.; Pike, D.G.; Desportes, G.; Øien, N.; Gunnlaugsson, T.; Bloch, D. Distribution and abundance of fin whales (*Balaenoptera physalus*) in the Northeast and Central Atlantic as inferred from the North Atlantic sightings surveys 1987–2001. *NAMMCO Sci. Publ.* **2013**, *7*, 49–72. [CrossRef]
52. Watkins, W.A. Activities and underwater sounds of fin whales. *Sci. Rep. Whales Res. Inst.* **1981**, *33*, 83–117.
53. Papale, E.; Pelagatti, M.; Pedrazzi, G.; Buscaino, G. Occurrence and patterns of fin whale songs reveal alternative migration strategies in Svalbard Islands, Norway. *Sci. Rep.* **2023**, *13*, 4436. [CrossRef]
54. Harris, P.; Sotirakopoulos, K.; Robinson, S.; Wang, L.; Livina, V. A statistical method for the evaluation of long term trends in underwater noise measurements. *J. Acoust. Soc. Am.* **2019**, *145*, 228–242. [CrossRef]
55. Zhang, Z.Y.; Gavrilov, A. Icebergs and storms in the Southern Ocean as sources of low frequency noise along Australian southern coasts. In Proceedings of the ACOUSTICS 2019, Cape Schanck, VIC, Australia, 10–13 November 2019.
56. Brewer, P.G.; Hester, K. Ocean acidification and the increasing transparency of the ocean to low-frequency sound. *Oceanography* **2009**, *22*, 86–93. [CrossRef]
57. McCauley, R. *Offshore Irish Noise Logger Program (March to September 2014): Analysis of Cetacean Presence, and Ambient and Anthropogenic Noise Sources*; Centre for Marine Science and Technology: Perth, Australia, 2015.
58. Weirathmueller, M.J.; Stafford, K.M.; Wilcock, W.S.D.; Hilmo, R.S.; Dziak, R.P.; Trehu, A.M. Spatial and temporal trends in fin whale vocalizations recorded in the NE Pacific Ocean between 2003–2013. *PLoS ONE* **2017**, *12*, e0186127. [CrossRef] [PubMed]

59. Shabangu, F.W.; Findlay, K.P.; Yemane, D.; Stafford, K.M.; Van den Berg, M.; Blows, B.; Andrew, R.K. Seasonal occurrence and diel calling behaviour of Antarctic blue whales and fin whales in relation to environmental conditions off the west coast of South Africa. *J. Mar. Syst.* **2019**, *190*, 25–39. [CrossRef]
60. Escajeda, E.; Stafford, K.M.; Woodgate, R.A.; Laidre, K.L. Variability in fin whale (*Balaenoptera physalus*) occurrence in the Bering Strait and southern Chukchi Sea in relation to environmental factors. *Deep Sea Res. (Part II Top. Stud. Oceanogr.)* **2020**, *177*, 104782. [CrossRef]
61. Tulloch, V.J.D.; Plagányi, É.E.; Brown, C.; Richardson, A.J.; Matear, R. Future recovery of baleen whales is imperiled by climate change. *Glob. Change Biol.* **2019**, *25*, 1263–1281. [CrossRef]
62. Kawaguchi, S.; Atkinson, A.; Bahlburg, D.; Bernard, K.S.; Cavan, E.L.; Cox, M.J.; Hill, S.L.; Meyer, B.; Veytia, D. Climate change impacts on Antarctic krill behaviour and population dynamics. *Nat. Rev. Earth Environ.* **2024**, *5*, 43–58. [CrossRef]
63. Huveneers, C.; Jaine, F.R.A.; Barnett, A.; Butcher, P.A.; Clarke, T.M.; Currey-Randall, L.M.; Dwyer, R.G.; Ferreira, L.C.; Gleiss, A.C.; Hoenner, X.; et al. The power of national acoustic tracking networks to assess the impacts of human activity on marine organisms during the COVID-19 pandemic. *Biol. Conserv.* **2021**, *256*, 108995. [CrossRef]
64. Coll, M. Environmental effects of the COVID-19 pandemic from a (marine) ecological perspective. *Ethics Sci. Environ. Politics* **2020**, *20*, 41–55. [CrossRef]
65. Castellote, M.; Clark, C.; Lammers, M. Acoustic and behavioural changes by fin whales (*Balaenoptera physalus*) in response to shipping and airgun noise. *Biol. Conserv.* **2012**, *147*, 115–122. [CrossRef]
66. Panigada, S.; Pesante, G.; Zanardelli, M.; Capoulade, F.; Gannier, A.; Weinrich, M.T. Mediterranean fin whales at risk from fatal ship strikes. *Mar. Pollut. Bull.* **2006**, *52*, 1287–1298. [CrossRef]

oceans

Article

A Kiss from the Wild: Tongue Nibbling in Free-Ranging Killer Whales (*Orcinus orca*)

Javier Almunia [1,*], Johnny van Vliet [2] and Debbie Bouma [2]

1 Loro Parque Fundación, Avda. Loro Parque s/n, 38400 Puerto de la Cruz, Spain
2 Wild-Encounters, Sint Vitusholt 7e Laan 21, 9674 AX Winschoten, The Netherlands; info@wild-encounters.com (J.v.V.); debbiebouma@hotmail.nl (D.B.)
* Correspondence: dir@loroparque-fundacion.org

Abstract: Tongue-nibbling is a rare and previously undocumented affiliative behaviour in free-ranging killer whales (*Orcinus orca*), until now seen only in individuals under human care. This study presents the first recorded observation of tongue-nibbling between two wild killer whales in the Kvænangen fjords, Norway. The interaction, captured opportunistically by citizen scientists during a snorkelling expedition, lasted nearly two minutes and involved repeated episodes of gentle, face-to-face oral contact. This behaviour closely resembles sequences observed and described in detail in zoological settings, suggesting that it forms part of the species' natural social repertoire. The observation also supports the interpretation of tongue-nibbling as a socially affiliative behaviour, likely involved in reinforcing social bonds, particularly among juveniles. The prolonged maintenance of this interaction in managed populations originating from geographically distinct Atlantic and Pacific lineages further indicates its behavioural conservation across contexts. This finding underscores the importance of underwater ethological observation in capturing cryptic social behaviours in cetaceans and illustrates the value of integrating citizen science into systematic behavioural documentation. The study also reinforces the relevance of managed populations in ethological research and highlights the ethical need for carefully regulated wildlife interaction protocols in marine tourism.

Keywords: *Orcinus orca*; tongue-nibbling; affiliative behaviour; cetacean sociality; underwater behaviour; citizen science; managed populations; ethology

1. Introduction

The study of social behaviour in animals is a cornerstone of behavioural ecology, offering crucial insights into the biological and ecological dynamics of different species. Social animals, ranging from insects to primates, exhibit complex behavioural repertoires that serve to facilitate communication, cooperation, and group cohesion [1,2]. In primates, for instance, research on social dynamics has elucidated behaviours related to hierarchy (dominance and submission), agonism, affiliation, and reconciliation [3]. Analogous dominance–submission patterns have also been described in species as diverse as gorillas [4], elephants [5], dolphins [6], and killer whales [7].

Despite the relevance of such research, cetacean social behaviour remains particularly challenging to study due to the aquatic environment in which most of their activity occurs, largely beneath the water's surface. This constraint has historically limited direct observation and led researchers to infer social interactions based on proximity at the surface. For example, Thomsen [8] categorised the surface behaviour of resident killer

https://doi.org/10.3390/oceans6020037

whales (*Orcinus orca*) near Vancouver Island into social and non-social contexts, depending on whether individuals were within or beyond one body length of each other—an approach grounded in earlier work by Ford [9]. This proximity-based framework has since been employed in various studies, including those by Gibson [10] on bottlenose dolphins (*Tursiops* sp.) in Shark Bay, Parsons [11] on killer whales, and Marley [12] on *Tursiops aduncus* in Western Australia. While useful, such methods often simplify the complexity of cetacean interactions, underscoring the need for more granular approaches supported by emerging technologies.

Efforts to refine our understanding of cetacean sociality have increasingly focused on describing specific patterns of synchronised behaviour, particularly among animals under human care. Clegg et al. [13], for instance, analysed synchronised swimming and tactile interactions in bottlenose dolphins (*Tursiops truncatus*) in managed environments. Serres et al. [14] extended these observations to multiple captive cetacean species. However, these studies remain largely constrained to behaviours visible at the surface, offering limited access to the full social repertoire displayed underwater.

To overcome these observational limitations, underwater studies have proven essential. Christiansen et al. [15] examined the surface-level social behaviour of Indo-Pacific bottlenose dolphins (*Tursiops aduncus*) off Zanzibar, noting that the presence of tourist vessels reduced resting and social time while increasing travel and foraging. In contrast, Dudzinski et al. [16] and Delfour [17] employed underwater methodologies to provide more detailed behavioural descriptions. More recently, Manitzas Hill [18] combined both surface and subsurface observations to document activity budgets in killer whales under human care, thereby offering a more comprehensive understanding of behavioural variation in managed populations.

A pivotal contribution to the ethology of killer whales was the ethogram compiled by Martínez and Klinghammer in 1978 [19], based on both captive and free-ranging individuals. Their work catalogued over 50 behaviours, including both surface and underwater actions. Among the behaviours described was a peculiar interaction termed "nibbling," in which one whale gently mouthed the tongue of another.

Decades later, Sánchez-Hernández et al. [7] revisited this behaviour under human care, describing it for the first time in detail and interpreting it as an affiliative interaction. The study found that nibbling occurred predominantly among females and juveniles and situated it within a broader analysis of reconciliation behaviours, suggesting its importance in promoting group cohesion.

However, in the absence of comparable reports from wild populations, the ethological validity of such behaviours has sometimes been questioned. Some could have hypothesised that such behaviours could represent stereotypy, aberrant behaviour or an ephemeral fad, akin to the placement of dead salmon atop the head—a behaviour previously reported in killer whales.

This study presents the first documented case of tongue-nibbling between two wild killer whales in Norway. This finding confirms that the behaviour, previously observed only under human care, also occurs in the wild, thereby supporting its interpretation as part of the natural social repertoire of the species.

2. Observation Context and Recording Conditions

The focal observation was made on 11 January 2024 at approximately 10:40 local time in the Kvænangen fjords, northern Norway. Specifically, the event occurred at the entrance of Tverrfjorden, a sheltered bay located on the northern side of Jøkelfjord (Figure 1). These fjords are known for hosting seasonal aggregations of herring and cetaceans, and their calm winter conditions are conducive to in-water observations. A small group of snorkellers, under the supervision of experienced expedition leaders, entered the water from a Zodiac support boat deployed from the main expedition vessel. The entry protocol followed standard cetacean observation practices aimed at minimising disturbance: animals were approached slowly and from the side, and snorkellers floated in a passive, horizontal position once in the water. The environmental conditions at the time were favourable, with light wind, low swell (less than one metre), and good underwater visibility (estimated at 12–15 m). The orcas were slowly travelling, and no feeding or evasive behaviours were observed. The focal behavioural interaction took place between two adult-sized killer whales diving under the snorkellers at an approximate distance of 10–15 m. The entire sequence was recorded using a handheld GoPro Hero 11 Black action camera operated by one of the snorkellers. The footage was later analysed to identify and characterise the behaviours observed. The video of the 2024 observation is publicly available via the institutional repository and can be accessed at (accessed on 5 June 2025): https://research-data.ull.es/datasets/xj2yxc3cwp/1.

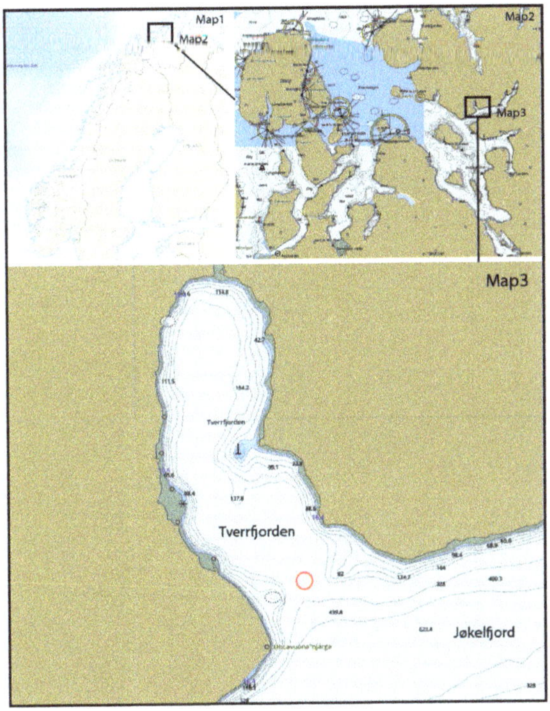

Figure 1. Map showing the location of the Kvænangen fjords, northern Norway, where the observed interaction between two wild killer whales (*Orcinus orca*) was recorded.

3. Results

On 31 October 2024, at approximately 13:30 local time, a behavioural interaction between two free-ranging killer whales (*Orcinus orca*) was recorded during a snorkelling expedition at the entrance of Tverrfjorden, a sheltered bay located on the northern side of Jøkelfjorden in the Kvænangen fjords, northern Norway. The weather conditions were calm and overcast, with light winds between 4 and 6 knots and mirror-like water surfaces within Tverrfjorden. These conditions provided a stark contrast to the adjacent fjord systems, where wind gusts exceeded 35 knots and wave heights reached approximately 50 cm in Jøkelfjorden. No other vessels or rigid inflatable boats (RIBs) were present in the area at the time of the observation.

The group of snorkellers operated two Zodiacs, each carrying six guests and an expedition leader. The main vessel, *Vestland Explorer*, remained anchored outside Jøkelfjorden and did not enter Tverrfjorden during the operations. Multiple groups of killer whales were present in the fjord throughout the day, with an estimated total of approximately 30 individuals. Observations began at around 11:00, with initial sightings focused on a group of approximately 12 individuals engaged in sub-surface apparent feeding behaviour on dispersed bait balls. Additional groups were observed displaying reduced locomotor activity near the coastline.

At the time of the focal event, the group of 12 guests was floating in a single line formation at the entrance of Tverrfjorden. Within this context, two killer whales were observed engaging in a prolonged mouth-to-mouth interaction (Figure 2) that lasted for a total of 1 min and 49 s. The individuals approached one another and maintained contact between the anterior portions of their heads. The interaction comprised three discrete episodes: the first lasting 10 s, the second 26 s, and the third 18 s. Following the final episode, the individuals separated and swam away (see the complete video in the Supplementary Materials).

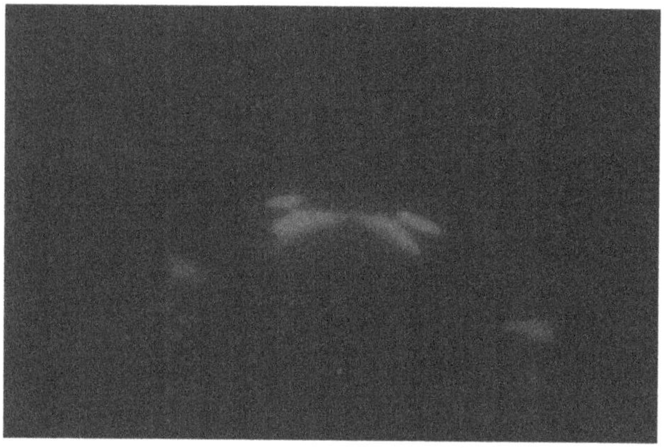

Figure 2. Still frame from video footage recorded in the Kvænangen fjords, Norway, in 2024, showing the tongue-nibbling interaction between two free-ranging killer whales. See the Supplementary Materials for the complete video sequence.

The observed behaviour is consistent with the affiliative tongue-nibbling interaction previously described in a 2013 study at Loro Parque involving two killer whales under human care (Figure 3). In that case, one individual protruded its tongue while the other made gentle nibbling movements. The behaviour occurred in three sequences, interrupted by the withdrawal and re-extension of the tongue, lasting a total of approximately 15 s (see the complete video in the Supplementary Materials).

To assess the rarity and broader context of this behaviour, the authors consulted three professional divers and underwater videographers with extensive experience documenting killer whales in different geographic locations. None of the individuals consulted reported having recorded or directly observed tongue-nibbling interactions. One diver did recall an incident in which several killer whales approached an RIB while the group was preparing to enter the water. At that time, some observers on board remarked that the animals appeared to be "kissing" beneath the boat—a description identical to that provided independently by the guests who recorded the event in Tverrfjorden. Additionally, the authors sought confirmation from the senior marine mammal trainers at Loro Parque, who affirmed their familiarity with the behaviour. They reported that tongue-nibbling was observed repeatedly among four individuals housed at the facility, although the behaviour had not been observed in subsequent years.

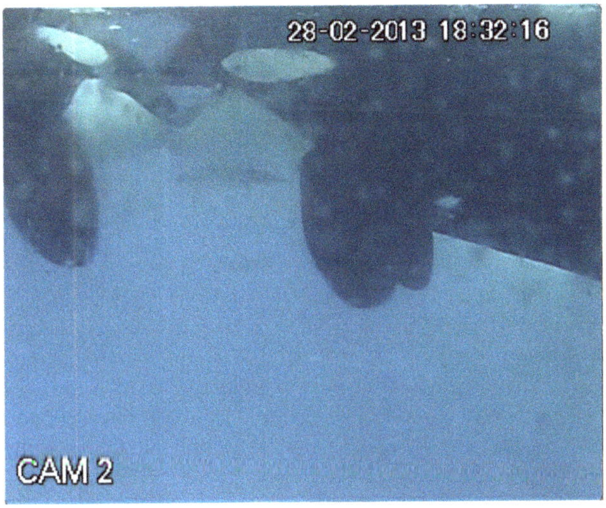

Figure 3. Still frame from video footage recorded at Loro Parque in 2013, illustrating the tongue-nibbling behaviour between two killer whales under human care. See the Supplementary Materials for the full video sequence.

4. Discussion

Although the video footage recorded in the Kvænangen fjords lacks the resolution required to discern fine-scale details of the tongue and mouth movements of the wild killer whales involved, the sustained frontal facial contact between the individuals, the prolonged duration of the episodes, and the overall behavioural pattern strongly support the interpretation that the observed interaction corresponds to the tongue-nibbling behaviour originally described by Martínez and Klinghammer in 1978 [19] and later documented in detail in a zoological context by Sánchez-Hernández et al. (2019) [7]. These new observations provide empirical support for the hypothesis that tongue-nibbling forms part of the natural behavioural repertoire of *Orcinus orca* and is not a behavioural artefact induced by captivity-related conditions such as stereotypy, aberrant expression or transient cultural novelty.

The reappearance of this behaviour more than three decades after its initial description under human care—first noted shortly after killer whale husbandry was established, and subsequently recorded in detail in 2013—suggests remarkable behavioural continuity across generations in zoological environments. This temporal consistency implies that tongue-nibbling may be a socially conserved behaviour. Furthermore, the fact that it has taken 47 years since its original ethological description to obtain comparable footage from a wild population highlights the considerable difficulty in documenting rare or cryptic behaviours in natural contexts, particularly in highly mobile marine species whose social interactions occur predominantly beneath the water's surface.

Notably, tongue-nibbling in killer whales exhibits significant parallels with recently documented mouth-to-mouth interactions in belugas (*Delphinapterus leucas*) under human care, as reported by Ham et al. (2023) [20]. In both species, these interactions primarily involve younger individuals and appear to serve a clearly affiliative function. As with the behaviour described by Martínez and Klinghammer [19] and later detailed by Sánchez-Hernández et al. [7], the beluga interactions are characterised by gentle, coordinated oral contact without any evident signs of aggression or dominance. This congruence supports the affiliative interpretation advanced by Sánchez-Hernández and colleagues and suggests that oral behaviours may represent socially meaningful interactions in odontocetes.

The cross-species resemblance reinforces the hypothesis that oral contact behaviours in toothed whales may contribute to the development of both social and motor skills during early ontogeny. In this context, tongue-nibbling and similar behaviours may serve as low-conflict mechanisms for strengthening social bonds among conspecifics not yet involved in adult roles such as mating or dominance competition. The recurrence of such behaviours in distantly related taxa, including killer whales and belugas, suggests that affiliative oral interactions may represent a phylogenetically conserved socio-developmental strategy in odontocetes.

The individual in which Martínez and Klinghammer [19] first recorded tongue-nibbling—Hugo—was a juvenile male of approximately 13 years, originating from the Southern Resident killer whale population off Vancouver Island in the Pacific Ocean [21]. In contrast, the individuals observed by Sánchez-Hernández et al. [7] at Loro Parque in 2013 were all born under human care and descended from a mix of lineages: some were of Icelandic (North Atlantic) and Canadian (North Pacific) origin, while one female, Morgan, was a rescued individual from the Norwegian (North Atlantic) population [22]. This diversity of geographic and genetic origins suggests that tongue-nibbling is not population-specific but rather a widely distributed behaviour within the species.

Historically, studies of cetacean social behaviour have relied heavily on surface-based observations from vessels or coastal vantage points [8,9], which has severely constrained the ability to describe the subtleties and complexity of social interactions. These methodological limitations have often led to the grouping of heterogeneous behaviours under broad, catch-all categories such as "socialising" [10–12], thereby obscuring distinctions between specific types of interaction. Only in recent years has progress been made in refining ethological classifications at the surface, with increased attention given to behavioural patterns such as synchronised swimming and physical contact in multiple cetacean species [13–15]. Simultaneously, the use of underwater observation techniques has yielded more detailed behavioural data, facilitating the identification of specific interactions such as petting, rubbing, contact swimming, nibbling, and nudging in dolphins [16,17] and more recently in killer whales [18].

That tongue-nibbling in killer whales—an interaction primarily expressed underwater—remained undescribed in the wild for nearly five decades underscores the indispensable role of subaquatic methodologies in cetacean behavioural research. Omitting the underwater dimension risks underestimating the richness of cetacean sociality and overlooking behaviours that may be central to group cohesion and individual bonding. Continued investment in underwater technologies and observation protocols is therefore vital to developing a more comprehensive and ecologically valid understanding of cetacean social systems.

Although killer whales under human care are unable to exhibit certain behaviours seen in wild populations, such as complex coordinated hunting strategies [23], the sustained occurrence of tongue-nibbling—an apparently affiliative behaviour—across multiple decades in zoological facilities calls into question generalised criticisms regarding the relevance of managed populations in the study of natural social behaviour in cetaceans [24]. It is important, however, to recognise that while the behaviour itself may occur in both settings, the triggers or underlying motivations may differ according to contextual variables. Behavioural responses are shaped by environmental and social factors, including the potential for stress, which is a natural physiological mechanism observable in both captive and wild cetaceans. Thus, behavioural comparisons must consider these contextual influences on expression and causation. These findings support the argument that killer whales, and potentially other cetaceans in human care, may serve as valuable models for investigating naturally occurring social dynamics under controlled, observable conditions.

While this case study provides limited evidence, the long-term preservation of affiliative interactions such as tongue-nibbling in managed settings suggests that narrow portrayals of captive environments—focused solely on aggression or stress-related behaviours—may not capture the full complexity of cetacean sociality [25]. This observation supports the need for further ethological research grounded in detailed behavioural descriptions and a thorough examination of behavioural triggers. A scientifically robust analysis of cetacean sociality in zoological settings must encompass the full behavioural spectrum—including affiliative, sexual, and reconciliatory interactions—as well as the dynamic processes that contribute to social stability [7,18].

Accordingly, evaluating the social complexity of killer whales in managed environments demands methodologically rigorous approaches, grounded in comprehensive ethograms and incorporating both surface and subaquatic observations. Such an integrative framework not only enhances the interpretive resolution of behavioural studies but also allows for a clearer distinction between the observable description of behaviour and the potential triggers behind it. This facilitates meaningful cross-context comparisons and contributes to a more nuanced understanding of cetacean socioecology.

The successful documentation of a rare and cryptic behaviour such as tongue-nibbling in wild killer whales was made possible in part through contributions from citizen science and the increasing global prevalence of recreational in-water cetacean encounters. While it is true that such interactions occasionally yield data of scientific relevance, it is well established that tourism-based activities—such as whale watching and swim-with-cetaceans programmes—may pose significant risks to wild populations. Numerous studies have documented potential adverse effects, including altered behavioural patterns, increased physiological stress, and disruptions to group cohesion [26–30].

It is therefore essential that all wildlife interaction activities comply strictly with established local and international regulations. Moreover, continuous welfare monitoring and robust impact assessments should be mandatory components of such programmes. Balancing the research potential of citizen-generated data with the ethical imperative to minimise anthropogenic disturbance is critical in ensuring both the integrity of behavioural science and the conservation of cetacean populations.

Supplementary Materials: The following supporting information can be downloaded at: https://www.mdpi.com/article/10.3390/oceans6020037/s1.

Author Contributions: J.A. conceptualized the study, wrote the manuscript, analyzed the data, and produced the figures. J.v.V. and D.B. run the citizen science program and participated in the review and editing of the manuscript. All authors have read and agreed to the published version of the manuscript.

Funding: This research received no external funding.

Institutional Review Board Statement: The research in Loro Parque has been approved by the Research Ethics and Animal Welfare Committee of the University of La Laguna. Reg. Num. (CEIBA2024-3515).

Data Availability Statement: The videos of the referenced behaviours are available as Supplementary Materials and can be downloaded from the University of la Laguna: Almunia Portolés, Javier (2025), "Tongue Nibbling in Killer Whales (Orcinus orca)", University of La Laguna, V1, doi (accessed on 5 June 2025): https://doi.org/10.17632/xj2yxc3cwp.1.

Acknowledgments: The authors would like to express their sincere gratitude to the individuals who captured the footage of the observed behaviour: Allison Kelly Estevez and Michael Estevez. We also extend our appreciation to the 95 guests who voluntarily shared their recordings during the Winter Whales of Norway Expeditions, operated by Waterproof Expeditions and led by Wild-Encounters under the direction of Johnny van Vliet and Debbie Bouma, between 22 October and 9 December 2024. Their generous contributions provided invaluable material for this study.

Conflicts of Interest: The authors declare no conflicts of interest.

References

1. Alexander, R.D. The Evolution of Social Behavior. *Annu. Rev. Ecol. Syst.* **1974**, *5*, 325–383. [CrossRef]
2. Eberhard, M.J.W. The Evolution of Social Behavior by Kin Selection. *Q. Rev. Biol.* **1975**, *50*, 1–33. [CrossRef]
3. de Waal, F.B.M.; Veenema, H.C., CHAPTER 2 The First Kiss: Foundations of Conflict Resolution Research in Animals. In *Natural Conflict Resolution*; University of California Press: Berkeley, CA, USA, 2023. [CrossRef]
4. Robbins, M.M.; Robbins, A.M.; Gerald-Steklis, N.; Steklis, H.D. Long-Term Dominance Relationships in Female Mountain Gorillas: Strength, Stability and Determinants of Rank. *Behaviour* **2005**, *142*, 779–809. [CrossRef]
5. Wittemyer, G.; Getz, W.M. Hierarchical dominance structure and social organization in African elephants, *Loxodonta africana*. *Anim. Behav.* **2007**, *73*, 671–681. [CrossRef]
6. Tamaki, N.; Morisaka, T.; Taki, M. Does body contact contribute towards repairing relationships? *Behav. Process.* **2006**, *73*, 209–215. [CrossRef]
7. Sánchez-Hernández, P.; Krasheninnikova, A.; Almunia, J.; Molina-Borja, M. Social interaction analysis in captive orcas (*Orcinus orca*). *Zoo Biol.* **2019**, *38*, 323–333. [CrossRef] [PubMed]
8. Thomsen, F.; Franck, D.; Ford, J.K.B. On the communicative significance of whistles in wild killer whales (*Orcinus orca*). *Naturwissenschaften* **2002**, *89*, 404–407. [CrossRef] [PubMed]
9. Ford, J.K.B. Acoustic behaviour of resident killer whales (*Orcinus orca*) off Vancouver Island, British Columbia. *Can. J. Zool.* **1989**, *67*, 727–745. [CrossRef]
10. Gibson, Q.A.; Mann, J. The size, composition and function of wild bottlenose dolphin (*Tursiops* sp.) mother-calf groups in Shark Bay, Australia. *Anim. Behav.* **2008**, *76*, 389–405. [CrossRef]
11. Parsons, K.M.; Balcomb, K.C.; Ford, J.K.B.; Durban, J.W. The social dynamics of southern resident killer whales and conservation implications for this endangered population. *Anim. Behav.* **2009**, *77*, 963–971. [CrossRef]
12. Marley, S.A.; Kent, C.P.S.; Erbe, C.; Parnum, I.M. Effects of vessel traffic and underwater noise on the movement, behaviour and vocalisations of bottlenose dolphins in an urbanised estuary. *Sci. Rep.* **2017**, *7*, 13437. [CrossRef] [PubMed]
13. Clegg, I.L.; Rödel, H.G.; Delfour, F. Bottlenose dolphins engaging in more social affiliative behaviour judge ambiguous cues more optimistically. *Behav. Brain Res.* **2017**, *322*, 115–122. [CrossRef]
14. Serres, A.; Hao, Y.; Wang, D. Swimming features in captive odontocetes: Indicative of animals' emotional state? *Behav. Process.* **2020**, *170*, 103998. [CrossRef] [PubMed]
15. Christiansen, F.; Lusseau, D.; Stensland, E.; Berggren, P. Effects of tourist boats on the behaviour of Indo-Pacific bottlenose dolphins off the south coast of Zanzibar. *Endanger. Species Res.* **2010**, *11*, 91–99. [CrossRef]
16. Dudzinski, K.M.; Gregg, J.D.; Ribic, C.A.; Kuczaj, S.A. A comparison of pectoral fin contact between two different wild dolphin populations. *Behav. Process.* **2009**, *80*, 182–190. [CrossRef] [PubMed]
17. Delfour, F.; Vaicekauskaite, R.; García-Párraga, D.; Pilenga, C.; Serres, A.; Brasseur, I.; Pascaud, A.; Perlado-Campos, E.; Sánchez-Contreras, G.J.; Baumgartner, K.; et al. Behavioural diversity study in bottlenose dolphin (*Tursiops truncatus*) groups and its implications for welfare assessments. *Animals* **2021**, *11*, 1715. [CrossRef]
18. Hill, H.M.M.; Themelin, M.; Dudzinski, K.M.; Felice, M.; Robeck, T. Individual variation in activity budgets of a stable population of killer whales in managed care across a year. *Behav. Process.* **2025**, *224*, 105135. [CrossRef]
19. Martinez, D.R.; Klinghammer, E. A partial ethogram of the killer whale (*Orcinus orca*). *Carnivore* **1978**, *1*, 13–27.
20. Ham, J.R.; Lilley, M.K.; Wincheski, R.J.; Miranda, J.; Ángel G. Velarde Dediós.; Kolodziej, K.; Pellis, S.M.; Hill, H.M.M. Playful mouth-to-mouth interactions of belugas (*Delphinapterus leucas*) in managed care. *Zoo Biol.* **2023**, *42*, 730–743. [CrossRef]
21. Wearing, S.; Buchmann, A.; Jobberns, C. Free Willy: The whale-watching legacy. *Worldw. Hosp. Tour. Themes* **2011**, *3*, 127–140. [CrossRef]
22. Kremers, D.; Lemasson, A.; Almunia, J.; Wanker, R. Vocal sharing and individual acoustic distinctiveness within a group of captive orcas (*Orcinus orca*). *J. Comp. Psychol.* **2012**, *126*, 433–445. [CrossRef]
23. Williams, R.; Lusseau, D. A killer whale social network is vulnerable to targeted removals. *Biol. Lett.* **2006**, *2*, 497–500. [CrossRef] [PubMed]
24. Corkeron, P. Marine Mammal Captivity, an Evolving Issue. In *Marine Mammals: the Evolving Human Factor. Ethology and Behavioral Ecology of Marine Mammals.*; Springer: Cham, Switzerland, 2022. [CrossRef]
25. Marino, L.; Rose, N.A.; Visser, I.N.; Rally, H.; Ferdowsian, H.; Slootsky, V. The harmful effects of captivity and chronic stress on the well-being of orcas (*Orcinus orca*). *J. Vet. Behav.* **2020**, *35*, 69–82. [CrossRef]
26. Erbe, C. Underwater noise of whale-watching boats and potential effects on killer whales (*Orcinus orca*), based on an acoustic impact model. *Mar. Mammal Sci.* **2002**, *18*, 394–418. [CrossRef]
27. Constantine, R.; Brunton, D.H.; Dennis, T. Dolphin-watching tour boats change bottlenose dolphin (*Tursiops truncatus*) behaviour. *Biol. Conserv.* **2004**, *117*, 299–307. [CrossRef]
28. Schaffar, A.; Madon, B.; Garrigue, C.; Constantine, R. Avoidance of whale watching boats by humpback whales in their main breeding ground in New Caledonia. *Int. Whal. Comm.* **2009**, *SC/61/WW6*, 9.

29. Christiansen, F.; Rasmussen, M.H.; Lusseau, D. Inferring energy expenditure from respiration rates in minke whales to measure the effects of whale watching boat interactions. *J. Exp. Mar. Biol. Ecol.* **2014**, *459*, 96–104. [CrossRef]
30. Oliveira, C.; Pérez-Jorge, S.; Prieto, R.; Cascão, I.; Wensveen, P.J.; Silva, M.A. Exposure to whale watching vessels affects dive ascents and resting behavior in sperm whales. *Front. Mar. Sci.* **2022**, *9*, 1–14. [CrossRef]

oceans

MDPI

Article

Vulnerability of Grey Seal Pups (*Halichoerus grypus*) to Storm Disturbances in the Context of Climate Change: A British Isles Case Study

Keely Saville [1,*], Laetitia Nunny [2], Daniel Jarvis [3], Sue Sayer [4], Laszlo Talas [1] and Mark P. Simmonds [1,2]

[1] Bristol Veterinary School, University of Bristol, Bristol BS40 5DU, UK; laszlo.talas@bristol.ac.uk (L.T.); mark.simmonds@sciencegyre.co.uk (M.P.S.)
[2] OceanCare, P.O. Box 372, CH-8820 Wädenswil, Switzerland; lnunny@oceancare.org
[3] British Divers Marine Life Rescue, Uckfield TN22 1DS, UK; dan@bdmlr.org.uk
[4] Seal Research Trust, Hayle TR27 5AD, UK; sue@sealresearchtrust.co.uk
* Correspondence: keels.saville@gmail.com; Tel.: +44-7895-825772

Abstract: Marine mammals may be particularly vulnerable to climate change. While some climate-change-induced impacts on these species have been recognised, the potential consequences of storminess have been less well-defined, and understanding of its significance largely relies on anecdotal evidence. To quantify the relationship between storminess and its possible impacts on grey seal pups (*Halichoerus grypus*), data from marine wildlife rescue databases and hospitalisation records (2015–2024) within the British Isles were examined (n = 20,686). Daily mean windspeed was used as a proxy for storminess. Significant relationships were found between storminess and pups presenting with malnourishment, head trauma, flipper injury, other wounds, and infections. This study provides the first empirical evidence of the explicit impacts of storminess on seal pup health, with all the presenting conditions increasing and higher rates of call-outs and pup admissions to rescue centres following stormy weather also recorded. Given the predicted increase in storm severity, these findings will aid rescue and rehabilitation planning by helping to predict when rescuers can anticipate increased admissions and the potential injuries that pups may experience following storms. The potential significance to the long-term conservation of the species is also highlighted.

Keywords: climate change; storm casualties; seal pups; pupping phenology; injuries; welfare; conservation

1. Introduction

The consequences of climate change are multifaceted and manifesting in ways that are detrimental to the health of marine ecosystems. Direct impacts of climate change within marine environments include rising sea surface temperature (SST) and sea surface height, and alterations in oceanic stratification [1]. These changes produce a cascade of effects for marine species, resulting in shifts in species abundance, distribution, predator–prey interactions, stranding frequency, and reproductive patterns, as well as increased exposure to algal blooms and disease [1]. Additional associated threats include increased seal disturbance caused by extreme weather events and increased coastal erosion and cliff falls [2].

The combined effects of long-term warming and extreme weather have emerged as a critical focus, particularly in recent Arctic research due to their profound influence on sea ice [3], and have been extensively investigated for various Arctic marine mammal species [4]. Regions other than the Arctic have also been assessed for climate change impacts, although the evidence often remains speculative, with anecdotal evidence and models used to infer effects [5,6]. However, it is now possible to gather empirical evidence of change which can validate and refine such predictions (e.g., [7,8]).

Recent changes in climate, including the impacts of the El Niño–Southern Oscillation (ENSO) and other episodic ocean anomalies, have produced well documented impacts on pinniped populations across the globe [5,9]. The reduction in oceanic upwellings drives changes in the body fat content, abundance, and distribution of seal prey, resulting in altered foraging behaviour and success [10,11]. In turn, observed declines in body condition have been documented in various seal species [5,12–14]. Climate change also impedes pinniped breeding dynamics, with associated shifts in reproductive cycles, shorter breeding seasons, and reduced reproductive outputs [5,15–19]. Cumulatively, these impacts affect survival and may ultimately manifest in declines in pinniped abundance [5,20].

One serious consequence of climate change is the escalating frequency and intensity of storm activity ('storminess') [21,22]. This is hypothesised to have serious implications for both the welfare and conservation of seal pups, although few cases in the published literature provide direct evidence. The early onset of strong autumnal storms in the Bering Sea has been tenuously linked to the alteration of departure-from-site phenology, direction of dispersal, and survival of Northern fur seal pups (*Callorhinus ursinus*) [23]. Similarly, severe storms, alongside the rapid acceleration of sea level rise, in subtropical US waters resulted in unforeseen changes in shoreline morphology with the nursing habitat for endangered Hawaiian monk seals (*Neomonachus schauinslandi*) being adversely affected by inundation [24]. This resulted in a considerable loss of pupping sites, increased pup drownings, and increased predation by Galapagos sharks (*Carcharhinus galapagensis*) [5].

Another species that may be at risk is the Atlantic grey seal (*Halichoerus grypus*), found along the UK's coastline, which hosts approximately 34% of the world's grey seal population [15,20,25,26]. These pinnipeds predominantly utilise beaches and rocky inlets as pupping sites, some of which flood during storms (Figure 1a,b) [27]. There is also evidence that some seal mothers exhibit considerable site fidelity to particular nursery sites [28–30], and this may be a complicating factor in restricting their ability to move to new sites when their original ones become unsuitable.

Since 2010, the northwestern coasts of Europe have experienced a series of powerful storms, with the number of major storms impacting the UK coastlines having tripled [31–34]. The roughening of the wave climate, as a result of increased windspeed from storms, has the potential to increase the severity of wave action on breeding sites, potentially increasing pup injuries and mortality [35–40]. This is supported by anecdotal evidence (Jarvis; Sayer; Howe; Bevington, pers comms/obs.).

Current responses by marine wildlife rescue organisations to injured or stranded pups involve admission to rescue centres or temporary holding facilities (Figure 1c). Such places are scarce across the British Isles. Projected increases in oceanic temperatures and sea levels will intensify storminess across the British Isles in coming decades [20,36,41]. A deeper understanding of the potential impacts of storminess on seal pups will better inform the rescue network and contribute to the protection and conservation of seal populations. Hence, this study aimed to quantify the broad hypothesis that storminess may be adversely impacting seal pup health and survival, and that this might be recognised by considering the frequency of casualties, the nature of these casualties, and the numbers of rescues. In addition, seal reproductive phenology (i.e., the timing of pupping) might be affected, and so this, too, was explored.

Figure 1. (**a**) Mother with her suckling whitecoat pup on Bardsey Island, Wales; (**b**) Grey seal mother with her whitecoat pup on a rocky shore on Bardsey Island, Wales; (**c**) Weaned grey seal pup admitted to Cornish Seal Sanctuary, Cornwall. Credit: (**a**,**b**): Mark P. Simmonds and (**c**): Keely Saville.

2. Materials and Methods

2.1. Data Collection and Processing

In the initial stages of this project, informal interviews were conducted with six relevant experts in seal rescue, rehabilitation, and pathology to help guide the research and identify potential datasets for investigation. Experts were identified from the research literature, online sources, and word of mouth. Representatives of the Seal Research Trust (SRT), Cornish Seal Sanctuary (CSS), British Divers Marine Life Rescue (BDMLR), the Royal Society for the Prevention of Cruelty to Animals (RSPCA), and Cornwall Marine Pathology Team were interviewed in person or online with, in some instances, follow-up questions by email. Where these experts are cited in the text, these quotes have been verified with them and permission was granted for their use. A total of 37 seal rehabilitation centres, organisations, and charities across the British Isles were subsequently contacted and, whilst twenty responses were received, only five were able to provide relevant data for this project. They provided 15,548 individual call-out records, 2644 admission records, and 2494 rescue records, all of which were considered in this investigation.

Grey seals are the most common seal species found around the British Isles [25], and harbour seals (*Phoca vitulina*) were only occasionally recorded in the data considered in this project. Initial data screening involved separating the two species and discarding harbour seal data due to their low representation in the dataset. To isolate species-specific and regional variations in pupping seasonality, the analysis was restricted to the primary pupping periods [42], with data from August to January considered. Based on previous studies, individuals were classified as 'pups' when they were up to 10 months old [40].

Records of 'juvenile' and 'adult' age classes, as well as any individual which was recorded to be older than 10 months, were subsequently discarded from the data. Animals of unknown age were also excluded. The remaining individuals, referred to as 'whitecoats', 'weaned', 'pre-moult', 'mid-moult', 'post-moult', or 'moulter' by rescue centres, are collectively referred to here as 'pups'.

These data were then organised into 4 strands. First, details of pup admissions into Cornish and Irish rescue centres were provided by BDMLR and CSS, and Seal Rescue Ireland (SRI), respectively. The individual 'reasons for rescue' were pooled together into 6 non-mutually exclusive categories (Table 1; Figure 2). When records indicated multiple reasons for an individual pup's admission to the facility, all reported reasons were included in the data analysis. Data on the number of rescued pups admitted into these facilities were subsequently extracted from the data sets, and this comprises the second strand of data. Pupping data makes up the third strand, wherein SRT provided monthly unique newborn pup counts (i.e., pups were only recorded on their first sighting) at sensitive seal sites across Cornwall.

Table 1. Classification of reasons for admission of seal pups to rescue centres where such data were available (i.e., in Cornwall, Ireland, and Suffolk).

Reason for Admission	Subcategories
Malnourishment	Starvation, underweight, lethargy, dehydration, unresponsive, hypoglycaemia, maternal separation.
Public Disturbance	Displaced by the public, public harassment, unadvised public lift, dog attack.
Head Trauma	Broken jaw, ocular ulcers, damaged teeth, mouth infection.
Flipper Injury	Broken limbs, nailbed swelling.
Other Wounds	Propeller wounds, bite wounds, entanglement, unspecified wounding (i.e., "wounds") or wounds on body parts other than head or flippers.
Infection	Gastrointestinal, respiratory, hypothermia, nasal discharge, high body temperature, pyrexia, seal pox, lungworm.

The final strand explored pup stranding data. This was extracted from the BDMLR national call-out database, the Environmental Records Centre for Cornwall and the Isles of Scilly (ERCCIS) database, and SRI hospitalisation records. Data obtained from the same regions across datasets were cross-checked against dates and locations of rescue to avoid any duplication of data. The 58 locations recorded in the call-out database were consolidated into 15 regions based on their geographic proximity.

Meteorological data was sourced by the National Meteorological Library and Archive—Met Office UK. Windspeed is a key factor in predicting storminess impacts and was used as a reliable proxy [22,36,43]. Due to the unavailability of county-wide daily windspeed data (Ball, pers comm.), the average daily windspeed from the nearest operational meteorological weather stations to the coastlines where pup stranding data was obtained was used. To correspond with the pupping data, the daily average windspeed in Cornwall was averaged to produce monthly windspeed values. As part of quality control, certain daily mean windspeeds were recorded as blank in the database (Ball, pers comm.) and these were subsequently excluded from data analysis.

Figure 2. Examples of pup presenting conditions to rescue centres following periods of stormy weather: (**a**) Maternal separation in storm-induced swell; (**b**) Blunt force head trauma; (**c**) Severe mandibular fracture with secondary infection; (**d**) Deep laceration wounding to chest cavity. Credit: (**a**): Sue Sayer, Seal Research Trust and (**b–d**): British Divers Marine Life Rescue.

2.2. Statistical Analyses

All data were recorded and organised using Excel® version 2405 and statistically analysed using RStudio® version 4.1.1.

To examine if time or windspeed affected how many pups were born, a linear regression analysis was performed. Logistic regression analysis was used to assess if wind speed affected why pups were being admitted to rescue facilities, with location as a control factor. Given the increased chance of detecting false positives (i.e., Type I error) due to multiple comparisons here, we applied the Bonferroni correction to p-values. The Bonferroni correction is a common method to adjust the significance threshold of multiple tests by dividing the alpha (here 0.05) by the number of comparisons [44]. To examine if windspeed affected how many pups were rescued, a logistic regression analysis was performed for each location. A similar analysis was also performed on the stranding data.

3. Results

3.1. Pupping

Fitting a linear regression with year and month as predictor variables and the number of pups being born in Cornwall as the outcome variable showed a significant effect (F_{11} = 47.058, p = 1.451 × 10^{-6}) (Figure 3). We found that the number of pups increased by 4.967 each year (SE = 2.2118, p = 0.0388). The effect of month was also significant and pairwise post-hoc analysis (Tukey test) indicated that the number of pups born in September was significantly higher than August (estimate = 73.8, SE = 7.34, p < 0.0001) and October (estimate = 74.2, SE = 7.34, p < 0.0001). There was no significant difference between the number of pups born in August and October (estimate = 0.4, SE = 7.34, p = 0.9984). Neither adding an interaction term between year and month (F_2 = 1.5484, p = 0.2643) nor including monthly average windspeed as a predictor variable (F_1 = 3.3802, p = 0.09583) improve the fit of the model significantly. Birthing data for other regions was not available.

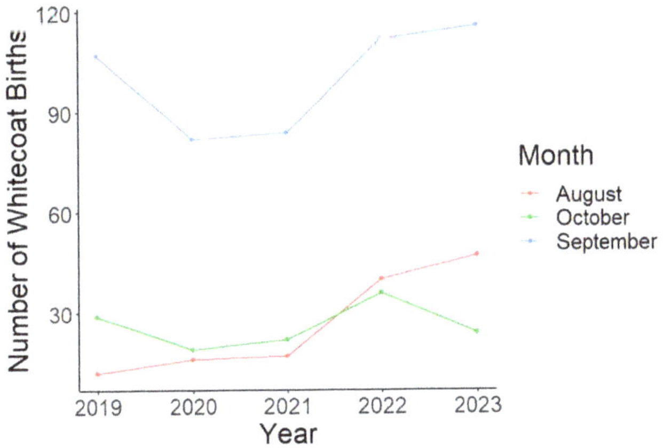

Figure 3. The number of unique whitecoat pups born during the pupping season over time in Cornwall.

3.2. Admission to Rescue Centres

Mixed effects logistic regression models with average windspeed as a fixed variable and location (Cornwall, Ireland, or Suffolk) as a random variable with a common slope and varying intercepts indicated significant associations between average windspeed and several presenting disorders between January 2015 and January 2024 (n = 1569). Pups were significantly more likely to be admitted to rescue centres with malnourishment, head trauma, flipper injury, other wounds, and/or infection as windspeed increased. To account for multiple comparisons, we applied the Bonferroni correction to the resulting p-values (Figure 4; Table 2).

Table 2. Results of mixed effects logistic regressions for each presenting condition of grey seal pups to rescue centres in Cornwall, Ireland, and Suffolk.

Reason for Admission	Estimate	p Value	% Increase per Unit of Windspeed (1 Knot)
Malnourishment	0.05367	1.14 × 10^{-8} *	5.5136
Public Disturbance	−0.009848	0.841	-
Head Trauma	0.03439	0.0482 *	3.4988
Flipper Injury	0.05778	0.000964 *	5.9482
Other Wounds	0.044621	6.05 × 10^{-6} *	4.5631
Infection	0.07317	4.10 × 10^{-9} *	7.5913

* Statistically significant.

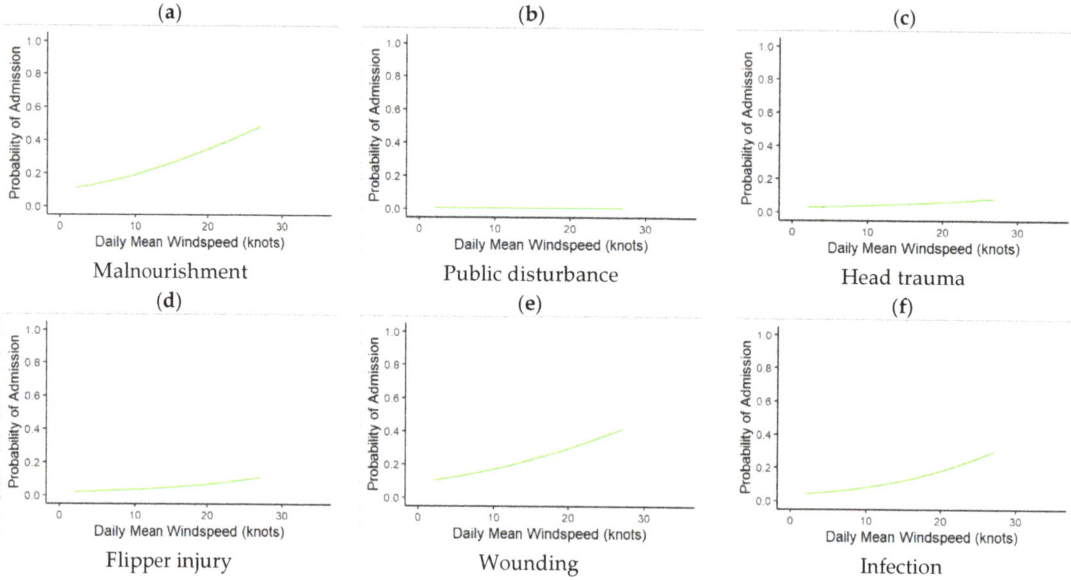

Figure 4. The relationship between storminess and the likelihood of pups presenting with (**a**) malnourishment; (**b**) public disturbance; (**c**) head trauma; (**d**) flipper injury; (**e**) wounding; or (**f**) infection, to rescue centres in Cornwall, Ireland, and Suffolk.

3.3. Rescue

To analyse the effect of windspeed on the number of rescues made for grey seal pups (n = 1511), we first fitted a mixed effects logistic regression model with a fixed effect of windspeed and a random effect of location with a common slope and different intercepts. We found that this model gave an estimate for the effect of windspeed on the number of rescues of 0.048687 (SEM = 0.006004), and this was significantly different from zero (Δdeviance = 65.259, df = 1, $p = 6.569 \times 10^{-16}$).

After fitting logistic regressions for each individual location, we found the windspeed had a significant effect on the chance of rescuing a pup in Cornwall, Ireland, Norfolk, Northumberland, and Suffolk (Figure 5; Table 3).

Table 3. Results of logistic regression models for grey seal pups needing to be rescued across the British Isles between January 2015 and January 2024 at different locations.

Location	Estimate	df	p Value	% Increase per Unit of Windspeed (1 Knot)
Cornwall	0.045904	2341	1.1×10^{-6} *	4.6974
Devon	0.05405	1114	0.265	-
Ireland	0.08352	1074	2.06×10^{-9} *	8.7107
Middlesbrough	0.05627	400	0.109	-
Newcastle	0.02735	1156	0.472	-
Norfolk	0.07678	942	0.00288 *	7.9804
Northumberland	0.07596	585	0.00149 *	7.8919
Scotland	−0.007995	1218	0.719	-
Suffolk	0.15730	1139	1.22×10^{-5} *	17.0347
Wales	−0.07928	1314	0.0911	-
Yorkshire	−0.01603	1327	0.51	-

* Statistically significant.

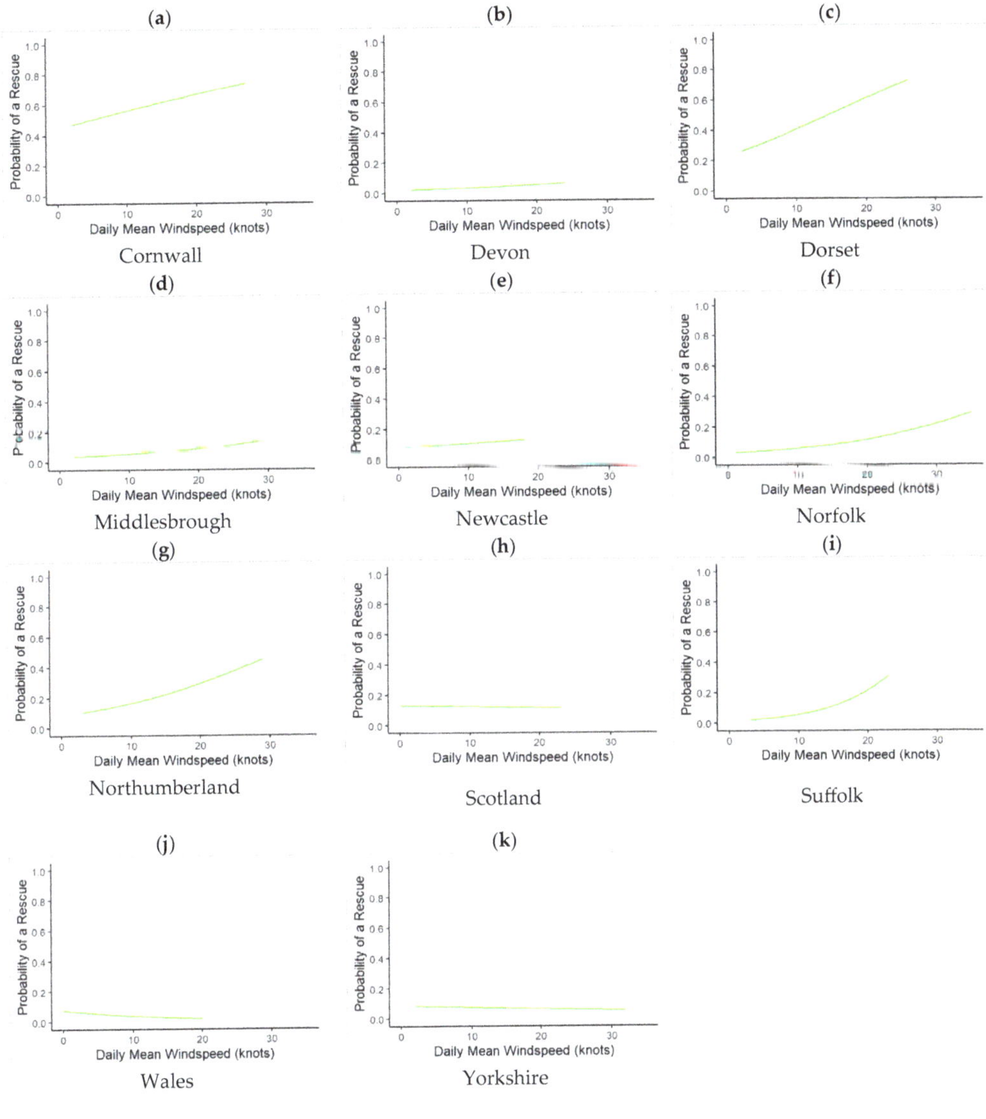

Figure 5. The relationship between storminess and the likelihood of a pup needing to be rescued in (**a**) Cornwall; (**b**) Devon; (**c**) Ireland; (**d**) Middlesbrough; (**e**) Newcastle; (**f**) Norfolk; (**g**) Northumberland; (**h**) Scotland; (**i**) Suffolk; (**j**) Wales; or (**k**) Yorkshire.

3.4. Call-Out

Like the previous analysis, to investigate the effect of windspeed on the number of call-outs made for grey seal pups (n = 10,215), we fitted a mixed effects logistic regression model with a fixed effect of windspeed and a random effect of location with a common slope and different intercepts. This model gave an estimate for the effect of windspeed on the number of call-outs of 0.0429 (SEM = 0.0042), and this was significantly different from zero (Δdeviance = 103.71, df = 1, $p < 2.2 \times 10^{-16}$).

After performing analyses for each location, we found the windspeed had a significant effect on the chance of getting a call-out in Cornwall, Ireland, Middlesbrough, Newcastle, Norfolk, and Northumberland (Figure 6, Table 4).

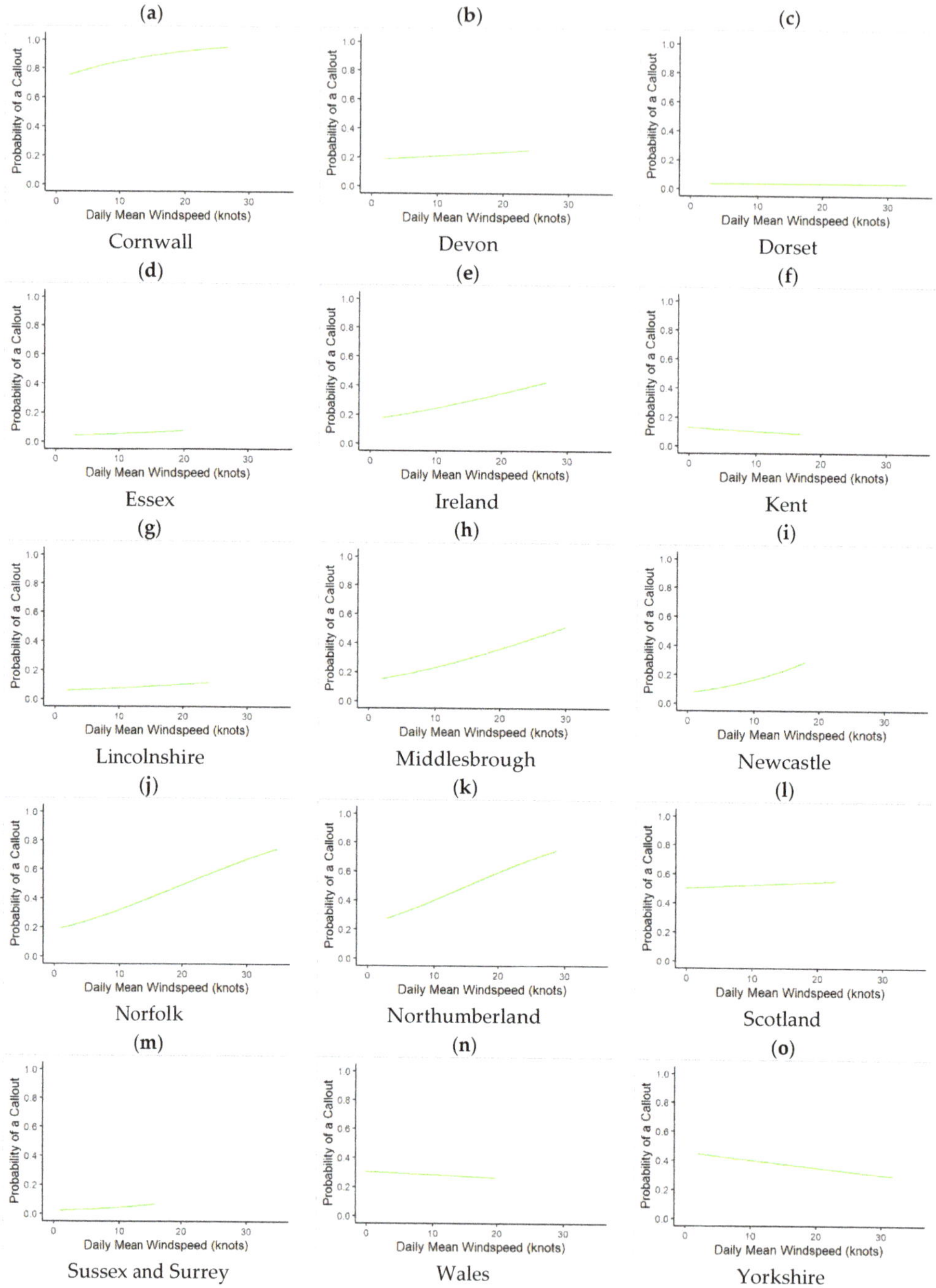

Figure 6. The relationship between storminess and the likelihood of a pup stranded in (**a**) Cornwall; (**b**) Devon; (**c**) Dorset; (**d**) Essex; (**e**) Ireland; (**f**) Kent; (**g**) Lincolnshire; (**h**) Middlesbrough; (**i**) Newcastle; (**j**) Norfolk; (**k**) Northumberland; (**l**) Scotland; (**m**) Sussex and Surrey; (**n**) Wales; or (**o**) Yorkshire.

Table 4. Results of logistic regression models for call-outs made for grey seal pups in locations across the British Isles between January 2015 and January 2024 at different locations.

Location	Estimate	df	p Value	% Increase per Unit of Windspeed (1 Knot)
Cornwall	0.07363	4122	1.37×10^{-12} *	7.6408
Devon	0.01622	1156	0.428	-
Dorset	−0.001522	765	0.961	-
Essex	0.03603	948	0.385	-
Ireland	0.4911	1822	1.82×10^{-5} *	5.0336
Kent	−0.02683	975	0.455	-
Lincolnshire	0.03286	958	0.307	-
Middlesbrough	0.06226	429	0.000974 *	6.4228
Newcastle	0.08986	1177	0.00419 *	9.4021
Norfolk	0.07325	1356	6.82×10^{-9} *	7.6
Northumberland	0.07908	1326	1.16×10^{-8} *	8.2291
Scotland	0.008445	1416	0.523	-
Sussex and Surrey	0.07151	948	0.236	-
Wales	−0.01004	1244	0.627	-
Yorkshire	−0.02199	1313	0.0917	-

* Statistically significant.

4. Discussion

4.1. Pupping

Pupping data indicated that September had a significantly higher whitecoat pup count in Cornwall, which is a month earlier than previous survey findings, and this supports the idea that pups may be being born earlier in the year than previously [45–49]. However, these results do not suggest that storminess is a reliable predictor of key parameters describing the timing of the pupping season. Hence, the data do not appear to support the hypothesis that storminess drives phenological advancements in pupping timing [46–49]. Previous studies have predominantly identified SST as the most influential proximate factor affecting pinniped reproductive phenology [47,50].

However, one previous study has noted that the pupping period of Australian fur seals (*Arctocephalus pusillus*) synchronises with winter zonal winds in the Bass Strait region [51], and a shift in peak birthing dates has also been described in harbour seals in the North Sea [49–51]. Peak pupping for grey seals in Cornwall was in October up until 2016 [45,52], but since then, has been in September, as shown in these data (Figure 3), as well as in the Report to Natural England 2020 [53]. The observed increase in births in August and reduced number of births in October, shown in Figure 3, support a change in the timing of pupping and suggest that continued monitoring is needed. This new situation also coincides with peak tourism time, creating increased overlap in interactions between humans and seals.

A trend of older females breeding earlier in the breeding season has been observed in pinnipeds [47]. Research has shown that mothers who give birth early in the season tend to nurse their pups longer, resulting in a higher likelihood of successful weaning [47,54,55]. It is plausible that older seal mothers are beginning to pup earlier as a response to mitigate the risk of pup loss or displacement during the latter part of the pupping season, which is characterised by a higher frequency of storms [39]. This shift in reproductive timing could have significant implications for population resilience in the context of climate change and, indeed, storms [5,47]. Therefore, maternal age should be considered in future analyses.

Female grey seals typically begin pupping at six years old and continue to reproduce annually for life, contingent upon sufficient body condition to sustain gestation [56,57]. One photo-identified grey seal along the Cornish coastline has reportedly given birth annually since 2003, with the exception of 2018. This interruption was suspected to be related to the extreme storm events, Brian and Ophelia, which occurred during the 2017 pupping season (Jarvis and Sayer, pers obs.). She lost her 2017 pup most likely as a result of the storms. One possibility was that her hormones were then out of sync for mating, resulting in a lack of oocyte fertilisation. Another possibility is that she had not regained condition at the time of implantation, so the 2018 fertilised oocyte did not implant.

4.2. *Admission to Rescue Centres*

Our results demonstrate that the likelihood of each presenting condition increases with windspeed, providing support for the notion that within the British Isles, the primary causes of mortality in seal pups are starvation, infection, and trauma [39], with storminess acting as a significant contributing factor. These findings further indicate an increased incidence of wounding during stormy weather.

Additionally, from our discussions with rescue centres, it became apparent that some were already overwhelmed at certain times of year by the numbers of pups needing treatment and that this may mean that some pups had to be euthanised.

4.2.1. Malnourishment

Storm surges have previously been correlated to high incidences of mother–pup separation [58–61], which prevents pups from feeding. This finding aligns with the observed relationship between increasing windspeed and malnourishment. This can lead to stranding due to malnutrition and starvation unless pups are rescued. Similarly, disturbances during nursing, such as those caused by storms, can disrupt necessary suckling periods, leading to insufficient body weight at weaning [62–64]. This can result in inadequate blubber insulation and a high surface-to-volume ratio, which increases metabolic demands [65].

One result of storminess is increased turbidity, and this may make it more difficult for the seals to locate prey using their senses of sight and 'wake-sensing' [66].

4.2.2. Public Disturbance

There is no apparent relationship between storminess and admissions related to human disturbance (Figure 4b). However, there has been a surge in public interest in marine wildlife and coastal land, sea, and air access in recent years that has increased pinniped disturbance, although the physiological consequences have not been defined [5,67,68]. Human disturbance may also be under-reported as it may not always be witnessed, and it may also be less prevalent in bad weather.

4.2.3. Head Trauma

The increased prevalence of head trauma in seal pups may be attributed to being thrown against underwater, offshore, or cliff rocks during storms [69]. Additionally, in the absence of their mothers, pups have been observed attempting to suckle on other seals or rocks (Sayer and Saville, pers obvs.). This latter behaviour could contribute to head trauma and subsequent dental damage [53].

4.2.4. Flipper Injury

Our findings, which show that flipper injuries increase with windspeed, support a previous study suggesting that the majority of flipper injuries and fractures are sustained during extreme weather [70]. Moreover, these datasets revealed numerous seal rescues occurred inland, possibly indicating that pups are seeking refuge from storm-inundated areas or being washed inland [36]. The increased incidence of flipper injuries during stormy weather may be attributed at least in part to prolonged movements across haul-outs on uneven land surface whilst searching for their mother, leading to injury and friction abrasions. Seals might also be forced to haul out on more marginal sites which offer more chance of harm, such as sharp rocky shores or public beaches.

4.2.5. Wounding

The influence of sea conditions later in the pupping season aligns with the potential for resource limitation (e.g., food and habitat resources) and intraspecific competition [47]. There is also an increased likelihood of seals overcrowding on haul-outs during high tides and stormy weather [68], which may help to explain the higher prevalence of bites and the significant relationship between storms and wounds observed in these results.

High windspeeds contribute to a rougher wave condition, often resulting in substantial coastal erosion [21,71]. The morphology and structural landscape of a pupping site can influence a pup's vulnerability to storm effects [36]. It can therefore be anticipated that physical injuries, such as wounds, will be more common along rocky shorelines during storms, as pups may be dashed against the rocks or get hit by stones falling from cliffs (Achilleos and Sayer, pers comms/obs.).

4.2.6. Infection

Poor nutrition may increase a pup's susceptibility to infection through decreased energy resources and exhaustion resulting in a weaker immune response [72]. This might result from altered food sources for weaned pups, as well as separation from their mothers during the nursing period for maternally dependent ones [60,72]. The observed significant positive relationship between windspeed and pups presenting with infection supports this hypothesis. For example, there has also been a marked increase in the prevalence of lungworm in pinnipeds examined in Cornwall in recent years (Barnett, pers comm.). While numerous factors may contribute to this trend, storminess could be a potential contributory factor perhaps by creating additional stress, thereby depressing the seals' immune systems, causing more serious infections (Barnett, pers comm.). Further research is necessary to quantify the impacts of storms on disease transmission in these populations.

4.3. Rescue and Call-Out

Anecdotally, a strong correlation exists between stormy weather and seal strandings (Jarvis; Thompson, pers comms.). The significant relationship found in our results between windspeed and number of rescues confirms this relationship. This is supported by the observed increase in serious injuries during stormy weather, indicating that as storminess increases, seals with more severe injuries are more likely to require rescuing.

The observed variation in call-outs between locations (Figure 6) might be partially explained by an increase in the number of pups along the coast (Howe, pers comm.). Future studies should incorporate abundance estimates to better understand the significance of trends. Additionally, standardising data collection methods across rescue centres would improve data consistency and comparability, enabling more robust analyses of injury patterns. Establishing universal protocols for recording injury type, severity, and environmental conditions at the time of rescue would help create a more comprehensive dataset, ultimately enhancing our ability to assess the impact of extreme weather events on seal populations. Refining storm impact modelling by integrating standardised rescue data with environmental variables could further improve predictions of weather-related injury risk.

4.4. Caveats

While windspeed is regarded as a reliable proxy for storminess [22,36,44], several authors have argued that a combination of strong winds and other climatic variables, such as low atmospheric pressure, is necessary to propagate storm surges [33,71]. The approach used here is limited to exploring a single variable (i.e., windspeed), and, in future research, incorporating multiple variables or developing a "storminess index" may provide a more accurate representation of stormy weather. From the perspective of the seal pup on the shore, the conditions that it is faced with will be a combination of factors including tidal height, wave and wind direction, and local topographical features.

In 2023, a particularly stormy year [73], a greater proportion of pups were observed swimming, exploring, and interacting with their mothers in the water [27]. This led to an alternative hypothesis suggesting that this exposure could potentially make them more resilient to poor wave conditions during storms and better able to survive. Generally speaking, though, younger pups are likely to be more vulnerable to being swept away.

Numerous studies have demonstrated that variations in the frequency of casualties can be attributed to differences in reporting effort [74,75]. Increased public awareness of seal rescue facilities is likely a contributing factor to improvements in data collection and reporting efforts, as indicated by a rise in public reports (Howe, pers comm.). However, this contradicts a recent report which recorded the lowest number of dead marine mammal strandings in August—a month associated with high tourist footfall [76]. This therefore suggests that the increased number of pup casualties was unlikely to be an artefact of increased awareness and reporting.

5. Conclusions

This project provides the first statistically supported evidence that storms are adversely impacting grey seal pups.

While data quality and recording limitations may have affected the ability to predict the influence of storminess on call-outs and pupping phenology, the results strongly suggest that grey seal pups are likely to experience increased welfare issues and reduced survival as a result of the anticipated larger magnitude storm events in British Isles waters in the coming decades. The site fidelity of mothers may be a significant factor contributing to the vulnerability of pups to severe weather changes.

Seals are exposed to the cumulative impacts of multiple stressors, some of which are increasing and being exacerbated by climate change. Our findings, given the predicted increase in seal pup casualties, reinforce the urgent need for an expansion of seal rehabilitation facilities in strategic locations with adequate long-term resourcing. This will complement the existing rescue network and alleviate the current and increasing strain on facilities, and will help improve the welfare of grey seals in the light of climate change.

Noting the importance of UK waters for this species, and the threat that increased storminess poses, conservation actions should be taken that will build resilience in the population—for example, ensuring that nursery sites are not affected by other negative impacts such as disturbance and plastic pollution. The research presented in this paper will hopefully serve as a catalyst for further investigation using a wider range of climatic variables to define storminess and encourage continued reporting efforts by rescue facilities. Continued research is essential to gain a comprehensive understanding of the vulnerability of grey seal pups in the British Isles, and seals more generally, to storm disturbances. In the long term, this may have significant conservation implications, particularly if storm-induced population losses exceed populations' capacity to maintain themselves.

Author Contributions: K.S.: conceptualisation, methodology, formal analysis, investigation, writing—original draft preparation, visualization; L.N.: conceptualisation, writing—review and editing; D.J. and S.S.: data curation, writing—review and editing; L.T.: formal analysis, writing—review and editing; M.P.S.: conceptualisation, methodology, writing—reviewing and editing, visualisation, supervision. All authors have read and agreed to the published version of the manuscript.

Funding: OceanCare provided support for data analysis and field work. Field work was also supported by a grant from IMarEST to K.S.

Institutional Review Board Statement: The study was conducted in accordance with the Declaration of Helsinki and approved by the Institutional Ethics Committee of University of Bristol Health Sciences under ethics approval code 18629 on 17 May 2024.

Informed Consent Statement: Informed consent was obtained from all subjects involved in the study. Written informed consent has also been obtained from the contributors to publish this paper.

Data Availability Statement: The raw data supporting the conclusions of this article will be made available by the authors on request.

Acknowledgments: The authors express their sincere gratitude to James Barnett, Anne VanDomburg, Ruth Williams, and Evangelos Achilleos for their generous contributions of time and expertise, which were instrumental in providing critical insights into this research. The authors also thank BDMLR and SRT for providing valuable photographic materials. Consultation with Duncan Ball at the Met Office was invaluable to the project and he, too, is thanked for his input and advice. Thanks to the Institute of Marine Engineering, Science and Technology (IMarEST) and OceanCare for their generous support in the form of grants, which funded travel expenses and writing up.

Conflicts of Interest: The authors declare no conflicts of interest. The funders had no role in the design of the study; in the collection, analyses, or interpretation of data; in the writing of the manuscript; or in the decision to publish the results.

References

1. Doney, S.C.; Ruckelshaus, M.; Emmett Duffy, J.; Barry, J.P.; Chan, F.; English, C.A.; Galindo, H.M.; Grebmeier, J.M.; Hollowed, A.B.; Knowlton, N.; et al. Climate Change Impacts on Marine Ecosystems. *Ann. Rev. Mar. Sci.* **2012**, *4*, 11–37. [CrossRef] [PubMed]
2. Hiscock, K.; Earll, B. *South-West Marine Ecosystems in 2022 (The State of South-West Seas)*; Marine Biological Association of the UK: Plymouth, UK, 2023; Volume 56.
3. Kovacs, K.M.; Lydersen, C.; Overland, J.E.; Moore, S.E. Impacts of changing sea-ice conditions on Arctic marine mammals. *Mar. Biodivers.* **2011**, *41*, 181–194. [CrossRef]
4. International Whaling Commission. *Report of the IWC Workshop on Impacts of Increased Marine Activities on Cetaceans in the Arctic*; International Whaling Commission: Anchorage, AK, USA, 2014.
5. Gulland, F.M.; Baker, J.D.; Howe, M.; LaBrecque, E.; Leach, L.; Moore, S.E.; Reeves, R.R.; Thomas, P.O. A review of climate change effects on marine mammals in United States waters: Past predictions, observed impacts, current research and conservation imperatives. *Clim. Change Ecol.* **2022**, *3*, 100054. [CrossRef]
6. Townhill, B.; Couce, E.; Lynam, C.; Pinnegar, J. *In Modelling Technical Report "Investigating Climate Change Resilience of Vulnerable Marine Species Around the UK [InCResiVul]"*; Centre for Environment Fisheries & Aquaculture Science (Cefas): Suffolk, UK, 2024.
7. Boveng, P.L.; Bengtson, J.L.; Cameron, M.F.; Dahle, S.P.; Logerwell, E.A.; London, J.M.; Overland, J.E.; Sterling, J.T.; Stevenson, D.E.; Taylor, B.L.; et al. Status review of the ribbon seal (*Histriophoca fasciata*). In *NOAA Technical Memorandum NMFS-AFSC*; U.S. Department of Commerce: Washington, DC, USA, 2013; Volume 255, p. 174.
8. Cameron, M.F.; Frost, K.J.; Ver Hoef, J.M.; Breed, G.A.; Whiting, A.V.; Goodwin, J.; Boveng, P.L. Habitat selection and seasonal movements of young bearded seals (*Erignathus barbatus*) in the Bering Sea. *PLoS ONE* **2018**, *13*, e0192743. [CrossRef]
9. Le Boeuf, B.J.; Crocker, D.E. Ocean climate and seal condition. *BMC Biol.* **2005**, *28*, 9. [CrossRef]
10. Pinnegar, J.; Townhill, B.; Lynam, C.; Chaniotis, P. *Investigating Climate Change Resilience of Vulnerable Marine Species Around the UK. Report of Project: BX041 Stakeholder Workshop held on 28th October 2020*; Centre for Environment, Fisheries and Aquaculture Science: Suffolk, UK, 2020.
11. Harwood, J. Marine mammals and their environment in the twenty-first century. *J. Mammal.* **2001**, *82*, 630–640. [CrossRef]
12. Ukkonen, P.; Aaris-Sørensen, K.; Arppe, L.; Daugnora, L.; Halkka, A.; Lõugas, L.; Oinonen, M.J.; Pilot, M.; Storå, J. An Arctic seal in temperate waters: History of the ringed seal (*Pusa hispida*) in the Baltic Sea and its adaptation to the changing environment. *Holocene* **2014**, *24*, 1694–1706. [CrossRef]
13. Schmölcke, U. Holocene environmental changes and the seal (Phocidae) fauna of the Baltic Sea: Coming, going and staying. *Mammal. Rev.* **2008**, *38*, 231–246. [CrossRef]
14. Boveng, P.L.; Ziel, H.L.; McClintock, B.T.; Cameron, M.F. Body condition of phocid seals during a period of rapid environmental change in the Bering Sea and Aleutian Islands, Alaska. *Deep. Sea Res. Part. II Top. Stud. Oceanogr.* **2020**, *181*, 104904. [CrossRef]

15. Blanchet, M.A.; Vincent, C.; Womble, J.N.; Steingass, S.M.; Desportes, G. Harbour seals: Population structure, status, and threats in a rapidly changing environment. *Oceans* **2021**, *2*, 41–63. [CrossRef]
16. Calandrino, G.; Pendleton, D.; Staudinger, M. *A Review of Literature for Gray and Harbor Seals*; Northeast Climate Adaptation Science Center, University of Massachusetts: Amherst, MA, USA, 2018.
17. Martin, E.; Banga, R.; Taylor, N.L. Climate change impacts on marine mammals around the UK and Ireland. *MCCIP Sci. Rev.* **2023**, *22*, 1–22.
18. Boehme, L.; Thompson, D.; Fedak, M.; Bowen, D.; Hammill, M.O.; Stenson, G.B. How many seals were there? The global shelf loss during the last glacial maximum and its effect on the size and distribution of grey seal populations. *PLoS ONE* **2012**, *7*, e53000. [CrossRef] [PubMed]
19. Abt, K. Shift of harbour seal (Phoca vitulina) pupping phenology: A result of climate change? In Proceedings of the European Cetacean Society, Istanbul, Turkey, 2–4 March 2009.
20. SCOS (Special Committee on Seals). *Scientific Advice on Matters Related to the Management of Seal Populations: 2022*; University of St Andrews: St. Andrews, UK; Sea Mammal Research Unit: St. Andrews, UK; Natural Environment Research Council: Swindon, UK, 2022.
21. Met Office. Recent Trends and Future Projections of UK Storm Activity. Available online: https://www.metoffice.gov.uk/research/news/2021/recent-trends-and-future-projections-of-uk-storm-activity (accessed on 11 January 2024).
22. Wood, M.; Haigh, I.D.; Le, Q.Q.; Nguyen, H.N.; Tran, H.B.; Darby, S.E.; Marsh, R.; Skliris, N.; Hirschi, J.J.M.; Nicholls, R.J.; et al. Climate-induced storminess forces major increases in future storm surge hazard in the South China Sea region. *Nat. Hazards Earth Syst. Sci.* **2023**, *23*, 2475–2504. [CrossRef]
23. Lea, M.A.; Johnson, D.; Ream, R.; Sterling, J.; Melin, S.; Gelatt, T. Extreme weather events influence dispersal of naive northern fur seals. *Biol. Lett.* **2009**, *5*, 252–257. [CrossRef]
24. Reynolds, M.H.; Berkowitz, P.; Courtot, K.N.; Krause, C.M. *Predicting Sea-Level Rise Vulnerability of Terrestrial Habitat and Wildlife of the Northwestern Hawaiian Islands*; (No. 2012-1182); US Geological Survey: Reston, VA, USA, 2012.
25. Thompson, D.; Duck, C.; Lonergan, M.E. The status of harbour seals (*Phoca vitulina*) in the United Kingdom. *NAMMCO Sci. Publ.* **2010**, *8*, 117–127. [CrossRef]
26. Carter, M.I.; Boehme, L.; Duck, C.D.; Grecian, J.; Hastie, G.D.; McConnell, B.J.; Miller, D.L.; Morris, C.; Moss, S.; Thompson, D.; et al. *Habitat-Based Predictions of at-Sea Distribution for Grey and Harbour Seals in the British Isles*; Report to BEIS, OESEA-16-7/OESEA-17-78; Sea Mammal Rescue Unit, University of St Andrews: St. Andrews, UK, 2020.
27. Büche, B.; Bond, S. *Grey Seal Breeding Census, Skomer Island 2023*; NRW Evidence Report No. 750; The Wildlife Trust of South and West Wales, Natural Resources Wales: Cardiff, UK, 2023.
28. Cordes, L.S.; Thompson, P.M. Mark-resight estimates of seasonal variation in harbor seal abundance and site fidelity. *Popul. Ecol.* **2015**, *57*, 467–472. [CrossRef]
29. Pomeroy, P.P.; Anderson, S.S.; Twiss, S.D.; McConnell, B.J. Dispersion and site fidelity of breeding female grey seals (*Halichoerus grypus*) on North Rona, Scotland. *J. Zool.* **1994**, *233*, 429–447. [CrossRef]
30. Pomeroy, P.P.; Twiss, S.D.; Redman, P. Philopatry, site fidelity and local kin associations within grey seal breeding colonies. *Ethology* **2000**, *106*, 899–919. [CrossRef]
31. Intergovernmental Panel on Climate Change (IPCC). Summary for Policymakers. In *IPCC Special Report on the Ocean and Cryosphere in a Changing Climate*; IPCC: Cambridge, UK, 2019.
32. Kew, S.; McCarthy, M.; Ryan, C.; Pirret, J.S.R.; Murtagh, E.; Vahlberg, M.; Amankona, A.; Pope, J.O.; Lott, F.; Claydon, O.; et al. *Autumn and Winter Storms over UK and Ireland are Becoming Wetter Due to Climate Change*; Imperial College London: London, UK, 2024.
33. Allan, R. Impacts of Climate Change on Storminess. In *Executive Summary*; Meteorological Office: Exeter, UK, 2021.
34. Met Office. *The Recent Storms and Floods in the UK*; Centre for Ecology and Hydrology: Exeter, UK, 2014.
35. Chatzimentor, A.; Doxa, A.; Katsanevakis, S.; Mazaris, A.D. Are Mediterranean marine threatened species at high risk by climate change? *Glob. Change Biol.* **2023**, *29*, 1809–1821. [CrossRef]
36. SCOS (Special Committee on Seals). *Scientific Advice on Matters Related to the Management of Seal Populations*; University of St Andrews: St. Andrews, UK; Natural Environment Research Council, Special Committee on Seals: Swindon, UK, 2021.
37. Büche, B.; Stubbings, E. *Grey Seal Breeding Census Skomer Island*; NRW Evidence Report No. 252; The Wildlife Trust of South and West Wales: Cardiff, UK, 2017.
38. Wilkie, N.; Zbijewska, S. *Grey Seal Breeding Census Skomer Island*; NRW Evidence Report No. 339; The Wildlife Trust of South and West Wales: Cardiff, UK, 2019.
39. Begbie, B.P.; Kenny, D.; Howe, L. *Calf of Man Seal Survey*; Manx Wildlife Trust: Isle of Man, UK, 2023.
40. Silpa, M.A.; Thornton, S.M.; Cooper, T.; Hedley, J. Prevalence of presenting conditions in grey seal pups (*Halichoerus grypus*) admitted for rehabilitation. *Vet. Sci.* **2015**, *2*, 1–11. [CrossRef]

41. Bricheno, L.M.; Amies, J.D.; Chowdhury, P.; Woolf, D.; Timmermans, B. Climate change impacts on storms and waves relevant to the UK and Ireland. MCCIP Science Review. *Storms Waves* **2023**, *2*, 2.
42. Russell, D.J.; Morris, C.D.; Duck, C.D.; Thompson, D.; Hiby, L. Monitoring long-term changes in UK grey seal pup production. *Aquat. Conserv. Mar. Freshw. Ecosyst.* **2019**, *29*, 24–39. [CrossRef]
43. Martínez-Alvarado, O.; Gray, S.L.; Hart, N.C.; Clark, P.A.; Hodges, K.; Roberts, M.J. Increased wind risk from sting-jet windstorms with climate change. *Environ. Res. Lett.* **2018**, *13*, 044002. [CrossRef]
44. Dunn, O.J. Multiple Comparisons Among Means. *J. Am. Stat. Assoc.* **1961**, *56*, 52–64. [CrossRef]
45. Sayer, S.; Witt, M. *Special Area of Conservation Condition Monitoring: Grey Seals (Halichoerus grypus) in the Isles of Scilly During the 2016 Pupping Season*; Report Commissioned for Natural England (NECR261); Natural England: York, UK, 2018; ISBN 978-1-78354-509-4.
46. Osinga, N.; Ferdous, M.S.; Morick, D.; Hartmann, M.G.; Ulloa, J.A.; Vedder, L.; de Haes, H.U.; Brakefield, P.M.; Osterhaus, A.D.M.E.; Kuiken, T. Patterns of stranding and mortality in common seals (*Phoca vitulina*) and grey seals (*Halichoerus grypus*) in The Netherlands between 1979 and 2008. *J. Comp. Pathol.* **2012**, *147*, 550–565. [CrossRef] [PubMed]
47. Bull, J.C.; Jones, O.R.; Börger, L.; Franconi, N.; Banga, R.; Lock, K.; Stringell, T.B. Climate causes shifts in grey seal phenology by modifying age structure. *Proc. R. Soc. B* **2021**, *288*, 20212284. [CrossRef] [PubMed]
48. Rogers, S.I.; Greenaway, B. A UK perspective on the development of marine ecosystem indicators. *Mar. Pollut. Bull.* **2005**, *50*, 9–19. [CrossRef]
49. Osinga, N.; Pen, I.; De Haes, H.U.; Brakefield, P.M. Evidence for a progressively earlier pupping season of the common seal (*Phoca vitulina*) in the Wadden Sea. *J. Mar. Biol. Assoc. UK* **2012**, *92*, 1663–1668. [CrossRef]
50. Boyd, I.L. Environmental and physiological factors controlling the reproductive cycles of pinnipeds. *Can. J. Zool.* **1991**, *69*, 1135–1148. [CrossRef]
51. Geeson, J.J.; Hobday, A.J.; Speakman, C.N.; Arnould, J.P. Environmental influences on breeding biology and pup production in Australian fur seals. *R. Soc. Open Sci.* **2022**, *9*, 211399. [CrossRef]
52. Sayer, S.; Hockley, K.; Witt, M.J.; Godley, B.J. *Monitoring Grey Seals (Halichoerus grypus) in the Isles of Scilly During the 2010 Pupping Season*; Natural England Commissioned Reports (No. 103); Natural England: York, UK, 2010.
53. Sayer, S.; Millward, S.; Witt, M. *Monitoring Grey Seal (Halichoerus grypus) Pupping Sites in Cornwall 2019/2020*; Natural England Commissioned Reports (PO 10070016704); Cornwall Seal Research Group Trust: York, UK; Natural England: York, UK; The University of Exeter: Exeter, UK, 2020.
54. Cordes, L.S.; Thompson, P.M. Variation in breeding phenology provides insights into drivers of long-term population change in harbour seals. *Proc. R. Soc. B Biol. Sci.* **2013**, *280*, 20130847.
55. Pomeroy, P.P.; Fedak, M.A.; Rothery, P.; Anderson, S. Consequences of maternal size for reproductive expenditure and pupping success of grey seals at North Rona, Scotland. *J. Anim. Ecol.* **1999**, *68*, 235–253.
56. Bowen, W.D.; Iverson, S.J.; McMillan, J.I.; Boness, D.J. Reproductive performance in grey seals: Age-related improvement and senescence in a capital breeder. *J. Anim. Ecol.* **2006**, *75*, 1340–1351. [CrossRef] [PubMed]
57. Bowen, W.D.; den Heyer, C.E.; Lang, S.L.; Lidgard, D.; Iverson, S.J. Exploring causal components of plasticity in grey seal birthdates: Effects of intrinsic traits, demography, and climate. *Ecol. Evol.* **2020**, *10*, 11507–11522. [PubMed]
58. Allen, S.J.; Bowen, W.D.; den Heyer, C.E. Birth-site habitat selection in gray seals (*Halichoerus grypus*): Effects of maternal age and parity and association with offspring weaning mass. *Mar. Mammal. Sci.* **2022**, *38*, 349–363.
59. Colegrove, K.M.; Greig, D.J.; Gulland, F.M. Causes of live strandings of northern elephant seals (*Mirounga angustirostris*) and Pacific harbor seals (*Phoca vitulina*) along the central California coast, 1992–2001. *Aquat. Mamm.* **2005**, *31*, 1–10.
60. Riedman, M.L.; Le Boeuf, B.J. Mother-pup separation and adoption in northern elephant seals. *Behav. Ecol. Sociobiol.* **1982**, *11*, 203–215.
61. Boness, D.J.; Bowen, D.; Iverson, S.J.; Oftedal, O.T. Influence of storms and maternal size on mother–pup separations and fostering in the harbor seal, Phoca vitulina. *Can. J. Zool.* **1992**, *70*, 1640–1644.
62. Osinga, N.; Hart, P. Harbour seals (*Phoca vitulina*) and rehabilitation. *NAMMCO Sci. Publ.* **2010**, *8*, 355–372.
63. Van der Zande, A.N.; Van Alphen, J.J.M.; Goodman, S.J.; Meijboom, F.L.B.; Stegeman, A.J.; Thompson, D.; Kuindersma, W.; Latour, J.B. *Advice of the Scientific Advisory Committee on Seal Rehabilitation in the Netherlands*; Wageningen Environmental Research; Wageningen University and Research: Wageningen, The Netherlands, 2018.
64. Sauve, C.C.; Van de Walle, J.; Hammill, M.O.; Arnould, J.P.; Beauplet, G. Stomach temperature records reveal nursing behaviour and transition to solid food consumption in an unweaned mammal, the harbour seal pup (*Phoca vitulina*). *PLoS ONE* **2014**, *9*, e90329.
65. Coltman, D.W.; Bowen, W.D.; Wright, J.M. Birth weight and neonatal survival of harbour seal pups are positively correlated with genetic variation measured by microsatellites. *Proc. R. Soc. Lond. Ser. B Biol. Sci.* **1998**, *265*, 803–809. [CrossRef] [PubMed]
66. Zheng, X.; Kamat, A.M.; Cao, M.; Kottapalli, A.G.P. Creating underwater vision through wavy whiskers: A review of the flow-sensing mechanisms and biometric potential of seal whiskers. *J. R. Soc. Interface* **2021**, *18*, 20210629. [CrossRef]

67. Islam, M.Z. Ocean wildlife and megafauna protection. In *The Blue Book: Smart Sustainable Coastal Cities and Blue Growth Strategies for Marine and Maritime Environments*; Springer International Publishing: Berlin/Heidelberg, Germany, 2024; pp. 15–33.
68. Kelly, C.; Glegg, G.A.; Speedie, C.D. Management of marine wildlife disturbance. *Ocean. Coast. Manag.* **2004**, *47*, 1–19. [CrossRef]
69. Baker, J.R. Mortality and morbidity in grey seal pups (*Halichoerus grypus*). Studies on its causes, effects of environment, the nature and sources of infectious agents and the immunological status of pups. *J. Zool.* **1984**, *203*, 23–48. [CrossRef]
70. Hespel, A.M.; Bernard, F.; Davies, N.J.; Huuskonen, V.; Skelly, C.; David, F. Surgical repair of a tibial fracture in a two-week-old grey seal (*Halichoerus grypus*). *Vet. Comp. Orthop. Traumatol.* **2013**, *26*, 82–87. [CrossRef]
71. Spencer, T.; Brooks, S.M.; Evans, B.R.; Tempest, J.A.; Möller, I. Southern North Sea storm surge event of 5 December 2013: Water levels, waves and coastal impacts. *Earth-Sci. Rev.* **2015**, *146*, 120–145. [CrossRef]
72. Gershwin, M.E.; Borchers, A.T.; Keen, C.L. Phenotypic and functional considerations in the evaluation of immunity in nutritionally compromised hosts. *J. Infect. Dis.* **2000**, *182*, S108–S114. [CrossRef]
73. Met Office. UK Storm Centre. Available online: https://www.metoffice.gov.uk/weather/warnings-and-advice/uk-storm-centre/index (accessed on 23 August 2024).
74. Haverkamp, H.; Chang, H.Y.; Newcomb, E.; Doughty, L.; Walk, D.; Seton, R.; Jones, L.S.; Todd, S.; Cammen, K.M. A retrospective socio-ecological analysis of seal strandings in the Gulf of Maine. *Mar. Mammal. Sci.* **2023**, *39*, 232–250. [CrossRef]
75. Olson, J.K.; Aschoff, J.; Goble, A.; Larson, S.; Gaydos, J.K. Maximizing surveillance through spatial characterization of marine mammal stranding hot spots. *Mar. Mammal. Sci.* **2020**, *36*, 1083–1096. [CrossRef]
76. Crosby, A.; Hawtrey-Collier, A.; Trew, S. *Marine Strandings in Cornwall and Isles of Scilly: 2022 Annual Report*; Cornwall Wildlife Trust: Truro, UK, 2022.

oceans

Article

Following the Food: Dynamic, Seasonal Changes in the Fine-Scale Distribution of Foraging Minke Whales Within a Scottish Marine Protected Area (MPA)

Duncan A. I. MacDougall [1,*] and Kevin P. Robinson [1,2,*]

1 Cetacean Research & Rescue Unit (CRRU), Banff AB45 3YQ, UK
2 Centre for Ecology & Conservation, University of Exeter, Cornwall TR10 9FE, UK
* Correspondence: duncan.macdougall@crru.org.uk (D.A.I.M.); kev.robinson@crru.org.uk (K.P.R.)

Abstract: Environmental heterogeneity is especially important in determining the distribution and spatial management of marine mammals. Intra-annual changes in distribution exhibited by highly mobile species such as baleen whales, however, present a challenge to traditional area-based management measures which should be accounted for in the designations, but these data are typically lacking. In the present study, we investigated the seasonal variables influencing the spatio-temporal distribution of feeding/foraging minke whales in the Southern Trench MPA in northeast Scotland. A presence–absence model was selected to determine the associations of feeding/foraging whales with areas of high prey density and other environmental determinants. Whale presence was strongly correlated with high burrowed sandeel density (BSD) in May and June and offshore thermal fronts (derived from the standard deviation of sea-surface temperature (SST SD)) from June to September. Both were concluded to be valuable proxies for the distribution of available prey and provided a compelling explanation for observed spatio-temporal shifts and high intraannual variability of whales from our long-term data. The present findings illustrate the value of prey data inclusion in habitat models for baleen whales on their feeding grounds, and advocate for a more dynamic, ecosystem-based approach to management for these highly mobile protected whales.

Keywords: presence–absence modelling; seasonal distribution; habitat preferences; *Balaenoptera acutorostrata*; predator–prey relationships; dynamic management

1. Introduction

Marine habitats are heterogenous over space and time leading to the non-random distribution of both predators and prey alike. This environmental heterogeneity is especially important when considering spatial management options for marine mammals, since the success of efforts to conserve protected populations relies on a clear understanding of the underlying factors driving their distributional changes [1].

In area-based management strategies for cetaceans, site selection is primarily structured upon fixed, physical features of the marine environment [2]. Highly sloped areas of the seabed, for example, promote upwellings in nutrient-rich waters, which heighten local productivity and aggregate targeted prey [3,4], whilst sea bottom sediments may provide integral habitats for benthic and demersal fish quarry [5]. Distribution in cetaceans, however, is also often linked to oceanographic variables, such as eddies, frontal systems, sea surface temperature (SST) and surface chlorophyll-*a* concentrations [6–8], which positively favour ecosystem richness and the availability of prey. Consequently, the relationship between predator–prey dynamics and the influence of the marine environment is highly complex and likely the result of compound interactions between topographic, oceanographic and even anthropogenic factors over multiple scales, which presents challenges to area-based management [9–11].

Whilst the mapping of important areas from spatial sightings is a crucial first step in the design of area-based management for cetaceans, temporal changes in the distribution of target species are rarely taken into account [12]. For example, many baleen whales exhibit such changes during their migrations from high latitude feeding grounds to low latitude breeding grounds [13], but also within feeding areas [14–18]. Thus, area-based management measures for such highly mobile species should account for these distributional changes that may limit the efficacy of traditional static MPAs, but these data are often lacking. There is subsequently a need to integrate the underlying behavioural factors driving these changes into spatial models in priority areas to better understand the distributional trends observed [19].

The minke whale (*Balaenoptera acutorostrata*) has recently been categorised as a 'Priority Marine Feature' by the Scottish Government [20] and is considered to be an important ecological component of the marine ecosystem in northeast Scotland [21,22]. The highly productive waters of the Moray Firth (57°41′ N, 2°40′ W) afford rich, inshore feeding grounds for these coastal baleen whales during the summer and autumnal months [21–24] where they are known to principally target sandeels (*Ammodytes marinus*), sprat (*Sprattus sprattus*) and herring (*Clupea harengus*) through both active corralling and more passive bird-associated feeding behaviour [21,22,24]. In 2020, the Southern Trench, along the southern coastline of the outer Moray Firth (Figure 1), was designated as a Marine Protected Area (MPA) for the protection of the species. Within the Southern Trench MPA, minke whales have previously been known to associate with ocean floor topography, water depth, seabed classification (according to sedimentary characteristics) and sea surface temperature [21,25]. Additionally, prey switching and seasonal dietary plasticity [22,26], as well as habitat partitioning by age-class [22], may explain the high inter-/intra-annual variability in observed sightings, but the likely changes in distribution as a result are not yet understood.

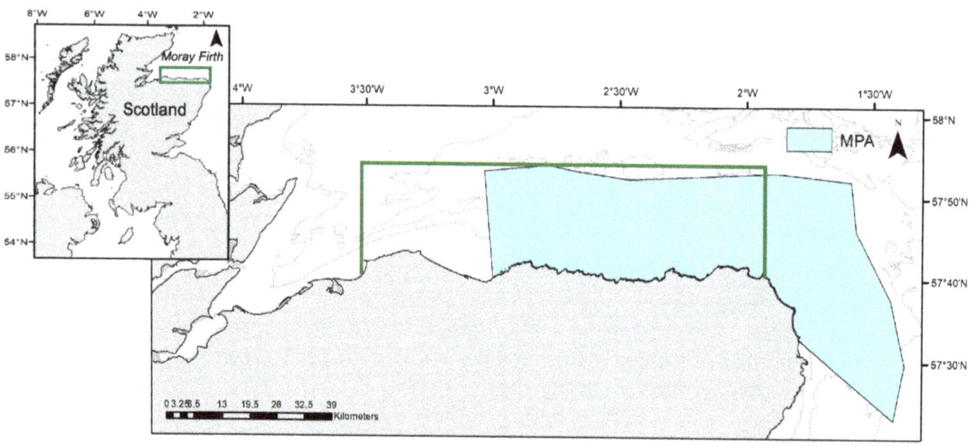

Figure 1. The location of the Moray Firth study area (green border) and the boundaries of the Southern Trench MPA along the southern coastline of the outer firth.

Studies addressing temporal shifts in the distribution of highly mobile species are necessary to inform improved area-based management of their important habitats (e.g., [27]), providing the biological rationale to proactively shift, extend or retract protective management boundaries. Thus, in the present study, we investigated the temporal changes influencing the seasonal distribution of feeding/foraging minke whales within the Southern Trench MPA, to better understand their intra-annual movements and identify seasonally important habitat features giving rise to feeding opportunities. A presence–absence model was employed to determine temporally dynamic associations of feeding/foraging whales with areas of high prey density and frontal systems, as well as associations with other environmental variables. The aim of the study was to identify features of biological importance within this recently established MPA, in order to inform and advance the adaptive management process for this and other Scottish MPAs also including minke whales as a priority feature.

2. Materials and Methods

2.1. Data Collection and Processing

Sightings data were collected during dedicated boat-based surveys within a 1980 km^2 area of the southern Moray Firth (Figure 1) between May and October, from 2009 to 2022 inclusive. Surveys were conducted using an 8 metre rigid-hulled inflatable boat (RHIB), operated at speeds of 7 to 10 knots during favourable sea conditions ($\leq$3 Beaufort Sea State, swells of less than 1 m and visibility $\geq$ 1.5 km) and with 4 to 8 trained observers (for full methodology, see Robinson et al. [21]). From a total of 24,640 km of survey effort, 570 whale sightings were recorded. Since the aim of the present study was to uncover the intra-annual associations by minke whales with biologically important environmental characteristics driven by feeding behaviour, only feeding/foraging animals (n = 458) were used in the following analyses. This is because non feeding/foraging animals that may be travelling between foraging habitats are not likely to be associated with the same conditions as actively feeding/foraging animals [28].

All spatial analyses were carried out using QGIS v3.28.1 [29]. Spatial layers were projected to the British National Grid Transverse Mercator coordinate reference system, so that spatial distances could be interpreted in metric units. A presence–absence approach was employed to produce a dependent variable for the probability of feeding/foraging whale presence. Since the study aimed to incorporate monthly and interannual variation, this required up to six presence–absence grids (one for every month surveyed) for each study year, 2009 to 2022, respectively, resulting in a total of 73 grids with a cell size of 2 km^2 set to capture fine-scale environmental variability. Therein, all grid cells containing positive search effort with at least one whale sighting were categorised as 'presence' cells, whereas those cells containing positive effort but with zero sightings were categorised as 'absence' cells.

In accordance with previous modelling studies of foraging cetaceans, where absence cells adjacent to presence cells are typically established as false-absences [30], each sighting location in the following analyses was assigned a 2 km buffer area (also in-line with the estimated foraging range typically utilised by minke whales in the study area of 1 to 2 km (KP Robinson, unpublished data [31])). In addition, survey track line layers were provided with a 1.5 km buffer area as a steadfast spotting distance for detecting whales during boat-based survey work.

A rectangular raster grid of the study area was created, measuring 553 by 159 cells, with a resolution of 200 m^2 per cell. Mean depth values were obtained using bathymetry data derived from Admiralty charts of the Moray Firth and the slope layer (in degrees) was derived from these depth data. Monthly composite images of sea surface temperature (SST) (°C) and chlorophyll-a concentration (CHL) (mg m^{-3}) were provided by the Natural Environment Research Council (NERC) Earth Observation Data Acquisition and Analysis Service (NEODAAS), at 1 km resolution (0.01 degrees).

Since frontal areas typically exhibit large SST gradients, the standard deviation of SST (SST SD) was used to quantify thermal fronts [32,33]. SST SD was calculated for each survey month using the *r.neighbors* tool from the GRASS 7 GIS core plugin, at the same resolution and coverage as SST. A 7 by 7 cell neighbourhood matrix was chosen to reflect the wider general area around the fronts, as well as the fronts themselves.

Direct variables for prey distribution and quality (e.g., prey density) are typically strong predictors of marine predator occurrence [34–39], but such data are usually scarce [35]. Sandeels, however, are highly spatially restricted by the availability of suitable burrowing sediment [40] and thus, their distribution, even at fine scales, is predictable [41]. For the present study, spatial estimates of burrowed lesser sandeel density (BSD) were available from Marine Scotland at a resolution of 200 m (under Open Government Licence v 3.0) [41].

All GIS data layers were subsequently overlayed with the corresponding cells in the presence–absence grid. The mean value for all variables in each cell was calculated for each presence–absence grid cell accordingly. The grids with data sampled for all environmental and prey layers were subsequently collated for the following analyses.

2.2. Statistical Analysis

All statistical analyses were carried out using RStudio v2023.12.1+402 [42]. The collinearity between explanatory variables, including physiographic (depth and slope), oceanographic (SST, SST SD and CHL), ecological (BSD) and temporal (month, year) variables, was examined using the 'pairs' function in RStudio.

A generalised additive model (GAM) was selected in the subsequent analyses since several of the relationships between presence/absence and the explanatory variables were expected to be non-linear. GAM functions were accessed using the *mgcv* package in RStudio [43] and low-rank thin plate smoothing splines were applied to depth, sea surface temperature (SST) and slope in the initial model, with the maximum number of smoothing basis functions (k) set at 4 for each of these smooth terms, to prevent overfitting. Tensor product interactions were included for both BSD and SST SD, interacting with month, with k set to 6 for these interactions. Smooth functions of the main effects (SST SD, BSD and month) were further included, with k set to 4 for SST SD and BSD and to 6 for month (May to October). Year was included as a linear predictor to capture interannual variation in the probability of minke whale presence.

The GAM family was set as 'binomial' using a logit link function. Restriction maximum likelihood (REML) was selected for smoothness selection to prevent overfitting, whilst concurvity amongst model smooth terms was assessed by looking at worst-case concurvity, as evaluated by the *concurvity* function in the *mgcv* package. Due to strong concurvity (concurvity > 0.8) being detected between SST SD and the tensor product interaction between SST SD and month, the smooth main effects for month and SST SD were removed from the model and a full tensor product smooth between SST SD and month was replaced. This may be preferable to simply discarding the main effects as the full tensor product smooth retains the main smooth effects within the term whilst using fewer smoothing parameters. However, a subsequent concurvity check indicated there was still high concurvity between the SST smooth and the full tensor product smooth interaction (>0.8). Since seasonal associations with fronts were expected to be more directly important for minke whales than SST, the SST smooth term was therefore removed from the model and replaced with a linear SST term. To check that the number of smoothing basis functions, k, was set at an appropriate level, checks were carried out using the *gam.check* function in *mgcv*. These indicated that k was set appropriately for all the smooth model terms except for depth. Subsequently, the number of basis functions for depth was increased to 8, with checks indicating this was an appropriate k value for depth as indicated by high p-values ($p > 0.05$) and effective degrees of freedom that were not close to the upper limits set by k. A final check indicated that there were no further issues of concurvity further to these changes (worst-case concurvity < 0.7).

2.3. Model Validation and Selection

To assess whether the model used met the assumptions of a GAM, model validation techniques were applied to examine the residuals. The assumption of homogeneity of residual variance was evaluated by plotting the mean residuals against all model explanatory variables. Due to the binary nature of the response variable (presence versus absence), however, the standard residual plots were difficult to interpret and so 'binned' residuals versus explanatory variables were created and examined using the *arm* package (version 1.12-2) [44] to display the mean residuals for discrete levels (bins) of the continuous explanatory variables. To assess whether the model had met the assumptions of temporal and spatial independence, mean residuals of the final model were plotted against the temporal variables (month and year) and spatial variables (latitude and longitude) x, respectively. Model selection was conducted with comparison of the Akaike's information criterion (AIC) between the full model and alternative models, where non-significant predictors were removed from the model until all remaining predictors were significant. The full model was selected unless differences in AIC between the full and any alternative models was greater than 2, in which case the model with the lowest AIC was selected accordingly.

3. Results

3.1. Sightings Data

The spatial distribution of feeding/foraging minke whales encountered in the study area between May and October 2009 to 2022 is shown in Figure 2. Whales were encountered throughout the study area but were more generally distributed towards the central and eastern areas of the study site, with a greater number of encounters offshore. Sightings were recorded from May to October inclusive (Figure 3a), but considerable variation was observed in the number of sightings from one month and from one year to the next (Figure 3a,b, respectively), with the highest number of sightings being recorded in 2018 and 2021 (Figure 3b).

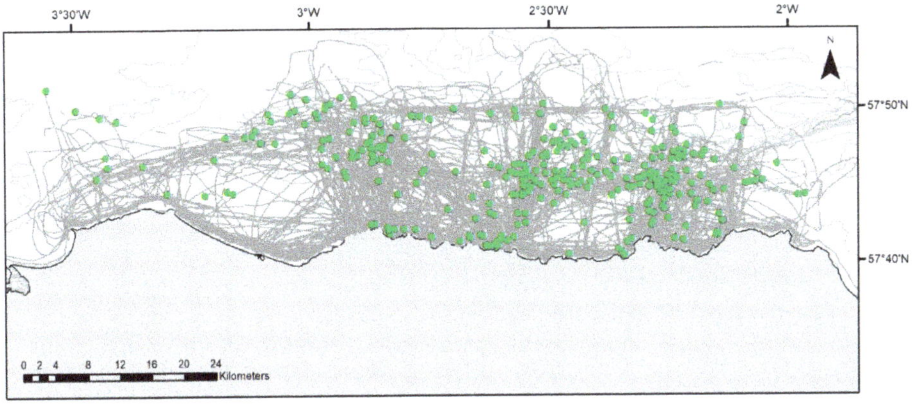

Figure 2. Showing the spatial distribution of feeding/foraging minke whales (n = 458) recorded during dedicated boat-based survey work (track lines in grey) along the southern coastline of the outer Moray Firth from May to October 2009 to 2022 inclusive.

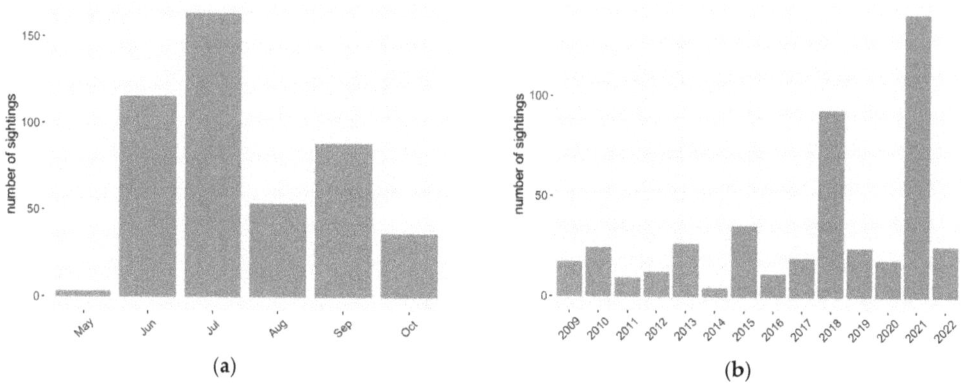

Figure 3. Showing (**a**) the pooled intra-annual and (**b**) inter-annual variability in the sightings of feeding/foraging minke whales using the southern Moray Firth study area from May to October 2009 to 2022 inclusive.

3.2. Model Selection and Outputs

Since differences between the full and alternative GAMs were minor (Δ AIC < 2; Table 1), the full model was selected for the following analyses. The selected model explained 21.6% of the deviance and the mean residuals displayed no notable pattern when plotted against temporal (i.e., month and year) or spatial (i.e., latitude and longitude) variables, indicating that the assumptions of temporal and spatial independence in the residuals had not been violated. Furthermore, there were no apparent patterns in the residuals when plotted against the explanatory variables and since residual variance was similar, the assumption of homogeneity of variance in the residuals by the model had also been adequately met.

Table 1. Comparison of the full and alternative generalised additive models using Akaike's information criterion (AIC). In the model equations, *n* denotes the probability of presence, *ti* denotes a tensor product interaction, *te* denotes a full tensor product smooth and *s* denotes a smooth function of a covariate.

Model	Equation	AIC
Full	$n \sim ti(\mathrm{BSD} : \mathrm{month}) + te(\mathrm{SSTSD} \times \mathrm{month}) + s(\mathrm{depth}) + s(\mathrm{CHL}) + s(\mathrm{BSD}) + s(\mathrm{slope}) + \mathrm{SST} + \mathrm{year}$	3104.575
Alternative 1	$n \sim ti(\mathrm{BSD} : \mathrm{month}) + te(\mathrm{SSTSD} \times \mathrm{month}) + s(\mathrm{depth}) + s(\mathrm{CHL}) + s(\mathrm{BSD}) + \mathrm{SST} + \mathrm{year}$	3102.965
Alternative 2	$n \sim ti(\mathrm{BSD} : \mathrm{month}) + te(\mathrm{SSTSD} \times \mathrm{month}) + s(\mathrm{depth}) + s(\mathrm{CHL}) + s(\mathrm{BSD}) + \mathrm{year}$	3103.725

The tensor product interaction (*ti*) between burrowed sandeel density (BSD) and month was found to have a strong significant effect (Table 2), with feeding/foraging minke whale presence being strongly associated with high BSD levels during May and June but declining in the consecutive months thereafter. From May to July, feeding/foraging minke whale presence was much lower at low sandeel densities than at high sandeel densities but was similar in August and considerably higher at low BSD than at high BSD in September and especially October (Table 2; Figure 4a). The tensor product smooth (*te*) between sea surface temperature standard deviation (SST SD) and month was also found to have a very strong significant effect, with feeding/foraging whale presence between June and August being strongly influenced by high SST SDs (i.e., thermal fronts), although apparently also to a lesser degree at low SST SDs during the same period. By September and October, feeding/foraging whale presence very steeply declined, particularly at high SST SDs (Table 2, Figure 4b).

Table 2. Test statistics of smooth and parametric predictor terms from the final GAM outputs for the presence of feeding/foraging minke whales.

Smooth Terms	Effective Degrees of Freedom	Reference Degrees of Freedom	Chi-Square	*p*
BSD	1.611	1.988	8.390	<0.05 *
ti (BSD, month)	4.448	5.376	25.177	<0.001 *
te (SST-SD, month)	14.050	16.666	78.470	<0.001 *
Depth	5.777	6.522	133.689	<0.001 *
CHL	1.020	1.039	10.070	<0.05 *
Slope	1.709	2.088	1.113	0.537
Parametric Terms	**Estimate**	**Standard Error**	**Z**	***p***
SST	−0.196	0.130	−1.504	0.133
Year (2009 as reference)				
2010	0.515	0.292	1.766	0.077
2011	0.130	0.377	0.035	0.972
2012	−0.235	0.357	−0.658	0.510
2013	0.651	0.278	2.341	<0.05 *
2014	−0.251	0.530	−0.474	0.636
2015	0.597	0.299	1.995	<0.05 *
2016	0.595	0.286	2.081	<0.05 *
2017	0.358	0.292	1.227	0.220
2018	1.962	0.252	7.793	<0.001 *
2019	0.143	0.304	0.472	0.637
2020	1.046	0.302	3.465	<0.001 *
2021	2.449	0.251	9.749	<0.001 *
2022	0.987	0.268	3.688	<0.001 *

* $p < 0.05$.

The main smooth effect of BSD had a weak but significant positive effect upon the presence of feeding/foraging minke whales (Table 2; Figure 5a). Other environmental variables, including depth and chlorophyll-a concentration (CHL), were also found to be significant predictors of the distribution of feeding/foraging whales (Table 2). Feeding/foraging minke whale presence was much more probable in water depths greater than 40 metres (Figure 5b). The smooth term for CHL showed a negative significant relationship with whale presence (Figure 5c), however the data were very limited at higher values and so interpretation of these results at these higher values should be considered conservatively. There was also significant interannual variation in feeding/foraging whale presence from year to year (Table 2). Conversely, the smooth terms for slope and the linear term for sea surface temperature (SST) had no significant effect upon minke whale presence (Table 2).

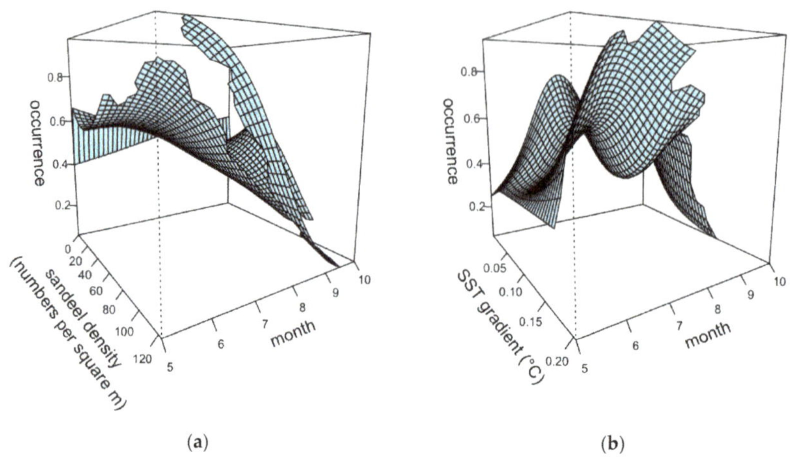

Figure 4. Partial effect plots showing the effect of (**a**) the tensor product interaction (*ti*) between BSD and month and (**b**) the tensor product smooth (*te*) between SST SD and month upon the presence of feeding/foraging minke whales.

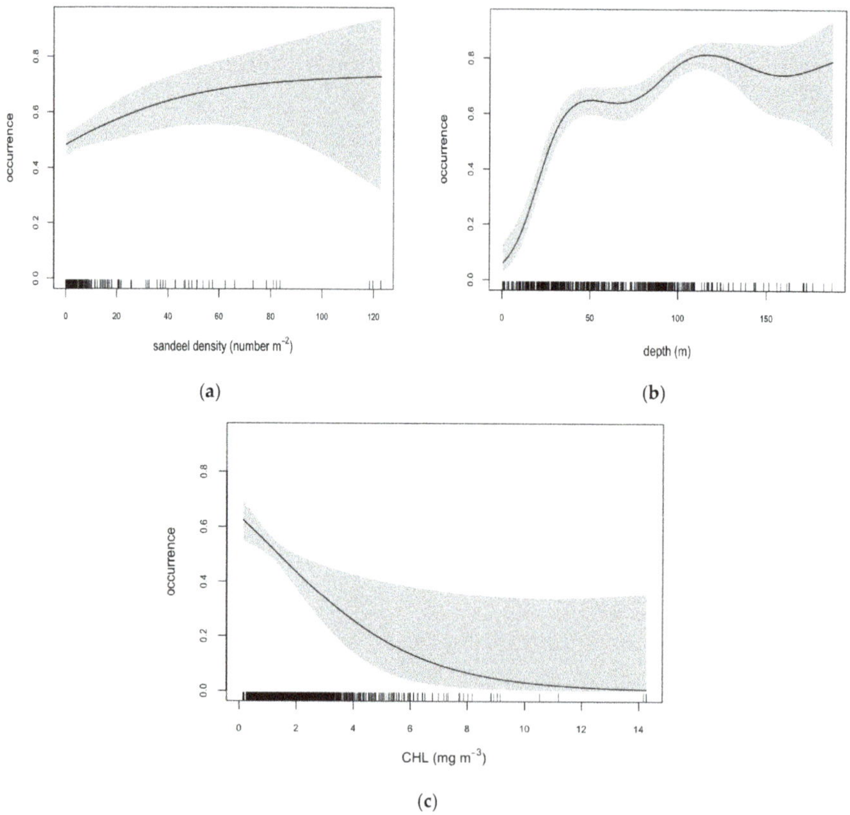

Figure 5. Partial effects plots from GAMs showing the component effect of the smooth terms for (**a**) BSD; (**b**) depth; (**c**) CHL. The confidence intervals take account of errors in the intercept estimation.

4. Discussion

The present results corroborate the importance of considering temporal changes in species distribution in the design of protected area frameworks for highly mobile species [9,11,27,45] such as minke whales. Whilst both physiographic and oceanographic environmental features can be informative for the spatial management of cetaceans, this study additionally illustrates the added importance of including prey data in habitat models for migratory whales on their feeding grounds. From May to July, feeding/foraging whales were found to show a strong preference for high-density sandeel burrowing habitat within the Moray Firth study area. However, from June through to August, the whales showed a marked proclivity for offshore frontal areas instead. In addition to environmental predictors such as depth and chlorophyll-a concentration (CHL), burrowed sandeel density (BSD) and sea surface temperature standard deviation (SST SD) were therefore concluded to provide valuable proxies for important minke whale feeding/foraging habitat and a compelling explanation for the observed spatio-temporal shifts in distribution over the summer feeding. These results further support previous studies suggesting that the temporal occurrence of these whales reflects the seasonal availability of their prey [22,25,46,47].

In Scottish waters, minke whales principally feed on sandeels, sprat and herring [48], which together contribute up to 86% of the total fish biomass in this North Sea coastal region [49]. Sandeels typically occur in the water column from April to July [50,51], but from July to August they enter their burrowing phase [50], thereby reducing predation opportunities for foraging minke whales. However, with the emergence of juvenile herring in the Moray Firth from July, followed by the inshore movements of pre-wintering sprat from August to October—as verified from monthly environmental DNA sampling in the study area [26]—whales are able to actively target these clupeid prey items with the concurrent decline in available sandeels.

The distributional changes observed in feeding/foraging minke whales in the present study—initially from high BSD areas then subsequently to frontal areas—subsequently supports seasonal prey switching by these coastal balaenopterids, as inferred from previous studies [22,26]. Predators may be flexible in their targeting of different prey species when their choices are limited or when a higher quality resource becomes more available [52] and this is assumed to occur when prey densities exceed a specific threshold that makes prey-switching more energetically efficient [22,46]. Nevertheless, the progressive decline observed herein in whale presence in high-density sandeel areas suggests that any dietary transition is gradual and more generalist foraging may occur throughout June and July when both sandeel and clupeid prey are equally available.

In contrast to sandeels, which by nature of their ecology remain bound to their habitat year-round [40], herring and sprat are highly migratory, pelagic species that are known to aggregate in areas of high zooplankton abundance [46,53,54], which proliferate along dynamic frontal systems and the transitional zones adjacent to these fronts. Frontal systems cause subtidal currents along predominant isobaths [55], serving as biological 'hot spots', where significant changes are observed in the pelagic food web due to enhanced vertical mixing [56]. Accordingly, the strong association of feeding/foraging whales with thermal fronts established in this study suggests the importance of these hotspots for clupeid prey and could partially explain the high interannual variability in the occurrence of whales observed within the Moray Firth study area.

The present results endorse the importance of utilising prey data for the interpretation of baleen whale occurrence on their feeding grounds. However, one obvious limitation to the use of BSD data is that these data do not necessarily reflect the sandeel densities in the water column. Although sandeels typically remain within a few kilometres of their burrows [40], the water column versus burrowed densities will be somewhat misaligned at fine spatial scales. With this in mind, the 2 km presence–absence cell size was selected to retain the fine spatial scale aspect of the study whilst providing a wider buffer, in comparison to a 1 km cell size for example, for extrapolation. Additionally, the sighting location buffers likely accounted for much of the wider area over which sighted whales were potentially foraging, and subsequently the wider area occupied by sandeels in the water column around their suitable burrowing habitat. For instance, a whale that was foraging for sandeels that was sighted within a low sandeel density grid cell adjacent to a high sandeel density grid cell would not necessarily be associated with low sandeel density in the model, as the buffered sighting locations result in adjacent cells containing higher values being considered as 'presence' cells. Thus, the results of this study indicate that BSD provides a convenient proxy for water column sandeel availability, supporting modelled prey data as a useful alternative to in situ prey data. However, given that the BSD main effect alone was a far less powerful predictor of whale presence than the tensor product interaction for BSD with month in the present study, the importance of incorporating a temporal component in such analyses may be essential to elucidate any seasonal trends.

In the absence of more accurate satellite-derived frontal data, SST gradients, such as SST SD, evidently provide a useful proxy for thermal fronts. However, SST SD and other gradient-based proxies may fail to capture detailed information about an animal's position in terms of its proximate distance or placement to the front. For example, areas directly adjacent to fronts may have heightened SST SDs, since the neighbourhood method used herein to quantify fronts might confound the boundaries between thermal fronts and adjacent mixed or stratified areas when applied at fine scales. However, transitional zones further from the fronts may have lower SST SDs as the neighbourhood matrix effect will be lower, resulting in reduced SST gradients. Thus, the peak in minke whale presence recorded during the summer months at low SST SDs might indicate an association with transitional zones (supporting prey resources) where SST gradients will be lower. Additionally, due to increases in winds and falling SSTs during September and October [57,58], SST gradients will likely be lower during these periods [59], which might explain the sheer decline in the association between presence and SST SD in this period. Thus, direct, satellite-derived frontal data would be preferable to more precisely identify oceanographic fronts and associated stratified and mixed zones.

Management measures for highly mobile species remain challenging, but insights into the dynamic spatial associations by species and their prey with environmental features are essential for the identification of important habitats that should be afforded protection against prevailing threats. Anthropogenic impacts such as overfishing, for example, may result in prey depletion and/or habitat loss [60]. A recent advancement in this area was the introduction of the Sandeel (Prohibition of Fishing) (Scotland) Order 2024, prohibiting sandeel fishing within the Scottish zone, which will not only be beneficial to minkes, but to the large number of other predators also dependent upon these baitfish. Of further concern to coastal minkes, however, is the imminent risk of entanglement in fishing gear [61]. Static creel pots, for example, are currently responsible for at least 30 fatalities a year in Scottish waters [62], although mitigation actions, such as sinking lead lines, are currently being sought to address this problem. The nearshore habitats utilised by minke whales further harbour impacts from export cable corridors and landfall sites for consented wind farms, and associated increases in vessel traffic [22]. Photo-identification studies demonstrate that some whales utilising the Southern Trench MPA may remain in the region, or return in subsequent years (Robinson, unpublished data [63]). This presents the added risk of potentially exposing the same individuals to recurring anthropogenic stressors over time. Consequently, there is a need to integrate temporal data into spatial management frameworks when identifying priority areas for protection and to take these outputs into consideration when assessing the impacts of detrimental anthropogenic activities.

5. Conclusions

Dynamic management—which incorporates flexible as opposed to strict boundaries in seasonally important areas for highly mobile cetaceans [11]—is becoming increasingly articulated as a more suitable approach for the management of protected species [10]. Very little protection in the form of restrictions upon potentially harmful activities is currently provided by the Southern Trench MPA besides a set of broad management objectives [64]. The present findings could be used to inform a more dynamic, targeted approach that focusses protective measures, such as restrictions on static fishing gear, in these seasonally important minke whale feeding habitats. Such an approach in minke whale feeding areas may benefit the much wider variety of marine predators that feed on sandeels [65–68] and around fronts [8,69–71], thus promoting consistency with the principles of ecosystem-based MPA management [11], whilst targeting measures in the most important areas and perhaps permitting activities in the less important areas of the MPA [9]. This study also illustrates the importance of including prey data, when available, in habitat models for the spatial management of baleen whales on their feeding grounds. Indeed, the systematic monitoring of protected species and their prey, influencing environmental factors and identified anthropogenic stressors are all crucial for modelling, without which the dynamism or adaptability of these models are compromised, since failure to consider such temporal changes may conceal pertinent information crucial for management.

Anthropogenic stressors will inevitably vary both spatially and temporally in and around the Southern Trench MPA. We know, for example, that the waters of the MPA provide important feeding grounds for minke whales in northeast Scotland, yet we know little about the site fidelity and movements of individual animals. Ongoing studies employing satellite-tracking devices will increase our understanding of the movements of these animals in and beyond the MPA. This information will be used to inform management decisions for this and other UK MPAs for the species (necessity for expansion, identification/mitigation of presenting threats) and will allow us to examine the regional policy decisions crucial for the future management of these and other protected cetaceans in our regional UK waters.

Author Contributions: Conceptualisation, K.P.R. and D.A.I.M.; methodology, D.A.I.M.; software, D.A.I.M.; validation, D.A.I.M.; formal analysis, D.A.I.M.; investigation, D.A.I.M. and K.P.R.; resources, K.P.R.; data curation, K.P.R.; writing—original draft preparation, D.A.I.M. and K.P.R.; writing—review and editing, K.P.R.; visualisation, K.P.R.; supervision, K.P.R.; project administration, K.P.R. All authors have read and agreed to the published version of the manuscript.

Funding: This research received no external funding.

Institutional Review Board Statement: This study was entirely observational, but all boat-based monitoring work was carried out under licence from NatureScot.

Informed Consent Statement: Not applicable.

Data Availability Statement: The raw data supporting the conclusions of this article will be made available by the authors on request due to privacy restrictions.

Acknowledgments: The authors would like to thank all CRRU interns and volunteers who generously gave their time and support with fieldwork. We thank Marine Scotland for providing the sandeel density data and NERC Earth Observation Data Acquisition and Analysis Service (NEODAAS) for supplying the oceanographic data. We also thank the editor and reviewers for their constructive feedback, which greatly improved the final manuscript.

Conflicts of Interest: The authors declare no conflicts of interest.

References

1. Bailey, H.; Thompson, P.M. Using Marine Mammal Habitat Modelling to Identify Priority Conservation Zones within a Marine Protected Area. *Mar. Ecol. Prog. Ser.* **2009**, *378*, 279–287. [CrossRef]
2. Cañadas, A.; Sagarminaga, R.; García-Tiscar, S. Cetacean Distribution Related with Depth and Slope in the Mediterranean Waters off Southern Spain. *Deep Sea Res. I Oceanogr. Res. Pap.* **2002**, *49*, 2053–2073. [CrossRef]
3. Franks, P. Sink or Swim, Accumulation of Biomass at Fronts. *Mar. Ecol. Prog. Ser.* **1992**, *82*, 1–12. [CrossRef]
4. Panigada, S.; Sciara, G.N.D.; Panigada, M.Z.; Airoldi, S.; Borsani, J.F.; Jahoda, M. Fin Whales (*Balaenoptera physalus*) Summering in the Ligurian Sea: Distribution, Encounter Rate, Mean Group Size and Relation to Physiographic Variables. *J. Cetacean Res. Manage.* **2005**, *7*, 137–145. [CrossRef]
5. Macleod, K.; Fairbairns, R.; Gill, A.; Fairbairns, B.; Gordon, J.; Blair-Myers, C.; Parsons, E. Seasonal Distribution of Minke Whales *Balaenoptera acutorostrata* in Relation to Physiography and Prey off the Isle of Mull, Scotland. *Mar. Ecol. Prog. Ser.* **2004**, *277*, 263–274. [CrossRef]
6. Kimura, S.; Kasai, A.; Nakata, H.; Sugimoto, T.; Simpson, J.H.; Cheok, J.V.S. Biological Productivity of Meso-Scale Eddies Caused by Frontal Disturbances in the Kuroshio. *ICES J. Mar. Sci.* **1997**, *54*, 179–192. [CrossRef]
7. Griffin, R.B. Sperm Whale Distributions and Community Ecology Associated with a Warm-Core Ring Off Georges Bank. *Mar. Mammal Sci.* **1999**, *15*, 33–51. [CrossRef]
8. Scales, K.L.; Miller, P.I.; Hawkes, L.A.; Ingram, S.N.; Sims, D.W.; Votier, S.C. REVIEW: On the Front Line: Frontal Zones as Priority at-Sea Conservation Areas for Mobile Marine Vertebrates. *J. Appl. Ecol.* **2014**, *51*, 1575–1583. [CrossRef]
9. Maxwell, S.M.; Hazen, E.L.; Lewison, R.L.; Dunn, D.C.; Bailey, H.; Bograd, S.J.; Briscoe, D.K.; Fossette, S.; Hobday, A.J.; Bennett, M.; et al. Dynamic Ocean Management: Defining and Conceptualizing Real-Time Management of the Ocean. *Mar. Policy* **2015**, *58*, 42–50. [CrossRef]
10. Maxwell, S.M.; Gjerde, K.M.; Conners, M.G.; Crowder, L.B. Mobile Protected Areas for Biodiversity on the High Seas. *Science* **2020**, *367*, 252–254. [CrossRef]
11. Hoyt, E. *Marine Protected Areas for Whales, Dolphins and Porpoises: A World Handbook for Cetacean Habitat Conservation*, 2nd ed.; Earthscan: Oxon, UK, 2011; ISBN 978-1-84407-763-2.
12. Camphuysen, K.C.; Shamoun-Baranes, J.; Bouten, W.; Garthe, S. Identifying Ecologically Important Marine Areas for Seabirds Using Behavioural Information in Combination with Distribution Patterns. *Biol. Conserv.* **2012**, *156*, 22–29. [CrossRef]
13. Horton, T.W.; Palacios, D.M.; Stafford, K.M.; Zerbini, A.N. Baleen Whale Migration. In *Ethology and Behavioral Ecology of Mysticetes*; Clark, C.W., Garland, E.C., Eds.; Springer International Publishing: Cham, Switzerland, 2022; pp. 71–104, ISBN 978-3-030-98449-6.
14. Lee, S.H.; Schell, D.M.; McDonald, T.L.; Richardson, W.J. Regional and Seasonal Feeding by Bowhead Whales Balaena Mysticetus as Indicated by Stable Isotope Ratios. *Mar. Ecol. Prog. Ser.* **2005**, *285*, 271–287. [CrossRef]
15. Konishi, K.; Tamura, T.; Isoda, T.; Okamoto, R.; Hakamada, T.; Kiwada, H.; Matsuoka, K. Feeding Strategies and Prey Consumption of Three Baleen Whale Species Within the Kuroshio-Current Extension. *J. Northwest Atl. Fish. Sci.* **2009**, *42*, 27–40. [CrossRef]
16. Bentaleb, I.; Martin, C.; Vrac, M.; Mate, B.; Mayzaud, P.; Siret, D.; de Stephanis, R.; Guinet, C. Foraging Ecology of Mediterranean Fin Whales in a Changing Environment Elucidated by Satellite tracking and Baleen Plate Stable Isotopes. *Mar. Ecol. Prog. Ser.* **2011**, *438*, 285–302. [CrossRef]
17. Mitani, Y.; Bando, T.; Takai, N.; Sakamoto, W. Patterns of Stable Carbon and Nitrogen Isotopes in the Baleen of Common Minke Whale Balaenoptera Acutorostrata from the Western North Pacific. *Fish. Sci.* **2006**, *72*, 69–76. [CrossRef]
18. Zhang, Y.; Sun, X.; Nong, Z.; Chen, M.; Hao, Y.; Wang, J.; Wang, K.; Wang, D.; Mei, Z. The First Baleen Whale Marine Protected Area Proposed for Bryde's Whales in the Beibu Gulf, China. *Mar. Mammal Sci.* **2024**, *40*, e13082. [CrossRef]
19. Ashe, E.; Noren, D.P.; Williams, R. Animal Behaviour and Marine Protected Areas: Incorporating Behavioural Data into the Selection of Marine Protected Areas for an Endangered Killer Whale Population. *Anim. Conserv.* **2010**, *13*, 196–203. [CrossRef]
20. NatureScot. Minke Whale. Available online: https://www.nature.scot/plants-animals-and-fungi/mammals/marine-mammals/minke-whale (accessed on 24 January 2025).
21. Robinson, K.P.; Tetley, M.J.; Mitchelson-Jacob, E.G. The Distribution and Habitat Preference of Coastally Occurring Minke Whales (*Balaenoptera acutorostrata*) in the Outer Southern Moray Firth, Northeast Scotland. *J. Coast. Conserv.* **2009**, *13*, 39–48. [CrossRef]
22. Robinson, K.P.; MacDougall, D.A.I.; Bamford, C.C.G.; Brown, W.J.; Dolan, C.J.; Hall, R.; Haskins, G.N.; Russell, G.; Sidiropoulos, T.; Sim, T.M.C.; et al. Ecological Habitat Partitioning and Feeding Specialisations of Coastal Minke Whales (*Balaenoptera acutorostrata*) Using a Recently Designated MPA in Northeast Scotland. *PLoS ONE* **2023**, *18*, e0246617. [CrossRef]
23. Paxton, C.; Scott-Hayward, L.; Rexstad, E. *Statistical Approaches to Aid the Identification of Marine Protected Areas for Minke Whale, Risso's Dolphin, White-Beaked Dolphin and Basking Shark*; Scottish Natural Heritage: Inverness, Scotland, 2014.
24. Robinson, K.P.; Tetley, M.J. Behavioural Observations of Foraging Minke Whales (*Balaenoptera acutorostrata*) in the Outer Moray Firth, North-East Scotland. *J. Mar. Biol. Assoc. UK* **2007**, *87*, 85–86. [CrossRef]

25. Tetley, M.J.; Mitchelson-Jacob, E.G.; Robinson, K.P. The Summer Distribution of Coastal Minke Whales (*Balaenoptera acutorostrata*) in the Southern Outer Moray Firth, NE Scotland, in Relation to Co-Occurring Mesoscale Oceanographic Features. *Remote Sens. Environ.* **2008**, *112*, 3449–3454. [CrossRef]
26. Boyse, E.; Robinson, K.P.; Beger, M.; Carr, I.M.; Taylor, M.; Valsecchi, E.; Goodman, S.J. Environmental DNA Reveals Fine-Scale Spatial and Temporal Variation of Marine Mammals and Their Prey Species in a Scottish Marine Protected Area. *Environ. DNA* **2024**, *6*, e587. [CrossRef]
27. Wilson, B.; Reid, R.J.; Grellier, K.; Thompson, P.M.; Hammond, P.S. Considering the Temporal When Managing the Spatial: A Population Range Expansion Impacts Protected Areas-Based Management for Bottlenose Dolphins. *Anim. Conserv. Forum* **2004**, *7*, 331–338. [CrossRef]
28. Hooker, S.K.; Cañadas, A.; Hyrenbach, K.D.; Corrigan, C.; Polovina, J.J.; Reeves, R.R. Making Protected Area Networks Effective for Marine Top Predators. *Endanger. Species Res.* **2011**, *13*, 203–218. [CrossRef]
29. QGIS Development Team. *QGIS Geographic Information System*; QGIS: London, UK, 2024.
30. Tepsich, P.; Rosso, M.; Halpin, P.; Moulins, A. Habitat Preferences of Two Deep-Diving Cetacean Species in the Northern Ligurian Sea. *Mar. Ecol. Prog. Ser.* **2014**, *508*, 247–260. [CrossRef]
31. Robinson, K.P. Minke Whale Sightings. *Cetacean Research and Rescue Unit*. 2025. Available online: https://crru.org.uk/research/minke-whales (accessed on 1 November 2024).
32. Forney, K.A.; Becker, E.A.; Foley, D.G.; Barlow, J.; Oleson, E.M. Habitat-Based Models of Cetacean Density and Distribution in the Central North Pacific. *Endanger. Species Res.* **2015**, *27*, 1–20. [CrossRef]
33. Gilles, A.; Viquerat, S.; Becker, E.A.; Forney, K.A.; Geelhoed, S.C.V.; Haelters, J.; Nabe-Nielsen, J.; Scheidat, M.; Siebert, U.; Sveegaard, S.; et al. Seasonal Habitat-Based Density Models for a Marine Top Predator, the Harbor Porpoise, in a Dynamic Environment. *Ecosphere* **2016**, *7*, e01367. [CrossRef]
34. Laidre, K.L.; Heide-Jørgensen, M.P.; Heagerty, P.; Cossio, A.; Bergström, B.; Simon, M. Spatial Associations between Large Baleen Whales and Their Prey in West Greenland. *Mar. Ecol. Prog. Ser.* **2010**, *402*, 269–284. [CrossRef]
35. Pendleton, D.E.; Holmes, E.E.; Redfern, J.; Zhang, J. Using Modelled Prey to Predict the Distribution of a Highly Mobile Marine Mammal. *Divers. Distrib.* **2020**, *26*, 1612–1626. [CrossRef]
36. Zerbini, A.N.; Friday, N.A.; Palacios, D.M.; Waite, J.M.; Ressler, P.H.; Rone, B.K.; Moore, S.E.; Clapham, P.J. Baleen Whale Abundance and Distribution in Relation to Environmental Variables and Prey Density in the Eastern Bering Sea. *Deep Sea Res. Part II Top. Stud. Oceanogr.* **2016**, *134*, 312–330. [CrossRef]
37. Pickens, B.; Taylor, J.; Campbell, M.; Driggers, W. Offshore Snapper and Shark Distributions Are Predicted by Prey and Area of Nearby Estuarine Environments in the Gulf of Mexico, USA. *Mar. Ecol. Prog. Ser.* **2022**, *682*, 169–189. [CrossRef]
38. Schick, R.S.; Lutcavage, M.E. Inclusion of Prey Data Improves Prediction of Bluefin Tuna (*Thunnus thynnus*) Distribution. *Fish. Oceanogr.* **2009**, *18*, 77–81. [CrossRef]
39. Friedlaender, A.S.; Halpin, P.N.; Qian, S.S.; Lawson, G.L.; Wiebe, P.H.; Thiele, D.; Read, A.J. Whale Distribution in Relation to Prey Abundance and Oceanographic Processes in Shelf Waters of the Western Antarctic Peninsula. *Mar. Ecol. Prog. Ser.* **2006**, *317*, 297–310. [CrossRef]
40. van der Kooij, J.; Scott, B.E.; Mackinson, S. The Effects of Environmental Factors on Daytime Sandeel Distribution and Abundance on the Dogger Bank. *J. Sea Res.* **2008**, *60*, 201–209. [CrossRef]
41. Langton, R.; Boulcott, P.; Wright, P.J. A Verified Distribution Model for the Lesser Sandeel Ammodytes Marinus. *Mar. Ecol. Prog. Ser.* **2021**, *667*, 145–159. [CrossRef]
42. RStudio Development Team. *RStudio: Integrated Development for R*; RStudio: Boston, MA, USA, 2020.
43. Wood, S.N. Fast Stable Restricted Maximum Likelihood and Marginal Likelihood Estimation of Semiparametric Generalized Linear Models. *J. R. Stat. Soc. Ser. B Stat. Methodol.* **2011**, *73*, 3–36. [CrossRef]
44. Gelman, A.; Su, Y.-S.; Yajima, M.; Hill, J.; Pittau, M.G.; Kerman, J.; Zheng, T.; Dorie, V. *Arm: Data Analysis Using Regression and Multilevel/Hierarchical Models*; Cambridge University Press: Cambridge, UK, 2021.
45. Conners, M.G.; Sisson, N.B.; Agamboue, P.D.; Atkinson, P.W.; Baylis, A.M.M.; Benson, S.R.; Block, B.A.; Bograd, S.J.; Bordino, P.; Bowen, W.D.; et al. Mismatches in Scale between Highly Mobile Marine Megafauna and Marine Protected Areas. *Front. Mar. Sci.* **2022**, *9*, 897104. [CrossRef]
46. Anderwald, P.; Evans, P.; Dyer, R.; Dale, A.; Wright, P.; Hoelzel, A. Spatial Scale and Environmental Determinants in Minke Whale Habitat Use and Foraging. *Mar. Ecol. Prog. Ser.* **2012**, *450*, 259–274. [CrossRef]
47. De Boer, M. Spring Distribution and Density of Minke Whale *Balaenoptera acutorostrata* along an Offshore Bank in the Central North Sea. *Mar. Ecol. Prog. Ser.* **2010**, *408*, 265–274. [CrossRef]
48. Pierce, G.J.; Santos, M.B.; Reid, R.J.; Patterson, I.A.P.; Ross, H.M. Diet of Minke Whales *Balaenoptera acutorostrata* in Scottish (UK) Waters with Notes on Strandings of This Species in Scotland 1992–2002. *J. Mar. Biol. Assoc. UK* **2004**, *84*, 1241–1244. [CrossRef]
49. Greenstreet, S.P.R.; McMillan, J.A.; Armstrong, E. Seasonal Variation in the Importance of Pelagic Fish in the Diet of Piscivorous Fish in the Moray Firth, NE Scotland: A Response to Variation in Prey Abundance? *ICES J. Mar. Sci.* **1998**, *55*, 121–133. [CrossRef]

50. Winslade, P. Behavioural Studies on the Lesser Sandeel *Ammodytes marinus* (Raitt) III. The Effect of Temperature on Activity and the Environmental Control of the Annual Cycle of Activity. *J. Fish Biol.* **1974**, *6*, 587–599. [CrossRef]
51. Pedersen, S.A.; Lewy, P.; Wright, P. Assessments of the Lesser Sandeel (*Ammodytes marinus*) in the North Sea Based on Revised Stock Divisions. *Fish. Res.* **1999**, *41*, 221–241. [CrossRef]
52. Robinson, B.W.; Wilson, D.S. Optimal Foraging, Specialization, and a Solution to Liem's Paradox. *Am. Nat.* **1998**, *151*, 223–235. [CrossRef] [PubMed]
53. Maravelias, C.; Reid, D. Identifying the Effects of Oceanographic Features and Zooplankton on Prespawning Herring Abundance Using Generalized Additive Models. *Mar. Ecol. Prog. Ser.* **1997**, *147*, 1–9. [CrossRef]
54. Maravelias, C.; Reid, D.; Swartzman, G. Seabed Substrate, Water Depth and Zooplankton as Determinants of the Prespawning Spatial Aggregation of North Atlantic Herring. *Mar. Ecol. Prog. Ser.* **2000**, *195*, 249–259. [CrossRef]
55. Van Haren, J.J.M.; Joordens, J.C.A. Observations of Physical and Biological Parameters at the Transition between the Southern and Central North Sea. *Neth. J. Sea Res.* **1990**, *25*, 351–364. [CrossRef]
56. Gissel, T.N.; Munk, P. Zooplankton Diversity and the Predatory Impact by Larval and Small Juvenile Fish at the Fisher Banks in the North Sea. *J. Plankton Res.* **1998**, *20*, 2313–2332. [CrossRef]
57. Elliott, A.J.; Clarke, T. Seasonal Stratification in the Northwest European Shelf Seas. *Cont. Shelf Res.* **1991**, *11*, 467–492. [CrossRef]
58. Van Haren, H.; Howarth, M.J.; Jones, K.; Ezzi, I. Autumnal Reduction of Stratification in the Northern North Sea and Its Impact. *Cont. Shelf Res.* **2003**, *23*, 177–191. [CrossRef]
59. Doniol-Valcroze, T.; Berteaux, D.; Larouche, P.; Sears, R. Influence of Thermal Fronts on Habitat Selection by Four Rorqual Whale Species in the Gulf of St. Lawrence. *Mar. Ecol. Prog. Ser.* **2007**, *335*, 207–216. [CrossRef]
60. Yan, H.F.; Kyne, P.M.; Jabado, R.W.; Leeney, R.H.; Davidson, L.N.K.; Derrick, D.H.; Finucci, B.; Freckleton, R.P.; Fordham, S.V.; Dulvy, N.K. Overfishing and Habitat Loss Drive Range Contraction of Iconic Marine Fishes to near Extinction. *Sci. Adv.* **2021**, *7*, eabb6026. [CrossRef] [PubMed]
61. Northridge, S.; Cargill, A.; Coram, A.; Mandleberg, L.; Calderan, S.; Reid, B. *Entanglement of Minke Whales in Scottish Waters; an Investigation into Occurrence, Causes and Mitigation*; Sea Mammal Research Unit: Fife, Scotland, 2010.
62. Leaper, R.; MacLennan, E.; Brownlow, A.; Calderan, S.; Dyke, K.; Evans, P.; Hartny-Mills, L.; Jarvis, D.; McWhinnie, L.; Philp, A.; et al. Estimates of Humpback and Minke Whale Entanglements in the Scottish Static Pot (Creel) Fishery. *Endanger. Species Res.* **2022**, *49*, 217–232. [CrossRef]
63. Robinson, K.P. Minke Whale Catalogue. *Cetacean Research and Rescue Unit.* 2025. Available online: https://crru.org.uk/research/catalogues?catalogue=minke-whale&page=1 (accessed on 1 November 2024).
64. NatureScot. *Conservation and Management Advice: Southern Trench MPA*; NatureScot: Inverness, Scotland, 2024.
65. Wright, P.J.; Begg, G.S. A Spatial Comparison of Common Guil lemots and Sandeels in Scottish Waters. *ICES J. Mar. Sci.* **1997**, *54*, 578–592. [CrossRef]
66. Carroll, M.J.; Bolton, M.; Owen, E.; Anderson, G.Q.A.; Mackley, E.K.; Dunn, E.K.; Furness, R.W. Kittiwake Breeding Success in the Southern North Sea Correlates with Prior Sandeel Fishing Mortality. *Aquat. Conserv. Mar. Freshw. Ecosyst.* **2017**, *27*, 1164–1175. [CrossRef]
67. MacLeod, C.D.; Santos, M.B.; Reid, R.J.; Scott, B.E.; Pierce, G.J. Linking Sandeel Consumption and the Likelihood of Starvation in Harbour Porpoises in the Scottish North Sea: Could Climate Change Mean More Starving Porpoises? *Biol. Lett.* **2007**, *3*, 185–188. [CrossRef]
68. Wanless, S.; Harris, M.P.; Greenstreet, S.P.R. Summer Sandeel Consumption by Seabirds Breeding in the Firth of Forth, South-East Scotland. *ICES J. Mar. Sci.* **1998**, *55*, 1141–1151. [CrossRef]
69. Pommier, M.; O'Donnell, C.; Barile, C.; McGill, R.; Berrow, S.; O'Brien, J. Exploring Environmental and Biological Drivers of Cetacean Occurrence in the Cross-Border Region of the Malin Shelf Using Data from a European Fishery Survey. *Front. Mar. Sci.* **2023**, *10*.
70. Scales, K.L.; Miller, P.I.; Embling, C.B.; Ingram, S.N.; Pirotta, E.; Votier, S.C. Mesoscale Fronts as Foraging Habitats: Composite Front Mapping Reveals Oceanographic Drivers of Habitat Use for a Pelagic Seabird. *J. R. Soc. Interface* **2014**, *11*, 20140679. [CrossRef]
71. Miller, P.I.; Scales, K.L.; Ingram, S.N.; Southall, E.J.; Sims, D.W. Basking Sharks and Oceanographic Fronts: Quantifying Associations in the North-East Atlantic. *Funct. Ecol.* **2015**, *29*, 1099–1109. [CrossRef]

oceans MDPI

Article

Resident Harbor Porpoises (*Phocoena phocoena* vomerina) in the Salish Sea: Photo-Identification Shows Long-Term Site Fidelity, Natal Philopatry, and Provides Insights into Longevity and Behavior

Cindy R. Elliser *, Katrina H. White and Maia C. Hansen

Pacific Mammal Research, Anacortes, WA 98221, USA; katrina.maciver@pacmam.org (K.H.W.); research@mhansen.com (M.C.H.)
* Correspondence: cindy.elliser@pacmam.org

Abstract: Harbor porpoises have been well studied in locations worldwide, but relatively little research has focused on site fidelity of individuals, which influences population structure and determines appropriate management and conservation measures. This study investigated the site fidelity and behavior of harbor porpoises through long-term, land-based photo-identification around Fidalgo Island, Washington, USA. Individuals were consistently re-sighted for up to 11 years, including natal philopatry. There was individual variation in seasonal site fidelity, with some individuals seen year-round, though general porpoise presence in the study area and individual re-sightings were greatly reduced during summer. Females had calves yearly for up to 3 consecutive years, with females as old as at least 14 successfully producing calves. There was some evidence for mother–calf associations post-weaning, but none long-term. Behavior was affected by tide, where porpoises traveled more and foraged and socialized less during ebb tide compared to other tidal states. Mating attempts with aerial behaviors were observed 84 times and seen year-round in every month and every season. This study provides documentation of a local, resident population of harbor porpoise, and emphasizes the need for more fine-scale studies like this to ensure the appropriate management and conservation of harbor porpoise populations and/or stocks worldwide.

Keywords: harbor porpoise; photo-ID; *Phocoena phocoena*; *Phocoena phocoena* vomerina; Salish Sea; site fidelity; seasonality

1. Introduction

Patterns of site fidelity of individual animals, or their tendency to return to, or remain in, a particular habitat, influence the social, genetic, and community structure of a population [1,2]. The movement of animals is driven by processes across multiple spatial and temporal scales, and influences the fate of individuals as well as the structure and dynamics of populations, communities, and ecosystems [2]. The harbor porpoise (*Phocoena phocoena*) is one of the most abundant coastal species of cetacean worldwide, with a wide geographic distribution spanning much of the northern hemisphere [3,4]; however, information on the site fidelity and behavior of individuals is lacking for this species. As a small, highly mobile, coastal marine predator, harbor porpoises live a large portion of their lives in close proximity to human activities, and knowledge of site fidelity and habitat use patterns is critical in assessing impacts from anthropogenic threats [5].

To date, research on harbor porpoises along the Western coast of the United States (USA) has been focused on large-scale abundance, stranded animals, and stock structure [6–15], or on mating behavior [16–18]. Similarly, the multitude of research in European waters has primarily focused on large-scale abundance, stranded animals, distribution, and the effects of pollutants and anthropogenic impacts (e.g., [19–23]), as well as mating behavior [18]. However, there is increasing evidence for fine-scale population structure in a variety of harbor porpoise populations worldwide. Research has shown that harbor porpoise behavior and distribution can be structured over varying spatial scales and across different temporal scales, including diel and seasonal differences [24–31]. Some harbor porpoise populations may also live in relatively small and restricted geographic areas [6,32,33], and significant genetic differences have been observed within geographic areas that were previously thought to be one genetic stock [14,15,34]. When population structure is unknown, it is difficult to define what effect the loss of individuals will have; if there are small or restricted porpoise populations, a localized conservation concern may exist [35]. Understanding fine-scale variations within and between populations is critical to ensure the appropriate management and conservation of harbor porpoise stocks.

Globally, the conservation status of the species differs between populations. In Atlantic and European waters there are harbor porpoise populations that are classified as either vulnerable, endangered, or critically endangered, all of which indicate that they are threatened with extinction [36,37]. Along the West Coast of the USA and the Salish Sea (a multinational body of water spanning between Washington (WA), USA, British Columbia, Canada, and many Indigenous Tribes and First Nations), harbor porpoises are in a somewhat better status [33], ranging from being a species of special concern in Canada under the Species At Risk Act [38], to the USA where it is a Priority Species under Washington Fish and Wildlife's Priority Habitat and Species Program [39] and a candidate for listing as a Washington State endangered, threatened, or sensitive species. In the Salish Sea, harbor porpoises are year-round residents with a high abundance and consistent presence [40–44], and in WA comprise the Washington inland waters stock (designated by the National Oceanographic and Atmospheric Association (NOAA)). This stock is of particular interest given their drastic decline in numbers by the 1970s [43,45,46] followed by a subsequent recovery [9,46,47]. Their status as a species of concern at various governmental levels in this region is based on the large amount of data gaps about many aspects of their biology, ecology, behavior, sociality, and regionally specific threats [33], combined with the fact that, due to their coastal habitat and prey preferences, harbor porpoises are vulnerable to anthropogenic impacts such as fishery interactions [35], noise disturbance [22,48], disease [49], and pollution [50–52]. They are also important as a prey species to mammal-eating Bigg's killer whales which frequent this region [53,54].

Dedicated long-term research is critical to help fill in the data gaps for harbor porpoises in this region and better understand fine-scale population structure, movements, site fidelity, seasonal variation, and life history that will inform effective management and conservation to ensure their continued success in the Salish Sea (see review [33]). The long-term tracking of individuals can provide valuable data on fine-scale site fidelity, as well as life history characteristics that are critical to understanding these population dynamics. Photographic identification (photo-ID) allows for the tracking of individual animals over long periods of time and can provide vital data on population dynamics and individual variation, offering insights into site fidelity, heredity, longevity, and ranging patterns, while also providing a unique opportunity to explore and observe behavioral patterns and anomalies as seen in other well-studied small cetacean populations (e.g., bottlenose dolphins, *Tursiops truncatus*) [55–59] and spotted dolphins, *Stenella frontalis* [60–62].

Harbor porpoises are known for being evasive and shy, making them a more difficult species to study. This behavior, along with their small size, cryptic coloration, and few readily observable distinguishing marks [63–67], has meant that they have historically been overlooked as candidates for photo-ID studies at the local population level. In spite of these qualities, pigmentation patterns on the side/flank of harbor porpoises have been known to be unique to individual harbor porpoises since the early 1970s [64,68]. Research utilizing these unique markings for identification has only recently been implemented in a small number of locations, demonstrating that these marks can be successfully used for photo-ID over shorter (months) and longer (years) periods of time in both USA and European waters [32,67,69–72]. Using photo-ID, our previous research successfully demonstrated site fidelity of Pacific harbor porpoises (*Phocoena phocoena* vomerina) over 3 years and year-round use of the waters off Fidalgo Island, Washington, USA, part of the Salish Sea [32].

This paper provides further data from the on-going, long-term, land-based, photo-ID study of harbor porpoises in Burrows Pass off Fidalgo Island, between January 2014 and October 2024, expanding on Elliser et al.'s [32] initial study. The objectives of this study included using photo-ID to determine the degree of individual site fidelity, along with related life history and behavioral observations. To our knowledge, this study is one of the longest consecutive datasets available on individual free-swimming harbor porpoises in the world and can provide invaluable insights into their behavior, site fidelity, and ecology.

2. Materials and Methods

2.1. Study Area

The land-based study site was located on Fidalgo Island, WA. Burrows Pass is a stretch of water flowing between Burrows and Fidalgo Islands, connecting Burrows Bay to Rosario Strait (Figure 1). The center of Burrows Pass is located at 48.49 N, 122.69 W. The depth varies from 12 to 40 m and the pass is approximately 1300 m long, with a maximum width of approximately 915 m at the western end opening to Rosario Strait and a minimum width of approximately 460 m at the eastern end heading into Burrows Bay. The observation point was situated about 6 m above sea level and provided a good vantage point of the entire Pass.

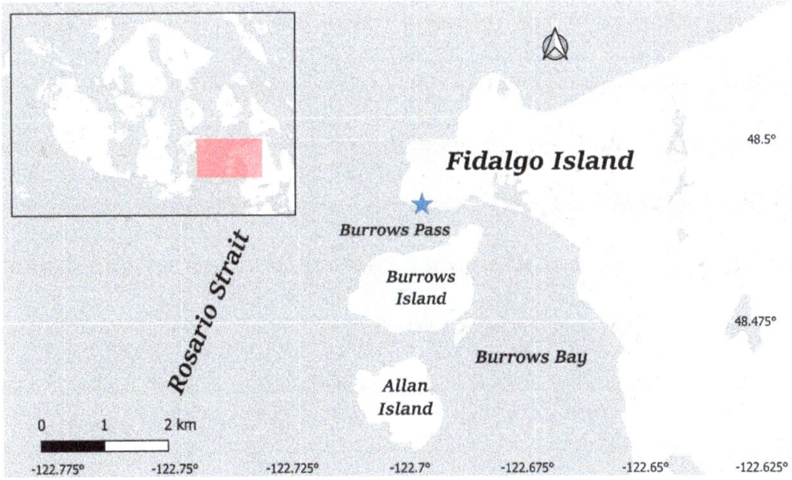

Figure 1. A map of the study area, Burrows Pass. The black square shows the land-based location where researchers conducted observations.

2.2. Data Collection

Photographic and observational data were collected between January 2014 and October 2024 using Canon EOS T3i, T5i, and 5D cameras fitted with either a 100–400 mm or 100–600 mm zoom lens. Data were collected year-round; however, effort was not evenly distributed throughout seasons due to poor weather (e.g., increased precipitation, wind, etc.), particularly during the winter months. For the first 6 months of the study (January 2014–June 2014), data were collected approximately once per month as part of a feasibility study [32], increasing to an average of 2–3 times per week.

Observation periods were defined as the total amount of time spent in the field on a given day, and the vast majority lasted 2 h; however, due to additional animal activity, poor weather, time constraints, etc., occasionally, sessions were shorter or longer (range = 1–3 h). However, over the entire study, the average time spent in an observation was approximately 2 h (average = 1.99 h). Observations were conducted on separate weekdays during daylight hours, though occasionally two observations were conducted on a single day when necessary. Observations were typically not conducted in conditions over a Beaufort 3, or in winds > 20 mph. During observations, environmental data were collected including temperature, wind direction and speed, Beaufort state, tidal rip/current, sun/cloud cover, lunar phase, glare, and tidal state, as well as the number of boats passing through.

Little is known about harbor porpoise sociality and group structure [32]; therefore, for the purpose of this study a group was defined based on behavior and proximity: all animals in sight and/or involved in the same activity [73]. Because of the small size and topography of Burrows Pass, all individual porpoises were likely within range of acoustic, if not visual, communication. Thus, all porpoises within the Pass were considered a group unless individuals remained spatially separated while in the Pass (i.e., individuals stayed at opposite ends of the Pass) and showed no observable interactions with one another, in which case they were considered as two separate groups [32]. Generally, the number of individuals in the study site was three or less and their behavior and relative cohesiveness made it clear they were interacting with one another. A sighting was defined as the duration of time in which a porpoise or group of porpoises were observed within the study area during an observation period. A sighting was considered over after 10 min of not seeing any individual in the group. The end time of the sighting was then set at 5 min after the last sighting of an individual or group. The vast majority of the time there was one sighting of harbor porpoises at a time, but occasionally two sightings (defined using the criteria above) would overlap in time.

A minimum, maximum, and best estimate for group size and number of calves (defined as the observers' best approximation based on the range of animals observed during a sighting) was recorded and updated after photographic analysis. Even if suitable identification photos were not obtained, photos could be used to confirm or update group size due to dorsal fin shape variability among individuals and small group sizes [32]. The final group size estimates presented are based on these best estimates. Calves were identified by visual estimation of body length (up to two-thirds the length of the adult [74]), along with typical calf behavior of swimming in echelon position. Calves included neonates and young (under 1 year old), while all other animals (including juveniles/subadults) that could not be visually distinguished from adult animals were classified as adults.

Sighting data (number of adults and calves, behavioral state (overall behavioral state that porpoises were engaged in during a sighting: probable foraging/foraging, traveling, socializing, unknown/other), behaviors observed (i.e., specific behaviors such as surface chases, fish chases, wake surfing, mating, aerials, porpoising, spy hops), and location within Burrows Pass) were recorded throughout the sighting, and environmental variables (see above) were recorded at the start of each sighting. Behavioral state was determined by observation of the following specific behaviors: probable foraging/foraging (surface chases, long dive intervals, fish chases/catches, and/or presence of bird mills/birds diving/catching fish in close proximity to porpoises), socializing (aerials, wake surfing, play, mating), travel (directional movement/porpoising of animals into or out of study area), and unknown (no clear behavioral state observed) [32]. All data were entered into FinBase, a customized Microsoft Access database used to store, manage, and visualize data from photo-ID surveys [75].

Seasons were defined as follows: winter (December–February), spring (March–May), summer (June–August), and fall (September–November). Effort was not evenly distributed across seasons; therefore, sightings were corrected for effort in order to compare sighting rates for harbor porpoise seasonal habitat use. Sightings per unit effort (SPUE) were calculated as the number of harbor porpoise sightings divided by total number of observations.

2.3. Data Analysis

Photos were processed and analyzed using the methodology described in Elliser et al. [32], with only photos of suitable quality (good clarity and focus, lighting, proximity of animal) and animals that were 'highly distinctive' (D1) and 'distinctive' (D2) being used for identification [67]. A matrix of 8 different identification categories was used to accurately identify individual harbor porpoises. Pigmentation and scars/lesions were the primary identification features used, alongside any secondary features (e.g., nicks, notches, or scars on the peduncle or dorsal fin trailing edge) that were present. Confirmation features (overall coloration, fin shape, fin size, fin base width) were used to confirm identifications if needed, but were not used as the primary method of identification [32].

Harbor porpoise markings are not bilaterally symmetrical; thus, there was a possibility that a single harbor porpoise could be identified as two different animals when only one side of the animal was photographed. To reduce this type of error, dorsal fin shape, secondary features, and/or other confirmation marks were used to help match the left and right sides of individual animals and reduce this potential bias. When both sides of an animal were photographed within one sighting, these markings could be used to confirm the left and right sides were of the same individual (this was aided by the small group sizes and variety in dorsal fin shapes, allowing researchers to match the left and right sides of a particular individual).

Once identified, individual porpoises were assigned a unique identifier (name) and added to a catalog of known individuals. Along with their unique identifier, each individual entry in the catalog included the following: (a) the best photo of that animal's left and/or right side, (b) when the individual was first observed and/or its birth year, (c) its distinctiveness category, (d) sex, if known, (e) calves and their birth years, if applicable, and (f) relationship to other porpoises in the catalog, if known (e.g., mothers and calves). The number of sightings was calculated for each individual. Sometimes an individual would be in more than one sighting during an observation period (in one day). Due to the close proximity of time and being in the same location, these subsequent sightings were not included in an individual's total sighting count, i.e., an individual was not counted as sighted more than once per day.

Porpoises seen with calves in echelon position were assumed to be females. Individuals involved in mating attempts were sometimes able to be sexed as well. Harbor porpoise mating behavior is highly lateralized and conserved around the world, with males exclusively coming up on the left-hand side of the female [17,18]. Thus, even if physical evidence (i.e., visible penis) was not visible to confirm the sex of an animal during a mating attempt, the animal on the left was presumed to be male and the animal on the right assumed to be female.

Data analyses were computed in Python using the latest stable release (0.14.4) of the statsmodels package [76]. Additionally, we used the latest stable release (0.11.2) of the scikit-posthocs package [77] for performing post hoc pairwise comparisons.

We tested for the multicollinearity of environmental variables by calculating the variance inflation factors (VIFs). VIFs > 10.0 are typically used to indicate multicollinearity [78]. The VIFs for our environmental variables were below this value (all VIFs < 2.5).

Individual binomial logistic regression analyses using a generalized linear model (GLM) were run to examine the predictive impact of the available environmental data for each of the recorded outcomes: presence/absence of porpoises as well as the observation of social, foraging, and/or travel behavior of observed porpoise groups. A full model including all three of the following environmental factors as predictors was set initially: season, tide state (slack low, flood, slack high, ebb), and rip tide strength (low, medium, high). To determine the relative significance of the predictor variables within the full model, the p-values from a Wald test were examined for each variable. Using Akaike's Information Criterion (AIC), and considering Cox and Snell's pseudo R^2 as a secondary measure of model fit, the original full model was compared to the reduced models containing only the predictors deemed significant ($p < 0.05$) for that outcome based on the Wald tests. In all cases, the reduced model showed a lower AIC value indicating a better fit, and was therefore selected for further analyses of that particular outcome.

For the analysis of seasonal differences in size of groups, a nonparametric Kruskal–Wallis test was used due to the skewed distribution of porpoise group sizes. Pairwise post hoc comparisons were performed using Dunn's test, with a Holm stepwise adjustment applied.

3. Results

3.1. Effort

Over 2290 h were spent on observations from Jan 2014 to October 2024. Effort was lowest in the winter due to poor weather conditions (e.g., wind, precipitation, etc.) that restricted time for fieldwork. Despite the fact that effort was highest in summer and that there were a relatively large number of sightings, that season had the lowest sightings per unit effort (SPUE = number of sightings/total number of observations), percentage of observations with sightings, and the lowest amount of time observing porpoises (Table 1).

Table 1. Total number of observations, sightings, percentages of amount of observations with sightings, and amount of time observing porpoises (during sightings). Sightings per unit effort (SPUE) is number of sightings divided by number of observations (average time spent per observation = 1.99 h). Italicized indicates lowest percentages and SPUE in summer.

Season	Number of Observations	Number of Sightings	Percentage (%) of Observations with Sightings	Percentage (%) of Time Observing Porpoises	SPUE
Winter	192	334	86.5%	42.1%	1.74
Spring	322	610	89.8%	53.9%	1.89
Summer	338	488	*75.7%*	*31.3%*	*1.44*
Fall	282	487	90.4%	53.2%	1.73
Total	1134	1919	85.19%	45.0%	1.69

3.2. Group Size

Group size ranged from 1 to 20 (x = 2.58 ± 1.72), with a median of 2, and was significantly different between seasons (Kruskal–Wallis, H = 81.436, $p < 0.001$). Group size was significantly smaller in summer (2.08 ± 1.40) compared to spring (2.82 ± 1.99), fall (2.8 ± 1.74), and winter (2.55 ± 1.42) (Dunn's test $p < 0.001$). The largest groups of 16 and 20 were seen in fall and spring, respectively, which also had the highest group size averages. The vast majority (78.9%) of sightings in all seasons had three or fewer individuals (fall 74.9%, spring 74.4%, summer 88.7%, winter 78.7%).

3.3. *Site Fidelity and Re-Sighting History*

Between January 2014 and October of 2024, 195 individual porpoises, including calves, were identified (Figure 2). Because the re-sightings for calves during the first year of life will be highly correlated with the mother, the first year of sightings and re-sightings for 31 calves were removed, leaving 164 individuals (28 calves were removed completely as they were not re-sighted/able to be matched after the first year; three calves remained, and their re-sightings post-weaning were counted). The majority (68.9%, $n = 113$) of these 164 individuals were re-sighted two or more times.

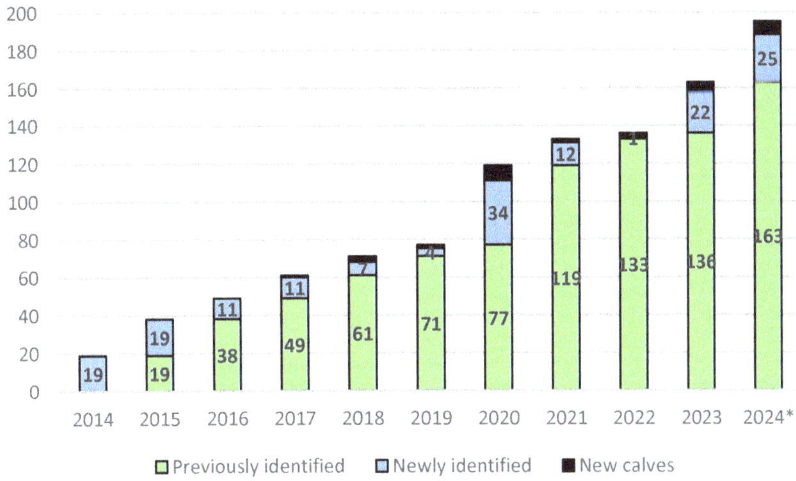

Figure 2. Discovery curve for period January 2014–October 2024 (* denotes this is not a full year) for individuals using Burrows Pass. New individuals (non-calf) and new calves are shown separately, above previously identified individuals.

Of these 113 re-sighted individuals, 57.5% were sighted in at least two consecutive or non-consecutive years ($n = 65$, Table 2). Sightings per calendar year for an individual ranged from 0 to 31.

Table 2. Sighting and re-sighting histories of non-calf harbor porpoises in Burrows Pass, WA, showing range of total sightings and mean re-sighting rates for all individuals, those seen at least two times, and those seen in at least 2 years.

	Number of Total Sightings	Mean Re-Sighting Rate
All non-calf individuals ($n = 164$)	1–125	5.4
Re-sighted at least twice ($n = 113$)	2–125	9.1
Re-sighted in at least 2 different years ($n = 65$)	2–125	12.8

The number of years an individual was re-sighted (not including missing years) ranged from 2 to 9 years, with an average of 2.9 years. If missing years are included (it being more likely we missed taking a photo of them vs. them not being present in the area), which may better indicate truer levels of site fidelity, then the range is 2 to 11 years, with an average of 4.0 years. This indicates at least 9 or 11 years of site fidelity for some individuals. Examples of long-term matches of individuals are provided in Figure 3.

Figure 3. Photographs of long-term matches of individual harbor porpoises. PIA, POI and RAI are known females. PHA (male), PIC (female), and PIZ (sex unknown) are all calves showing long-term site fidelity.

3.4. Seasonality

Logistic regression analysis revealed that season (Wald $\chi^2 = 32.102$, $p < 0.001$) as well as rip tide strength (Wald $\chi^2 = 9.218$, $p = 0.010$) both significantly affected harbor porpoise presence. Porpoises were more likely to be seen in fall, winter, or spring compared to summer ($p < 0.001$, $p = 0.022$, and $p < 0.001$, respectively). Additionally, harbor porpoises were more likely to be seen in low ($p = 0.009$) and medium ($p = 0.005$) rip tides when compared to high rip tides. Tide level (Wald $\chi^2 = 0.805$, $p = 0.848$) was not found to have a statistically significant impact on porpoise presence. The VIFs of all three environmental variables fell below the cutoff value used to indicate multicollinearity (all VIF values < 2.5).

Over half (55.4%, $n = 36$) of the 65 individuals seen in multiple years (not including a calf's first year) were also seen in more than one season within a calendar year (66.2%, n = 43, if calves are included). Fourteen of these individuals (38.9%) were seen in every season, some with all four season sightings coming within a calendar year and others across multiple years. Some individuals were seen in multiple seasons across years (e.g., spring in one year, fall in another), while others were seen in only one season across years (e.g., only fall of multiple years).

To look at seasonal trends within animals showing a stronger degree of site fidelity, re-sightings were calculated for individuals with 10 or more sightings across multiple years (Figure 4). There was a strong seasonal trend, with the highest re-sightings in spring ($n = 253$) and fall ($n = 170$). Summer ($n = 86$) and winter ($n = 62$) had about half or less compared to spring and fall. Six of these individuals were seen in every season within a calendar year (Figure 4).

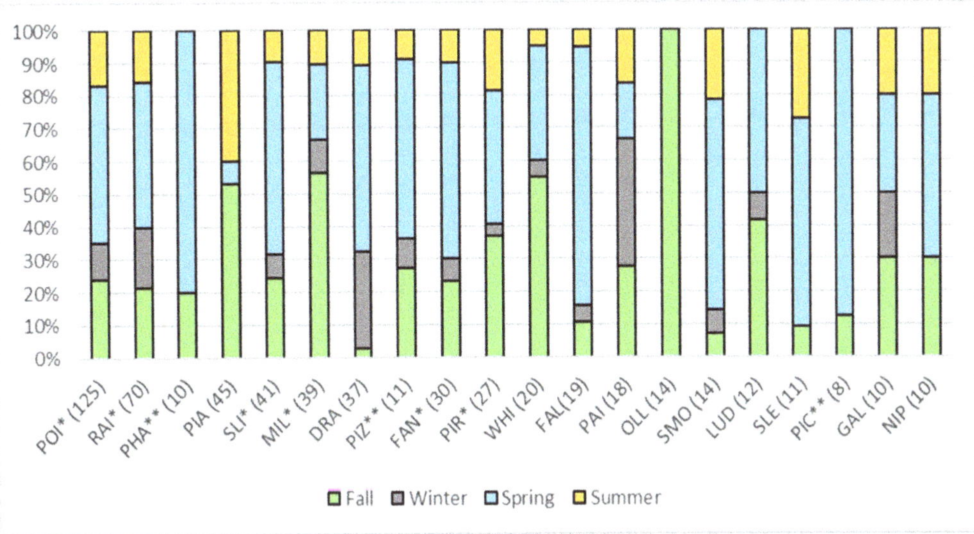

Figure 4. The percentage of seasonal site fidelity of individuals seen 10 or more times, with total sightings given in parentheses next to the individual's three-letter identifier. * indicates an individual seen in all seasons during one calendar year. ** indicates an individual first sighted as a calf; only re-sightings after their first year are included in this graph, so the total sightings of these individuals are higher than what is in parentheses.

3.5. Mothers and Calves

There were 19 presumed females, 18 documented by presence of calves consistently in echelon position and 1 inferred female from an observed mating attempt (Table 3). Each of the eighteen females had one to six calves over the study period. Four females had calves in a set of consecutive years, with one female having five calves in six years. Some females were not seen every year; thus, it is possible that they had calves that could not be documented.

Three of the most re-sighted females, PIA, POI, and RAI (Figure 3), were likely adults when they were first observed (2014, 2018, and 2018, respectively). PIA was sighted with a calf in 2014, indicating she was at least 3 years old. She had five confirmed calves, including the most recent in 2024 (Table 3). Conservatively, this means she is at least 14 and still having calves. Estimating that POI (4 calves) and RAI (6 calves) were at least 3 when first sighted in 2018 (each had a calf in 2019), this would make them at least 10 years old and still having calves (each had one in 2024, Table 3).

At least three calves have shown site fidelity long past weaning. PHA is a known male and son of POI, with 47 sightings over 5 years. He has been seen in multiple sightings with POI since weaning (once in 2021, once in 2023, and four in 2024), including a mating attempt with his mother on 5/17/24 (Figure 5). PIC is a daughter of PIA with 10 sightings over 8 years, including 5 years not seen. In addition, PIC was sighted in 2023 with calf PIP, making this a three-generation family (PIA, PIC, PIP), with the grandmother PIA also having a calf in 2023. PIZ is of unknown sex, the offspring of PIA, with 31 sightings over 5 years, including 1 year not seen. In addition, two calves showed close association (similar to echelon position) with their mothers during sightings over a year after birth, PIZ (two sightings in November) and an unnamed (UN) 2020 calf of RAI (three sightings in October and November), indicating that they were still associated with their mother at over a year old. Both mothers, PIA and RAI, were not observed with new calves that year (2021).

Table 3. Known females and their calving history, including the number or re-sightings of the mother, the number of calves she was seen with, the years the female was seen (gray boxes), and the years she had calves (letters in boxes). UN means that the calf was not named (due to lack of good photos or identifying marks), other letters are the three letter abbreviation of their name. Individuals with a * were of the three-generation family. Note that the lack of a calf in a given year is not necessarily indicative of no calf that year; the mother may have only been sighted before having a calf, or the calf did not survive, or the female was not seen enough times to determine. ** indicates that the previous year's calf was still with the female through November of the current year (indicating the calf was over a year old and still associated with the mother).

Mom	Re-Sights	# Calves	2014	2015	2016	2017	2018	2019	2020	2021	2022	2023	2024
ARC	2	1											UN
AUR	4	2							UN			AUS	
DOL	1	1										X	
DRA	2	2							UN				DAM
FAL	19	1						UN					
FLI	7	2							UN			UN	
MIL	39	2						MIN	MEL				
PEA	2	2		UN					PAP				
PIA *	45	4	UN			PIC *	UN		PIZ	**		PER	PIO
POI	124	4						PHA		PHS	UN		UN
PIC *	10	1				born	NA					PIP *	
POL	2	1											UN
RAI	71	6				UN		RAN	UN	**	RAB	UN	UN
SLE	11	1							UN				
SIC	4	1						UN					
SLI	41	2							UN				UN
SMO	14	1										UN	
SLO	2	1										UN	

Figure 5. Mating attempt on 5/17/24 by 5 year old male PHA towards his mother POI (at least 10 years old).

3.6. Behavior

Harbor porpoises are often cryptic in their behavior at the surface, such that it is often unclear in what overall behavioral state they are in (listed as "Unknown" in Table 4). Behavior, categorized as travel, foraging, social, or other, was examined in relation to season, tide, and rip tide (Table 4).

Table 4. The percentage of porpoise time in behavioral states by season and tide (number of sightings with the given behavior or tide divided by the total number of sightings). There can be more than one behavior observed during a sighting (i.e., foraging and social); thus, the overall totals may be over 100% as those sightings were counted for each behavioral state in the calculations. Bold/italicized indicates significantly higher levels, as described in text.

	Travel	Foraging	Social	Unknown
Winter	***35.9% (n = 120)***	32.9% (n = 93)	5.10% (n = 17)	34.1% (n = 114)
Spring	23.6% (n = 144)	***36.9% (n = 225)***	6.10% (n = 37)	42.8% (n = 261)
Summer	26.2% (n = 128)	23.4% (n = 114)	7.20% (n = 35)	48.2% (n = 235)
Fall	24.4% (n = 119)	***38.4% (n = 187)***	9.90% (n = 48)	40.7% (n = 198)
Slack low	***31.28% (n = 56)***	***35.19% (n = 63)***	***6.15% (n = 11)***	34.08% (n = 61)
Flood	18.31% (n = 156)	***40.02% (n = 341)***	***11.62% (n = 99)***	41.31% (n = 352)
Slack High	19.67% (n = 24)	***46.72% (n = 57)***	***9.02% (n = 11)***	40.98% (n = 50)
Ebb	***36.57% (n = 325)***	20.88% (n = 157)	2.13% (n = 16)	44.15% (n = 332)

Travel behavior was influenced by season (Wald χ^2 = 12.849, p = 0.005), tide (Wald χ^2 = 37.238, p < 0.001), and rip tide (Wald χ^2 = 7.333, p = 0.026). Travel was more likely to be observed in winter than any other season (vs. spring p < 0.001, vs. summer p = 0.026, and vs. fall p = 0.009). Travel was also more likely to be seen during ebb tide vs. flood or slack high (p < 0.001), as well as being more likely to be seen during slack low vs. flood (p = 0.008) and slack low vs. slack high (p = 0.019). Travel was observed less frequently during medium rip tides as compared to low rip tides (p = 0.01) (Table 4).

Foraging behavior was influenced by season (Wald χ^2 = 29.137, p = 0.005) and tide (Wald χ^2 = 72.386, p < 0.001). Rip tide was found not to be statistically significant (Wald χ^2 = 0.779, p = 0.677). Foraging behavior was more likely to be observed in fall and spring than it was in either winter (p = 0.019 and p = 0.001, respectively) or summer (p < 0.001 and p < 0.001, respectively). Additionally, foraging behavior was less likely in ebb tide as compared to all other tide states (flood, slack high, and slack low (p < 0.001). Observed foraging differences across non-ebb tide states were not statistically significant (Table 4).

Social behavior was influenced by tide (Wald χ^2 = 32.870, p < 0.001) and rip tide (Wald χ^2 = 10.536, p < 0.005). Season was found not to be statistically significant (Wald χ^2 = 3.510, p = 0.320). Observed social behavior was significantly less likely during ebb tide as compared to all other tide states (ebb vs. flood (p < 0.001), vs. slack high (p < 0.001), and vs. slack low (p = 0.005)). Social behavior was more likely to be observed during medium rip tides vs. high (p = 0.048) and vs. low (p = 0.002) rip tides (Table 4).

Mating attempts involving aerial behavior were observed during 84 sightings between 2015 and 2024. Typically, there was one attempt per sighting, but occasionally there would be multiple (2–4) attempts observed. Mating attempts were documented year-round. Attempts were seen in every month across all years, although almost, but not all, in one concurrent year (Table 5). Mating attempts were observed in all seasons across years, and in all seasons during a calendar year in five years (Table 5). All mating attempts were confirmed visually, and when possible with photographic documentation.

Table 5. The number of mating attempts observed between 2015 and 2024, and what months and seasons they were documented in. Confirmed visually and with photo documentation when possible.

Year	Number of Mating Attempts	Number of Months	Seasons
2015	1	1	Sp
2017	4	3	Sp, F, W
2018	4	4	Sp, F, W
2019	9	5	Sp, Su, F, W
2020	17	8	Sp, Su, F, W
2021	17	9	Sp, Su, F, W
2022	3	3	Su, F
2023	11	7	Sp, Su, F, W
2024	18	8	Sp, Su, F, W
Total	84	Seen in every month	In all seasons

4. Discussion

This study has successfully demonstrated that harbor porpoises in the Salish Sea can be identified over long periods of time (over a decade) using photo-ID, and that these data can provide unique insights into site fidelity, life history characteristics, behavior, habitat use, and population structure. We have demonstrated consistent long-term site fidelity of up to 11 years and consistent year-round, multi-year use of a habitat, including natal philopatry, by individually identified free-ranging harbor porpoises, indicating a resident, local population. Understanding this type of fine-scale population structure is vitally important in order to consider the appropriate biological context for the animals [33,79] and make biologically meaningful conservation decisions.

4.1. Group Size

Trends in group size were consistent with previous work, including summer having the smallest average group size, though the overall average of x = 2.58 was slightly higher than previously observed (x = 2.32) [32]. Harbor porpoises are most commonly observed in group sizes of 1–3 animals throughout their range [7,9,16,32,80,81]. However, large aggregations of harbor porpoises have been seen along the west coast of Canada in the Port of Prince Rupert (ranging from 30 to 1000 individuals) [30] and in the US waters of the Salish Sea (ranging from 20 to >300 individuals), and are more common than previously thought [82]. We did not observe any large aggregations in our local study area, likely due to the topography of the habitat that is not conducive to large groupings. In previous work north of our study site in the San Juan Islands area of the Salish Sea, group sizes were lower, with an average 1.87 individuals [43]. With a mean group size of 2.58, our findings are on the higher end of the local and global ranges, demonstrating the potential variability in harbor porpoise group sizes in different parts of the world, and even different locations within a population range.

4.2. Site Fidelity and Re-Sighting History

The regular and long-term site fidelity documented here indicates that there is a local, resident population of harbor porpoises that regularly utilize Burrows Pass. Almost 70% of non-calf harbor porpoises were re-sighted in Burrows Pass at least twice, with 57.5% of these animals showing site fidelity over multiple years. Long-term site fidelity within these re-sighted individuals ranged from 2 to 11 years, with as many as 31 sightings in a calendar year. It is important to note that these re-sightings are likely an under-representation of the true re-sighting rates. Usable photos of the animals are not always obtainable, due to behavior and/or distance to the photographer; thus, we are likely to have missed re-sightings of individuals (i.e., the porpoises were present, but we could not identify the individuals during a sighting). The local population of harbor porpoises that are resident and utilize Burrows Pass regularly is therefore likely larger than that documented here.

Until recently, there has been very little published information on site fidelity in individual harbor porpoises. Recent research has demonstrated varying degrees of site fidelity in harbor porpoises, ranging from weeks to years [32,33,69–72,83]; however, to our knowledge, only one other study has long-term photo-ID data similar to that shown here. Stichting Rugvin has been conducting photo-ID in the Eastern Scheldt, Netherlands, since 2009 [71]. They have documented harbor porpoise site fidelity of up to 14 years [F. Zanderink, pers. comm.].

Looking beyond photo-ID, other recent research supports the suggestion that long-term site fidelity over multiple years may be more widespread within this species than has previously been believed. At the population level, environmental DNA (eDNA) research has indicated that harbor porpoises in coastal and inland waters of Alaska show natal philopatry at multiple geographic scales, and have limited gene flow/dispersal between neighboring areas [14,15]. Two regions in the Southeast Alaska stock are suggested to have demographically independent populations (DIPs) based on genetic data, trends in abundance, and distribution, and it is thought that multiple other DIPs exist within the currently recognized harbor porpoise stock [84,85]. Along the California, (CA), WA, and Oregon (OR) coast, recent genetic studies have proposed creating a Central OR stock from the Northern CA/Southern OR stock, indicating more genetic differentiation within the stock than previously thought [13].

While some harbor porpoise populations may be more fluid and migrate seasonally [24,25,86,87], it is evident that some harbor porpoises do remain resident to specific areas, showing strong site fidelity and genetic differentiation, and that this is not isolated to one region or even one part of the world. Clearly, harbor porpoise populations and movement patterns are not universal across their range, and should be investigated for each population. Further research with photo-ID and genetics in our study area and the surrounding waters is necessary to determine whether there are other smaller, genetically or socially distinct populations or communities within the larger Salish Sea population. Understanding these dynamics is critical for the conservation of this species and redefining stock assessments, supporting the importance of fine-scale data in effectively managing and conserving this small coastal species [15,26,31,33].

4.3. Seasonality

Season and rip tide strength affected the presence/absence of harbor porpoises in Burrows Pass, with summer lower than every other season and porpoises more often seen in low and medium rip compared to high, consistent with our previous work [32]. The lower presence during high rip may indicate that the rip strength in Burrows Pass is above the upper limit for adult harbor porpoises in tidal stream habitats [32]. While harbor porpoises were observed year round, over all 11 years of the study we consistently observed that, despite increased effort, harbor porpoise presence was lower during summer in the months of June–August, with a significant drop in presence typically occurring in mid-late July. Seasonal variations in harbor porpoise presence and relative densities are well documented around the world, with different locations seeing peaks in harbor porpoises across different seasons [24,25,88–90], indicating that seasonal variations in abundance and density are likely population-specific and may be due to location-specific drivers, such as environmental conditions or prey availability.

Burrows Pass has the characteristics of a tidal stream habitat and is an important foraging location for harbor porpoises [32], with up to 34% of sightings in the current study including foraging behavior. This is likely an under-representation, as we could only document foraging behavior when there were observable surface behaviors (such as extended dive times, surface chases, fish catches, bird presence/diving). Harbor porpoises in Burrows Pass have been observed feeding on both small and large fish species (Elliser unpublished data, [91]). Worldwide, tidal stream habitats have been shown to have conditions that likely improve foraging opportunities for porpoises [26]. Feeding in distinctive areas, and thus reducing their need for movement between foraging patches, has been hypothesized to be a beneficial strategy for harbor porpoises [31,92], and is likely driven by prey availability in different locations. Prey movements and seasonal peaks in specific prey types have been documented as influences on harbor porpoise presence and seasonal movements [24,25,31,93]; thus, it is possible that a similar effect is occurring here and driving the reduced sightings in summer. Additionally, females are commonly observed returning to the Pass with new calves in August to early September; thus, it is possible that at least some of the animals showing strong site fidelity spend more time elsewhere to have their calves before returning to Burrows Pass.

Our study site was located adjacent to a busy marina, with boats traveling through the Pass as the main thoroughfare to Rosario Strait and the San Juan Islands. The number of boats in the area was higher during the summer than in other seasons; however, throughout the duration of the study, porpoises seem to have little or no direct behavioral response to vessel traffic in the Pass [32], a trend which is also seen in the southern Salish Sea (Puget Sound, D. Anderson pers. comm.). To the north, in the Port of Prince Rupert, Canada, harbor porpoises are consistently present despite the presence of large vessels in this very busy port, suggesting some level of acclimatization to localized disturbance [30], which has also been documented in European waters [94]. While harbor porpoises in the Salish Sea may appear to be more acclimated to boats than in other areas of their range, it is unknown whether boat presence may affect harbor porpoise in less overt and/or indirect ways in Burrows Pass which are not immediately visible through direct behavioral responses and might influence their seasonal presence [32]. For example, harbor porpoises with DTAGS have shown cessation of echolocation that was correlated with high boat-noise levels, leading to significantly fewer prey captures [95], and in another study, although little habitat displacement occurred, the distribution of harbor porpoises was significantly affected by shipping-noise levels [94]. It is important to note that some regularly re-sighted individuals have also been re-sighted during summer months when porpoise sightings are low, so while they are not completely absent from the area they are, in general, greatly reduced in time, presence, and number of re-sightings.

The lower re-sighting rate of individuals observed in winter is likely more of an artifact from difficulties in data collection, rather than the absence of animals. While effort was reduced due to poor weather conditions, the SPUE and percentage of observations with sightings were similar to spring and fall. However, the time spent observing porpoises was about 10% lower in winter, corresponding with travel being seen significantly more in winter than in any other season. While traveling, porpoises are typically directional and moving quickly through the area. While we can document this behavior, it is difficult to obtain photos that are usable for ID because the porpoises spend less time in the pass, providing few opportunities for good ID photos. The combination of these factors, along with less-ideal environmental conditions (such as higher Beaufort, glare, and low-light levels), make obtaining good ID photos more difficult in winter. It is likely that individuals with high re-sighting rates in spring and fall are also around more in the winter than we were able to document.

4.4. Mothers and Calves

Our study shows that females that utilize Burrows Pass are capable of successfully reproducing yearly for at least 3 consecutive years. It is likely that this is similar for many parts of the Salish Sea, as previous work on stranded individuals in the region suggested harbor porpoises may be able to reproduce yearly [10]. It is understood that harbor porpoises vary between populations on yearly or bi-yearly birthing patterns. Atlantic harbor porpoises seem to reproduce yearly, whereas it was thought that in the Pacific they reproduced bi-yearly [96]. It is unclear if the trend for yearly birthing seen in the Salish Sea extends to the coast and Pacific Ocean.

In Burrows Pass, new calves are most often first sighted in August and September, though one was observed as early as April [32]. This corresponds with stranding data for the area, which shows the beginning of calving season being May–late July and gestation being around 10.8 months [10]. Calves in Burrows Pass were consistently seen in close proximity (often echelon position) with their mothers through spring of the following year. This supports the suggestion that calves should stay in close proximity during the 8–9 months of lactation [97,98]; however, not much is known about their associations beyond that time frame.

Our study is the first to show site fidelity in harbor porpoise calves of at least 5 years, indicating natal philopatry in this area. This supports recent genetic evidence from southeast Alaska, which suggests levels of natal philopatry and limited dispersal in specific areas [14,15]. Additionally, in our study there were a few instances where calves were sighted with their mothers more than a year after their birth, and thus after weaning occurred. Those five sightings occurred in fall of the following year, and the mother and calf were observed in close proximity, similar to echelon position, not just sighted within the same group. Interestingly, in these years the mother had not been seen with a new calf, which may influence how long a calf will remain associated with the mother. In one unique case, the male porpoise PHA was re-sighted in groups with his mother six times after he was older than one year. He was observed attempting to mate with his mother in one sighting; thus, these associations may be more related to reproduction than a true social association. No other associations between mothers and grown calves were observed, though the sample size was small. The potential lack of these associations may be linked to the unknown social structure of harbor porpoises, as longer-term mother–calf associations have been documented in highly social small cetacean species such as bottlenose [57] and spotted dolphins [62]. In this population, it seems that calves may remain in their natal range; however, this does not equate to continued associations with their mothers. This may derive from a relatively weak mother–calf bond compared to other odontocete species.

Longitudinal drone work in Denmark that followed an individual female and her calves over nine years [72] has shown there is considerable separation between the mother and calf starting as early as even a month after birth, and the mother–calf bond seems to be maintained by the calves more so than the mother [72]. If the bond between mother and calf is not as strong in the early months of life, it seems unlikely that it would extend beyond the weaning process. While current evidence suggests that there may be little association between mothers and calves post-weaning in harbor porpoises, our data shows that it cannot be completely dismissed, and may depend on the individuals, inter-birth intervals, and other variables.

Even less is known about harbor porpoise calf associations with individuals other than their mother. In the Salish Sea, harbor porpoise calves are seen during large aggregation events that last from hours to months [82], and group size with calves is larger in Burrows Pass [32], indicating that calves do have some interaction with other porpoises, either directly or indirectly. Hamel et al. [72] observed no social associations between calves and other porpoises other than their mother, suggesting that the mother may be the only individual that interacts with the calf prior to weaning. An important factor to consider, however, is that the Hamel et al. [72] study focused on only one individual female and her offspring over nine years, and, thus, may not be representative of typical harbor porpoise mother–calf relationships. In addition, this study was conducted between April and November, leaving a four-month period over the winter where it is possible more interactions could have occurred. Differing maternal styles have been shown in bottlenose dolphins, where different behavioral patterns including proximity maintenance, discipline, pectoral fin contact, and initiation of separations and reunions with calves are exhibited [99–101]. It is possible such individual variations exist in harbor porpoise females as well, which could account for some of the differences in mother–calf relationships or associations observed.

Together, these data indicate that calves in Burrows Pass consistently associate with their mother and likely do associate to some unknown degree with other porpoises during the lactation period, whereas long-term associations with their mother post-weaning are uncommon and may vary by individual. While these data help to provide some evidence for association patterns (and possible differences between populations), the social structure of harbor porpoises is still poorly understood. More research is needed to fully understand this aspect of harbor porpoise society, which may vary depending on the population, habitat, and other spatiotemporal factors.

4.5. Behavior

Tide was not a significant factor for overall presence/absence, similar to our previous work [32]; however, when looking at what the animals were doing while present in the Pass, tide greatly influenced behavior. There was a strong correlation with behavior and ebb tide. The porpoises were significantly more likely to be traveling during ebb and slack low, and significantly less likely to be foraging or socializing, correlating with the much lower number of sightings that included behaviors like mating, porpoising, surface chases, and wake surfing. In addition, the presence of calves was found to be significant in relation to tide in Burrows Pass, with calves more likely to be present during slack high tides [32]. This could affect behavioral trends, as mothers may restrict their movements when they have young calves that may find it more difficult to swim in various tidal flows.

Trends for harbor porpoise behavior in relation to tide seem to be site-specific throughout their worldwide range. In San Francisco Bay, harbor porpoises were seen typically traveling out of the bay in response to ebb currents [81]. Buzzing detections (and potential foraging activity) in the Wadden Sea showed site-specific diel and tidal patterns, even at small spatial scales [102]. In Oregon, USA, porpoises were seen foraging at a rocky reef site during ebb phase [103]. In southwest Britain, foraging occurred at high tide, with travel occurring throughout the tidal states in one location, whereas behavioral activity did not vary by tidal state in another close study site [104]. General conclusions about marine top predator distribution need to take into account local differences in topography and tides that can have various effects at different sites [102,105] and can affect behavior and habitat use. It has also been shown that significant changes in porpoise behavior can be found in short and long-term observations [102]. Our data supports the increasing evidence that trends in harbor porpoise behavior are highly variable at small spatiotemporal scales, and that long-term datasets are necessary to tease apart the strong seasonal and spatial variation in diel and tidal effects that are likely site-specific [102].

Social behavior was observed the least in Burrows Pass compared to foraging and travel, but that is more likely because harbor porpoises are relatively less active at the surface, making it difficult to identify social behaviors. Recent work with drones in our study area has revealed that social interactions (like mating attempts) were occurring underwater, when from the surface they looked like regular surfacing behaviors. Thus, the amount of social activity presented here is likely a large under-representation. For this study, most of the social behavior documented was mating attempts, as those are often the most conspicuous. Male harbor porpoises often, but not always, leap out of the water when mating, and it is a highly lateralized behavior worldwide, where males approach the left side of the female 100% of the time [16,18]. Other less-frequently observed social behaviors included wake surfing, spy hops, surface splashing (not related to foraging), and play. Social behavior was observed year-round, including mating attempts occurring in every month and every season, which has been observed in San Francisco, CA, USA (except in June due to fog restricting visibility) [16]. Due to the fast, high-intensity maneuvers required by males during mating, year-round practice of female approaches with correct timing may be important for immature as well as mature males [16]. Practice may start before weaning. In Denmark, mother–calf sexual interactions have been documented between a 9–10 month old unweaned calf and its mother [18]. We provide evidence of an adult male son (PHA) attempting to mate with his mother (POI). It is unknown if the male knew she was his mother, or if this is a common occurrence in harbor porpoises. Mating attempts in Burrows Pass appeared to be observed the most in fall, although this was not statistically significant. However, the estimated conception dates for Salish Sea harbor porpoise range between mid-August and the end of December [10], supporting the idea that mating attempts may be more prevalent during fall.

4.6. Longevity

Long-term photo-ID data can also provide important insights into the longevity of harbor porpoises, which is currently estimated to be anywhere from 8 to 25 years of age, with most averages given as 8–12 years [80,106]. By tracking the calving history of known females and correlating this to the average age of sexual maturity in female harbor porpoises of 3–4 years [80,106,107], it is estimated that several of the animals in our study site are conservatively 10–14 years old and still successfully reproducing. Similarly, there has been a female harbor porpoise in Denmark that was at least 12 years old [40], as well as an individual that is at least 13 years old in the Netherlands [F. Zanderdink, pers. comm.] that were still successfully reproducing. We have documented an individual that is at least 14 years old in the wild, which has also been documented in the Netherlands [F. Zanderink, pers. comm.]. These data have profound implications for the life history of free-swimming harbor porpoises, and suggest that a minimum life expectancy of 8–12 years may be an underestimate for otherwise healthy animals in at least some populations.

5. Conclusions

Through long-term photo-ID, we have identified a local population of harbor porpoises in the Salish Sea that show up to over a decade of consistent site fidelity and at least five years of natal philopatry, and provide unique insights into their behavior, associations, reproduction, and longevity. This long-term dataset provides the opportunity to explore community, social, and population structure through analyses of associations and mark–recapture data that will be a focus of future research. Photo-ID on harbor porpoises has long been overlooked as a viable strategy for tracking harbor porpoises; however, this study shows that in at least some locations it can provide a wealth of data on this cryptic species. Other populations around the world have shown the potential for this type of long-term work, including in San Francisco Bay, USA [69,81], Denmark [18,72], the Netherlands [71], and British Columbia, Canada [33], and there are likely many more.

These results are especially important because of the pronounced variability and flexibility of harbor porpoises between and within population boundaries in their distribution, habitat use, and behavior, even at small spatiotemporal scales. Not all harbor porpoise populations are the same, and attempting to utilize uniform conservation strategies, without appropriate biological context for the animals [33,79], may not effectively protect a population, particularly where a localized conservation concern (small community/population) may exist for this species [35]. From a variety of research techniques, it is increasingly evident that many harbor porpoise populations along the west coast of the USA are structured at a finer scale than previously thought [6,14,15,32–34].

The long-term site fidelity, natal philopatry, and fine-scale variability in habitat use observed in this study support the idea that there are smaller communities or populations of harbor porpoises that may be socially or genetically distinct within the larger Salish Sea population. Cetaceans in USA waters are federally managed and protected as stocks delineated by NOAA. This research highlights the vital need for more research and the possible reassessment of stock structure in the Salish Sea [33]. The site fidelity, behaviors observed, and habitat use documented here also continue to show that Burrows Pass is likely a biologically important area [32,108], and we suggest that areas like these be identified elsewhere in the Salish Sea and provided additional protections for the continued conservation of this species. Long-term research such as this study provides the data needed to make these decisions and is vital for the survival of populations where large data gaps exist, in order to prevent potential future population declines [33]. Finally, as this is a multinational body of water, collaboration between researchers across borders is necessary to provide accurate context for the biologically meaningful conservation of Salish Sea harbor porpoises [33], similar to what has been accomplished between countries in Europe [69]. Future research on harbor porpoises should be undertaken with these conservation goals in mind, incorporating photo-ID where possible as a powerful tool to better understand site-specific site fidelity, population, and community structures in order to provide biologically meaningful measures for their protection.

Author Contributions: Conceptualization, C.R.E. and K.H.W.; methodology, C.R.E. and K.H.W.; software, C.R.E., K.H.W. and M.C.H.; validation, C.R.E., K.H.W. and M.C.H.; formal analysis, M.C.H.; investigation, C.R.E. and K.H.W.; data processing and analysis, C.R.E., K.H.W. and M.C.H.; writing—original draft preparation, C.R.E. and K.H.W.; writing—review and editing, C.R.E., K.H.W. and M.C.H. All authors have read and agreed to the published version of the manuscript.

Funding: This research was funded by the Marathon Petroleum Foundation, Gary Milgard Family Foundation, Melinda Gray Ardia Environmental Foundation, Orange County Community Foundation, Norcross Wildlife Foundation, Andeavour Foundation, and private donations.

Institutional Review Board Statement: Not applicable.

Informed Consent Statement: Not applicable.

Data Availability Statement: The data presented in this study are available on request from the corresponding author. Data are not publicly available as they are still in use by the authors.

Acknowledgments: We would like to thank all the staff, Board of Directors, interns, volunteers, and supporters of Pacific Mammal Research. Special thanks to S. Elliser, L. Rogers, and many others for their generous financial and technical support.

Conflicts of Interest: The authors declare no conflicts of interest. The funders had no role in the design of the study; in the collection, analyses, or interpretation of data; in the writing of the manuscript; or in the decision to publish the results.

References

1. Switzer, P.V. Site fidelity in predictable and unpredictable habitats. *Evol. Ecol.* **1993**, *7*, 533–555. [CrossRef]
2. Nathan, R.; Getz, W.M.; Revilla, E.; Holyoak, M.; Kadmon, R.; Saltz, D.; Smouse, P.E.A. Movement ecology paradigm for unifying organismal movement research. *Proc. Natl. Acad. Sci. USA* **2008**, *105*, 19052–19059. [CrossRef] [PubMed]
3. Gaskin, D.E. The harbor porpoise *Phocoena phocoena*: Regional populations, status, and information on direct and indirect catches. *Rep. Int. Whal. Comm.* **1984**, *34*, 569–586.
4. Fontaine, M.C.; Thatcher, O.; Ray, N.; Piry, S.; Brownlow, A.; Davison, N.J.; Jepson, P.; Deaville, R.; Goodman, S.J. Mixing of porpoise ecotypes in southwestern UK waters revealed by genetic profiling. *R. Soc. Open Sci.* **2017**, *4*, 160992. [CrossRef]
5. Passadore, C.; Möller, L.; Diaz-Aguirre, F.; Parra, G.J. High site fidelity and restricted ranging patterns in southern Australian bottlenose dolphins. *Ecol. Evol.* **2017**, *8*, 242–256. [CrossRef] [PubMed]
6. Chivers, S.J.; Hanson, B.; Laake, J.; Gearin, P.; Muto, M.M.; Calambokidis, J.; Duffield, D.; McGuire, T.; Hodder, J.; Greig, D.; et al. *Additional Genetic Evidence for Population Structure of Phocoena phocoena off the Coasts of California, Oregon, and Washington*; Southwest Fisheries Science Center: La Jolla, CA, USA, 2007.
7. Hobbs, R.C.; Waite, J.M. Abundance of harbor porpoise (*Phocoena phocoena*) in three Alaskan regions, corrected for observer errors due to perception bias and species misidentification, and corrected for animals submerged from view. *Fish. Bull.* **2010**, *108*, 251–267.
8. Huggins, J.L.; Raverty, S.A.; Norman, S.A.; Calambokidis, J.; Gaydos, J.K.; Duffield, D.A.; Lambourn, D.M.; Rice, J.M.; Hanson, B.; Wilkinson, K.; et al. Increased harbor porpoise mortality in the Pacific Northwest, USA: Understanding when higher levels may be normal. *Dis. Aquat. Org.* **2015**, *115*, 93–102. [CrossRef]
9. Jefferson, T.A.; Smultea, M.A.; Courbis, S.S.; Campbell, G.S. Harbor porpoise (*Phocoena phocoena*) recovery in the inland waters of Washington: Estimates of density and abundance from aerial surveys, 2013–2015. *Can. J. Zool.* **2016**, *94*, 505–515. [CrossRef]
10. Norman, S.A.; Hanson, M.B.; Huggins, J.; Lambourn, D.; Calambokidis, J.; Cottrell, P.; Greene, A.; Raverty, S.; Berta, S.; Dubpernell, S.; et al. Conception, fetal growth, and calving seasonality of harbor porpoise (*Phocoena phocoena*) in the Salish Sea waters of Washington, USA, and southern British Columbia, Canada. *Can. J. Zool.* **2018**, *96*, 566–575. [CrossRef]
11. Forney, K.A.; Carretta, J.V.; Benson, S.R. *Preliminary Estimates of Harbor Porpoise Abundance in Pacific Coast Waters of California, Oregon, and Washington, 2007–2012*; NOAA Technical Memorandum NMFS; Southwest Fisheries Science Center: La Jolla, CA, USA, 2014.
12. Forney, K.A.; Moore, J.E.; Barlow, J.; Carretta, J.V.; Benson, S.R. A multidecadal Bayesian trend analysis of harbor porpoise (*Phocoena phocoena*) populations off California relative to past fishery bycatch. *Mar. Mamm. Sci.* **2021**, *37*, 546–560. [CrossRef]
13. Forney, K.A.; Benson, S.R.; Becker, E.A. *A Habitat-Based Spatial Density Model for Harbor Porpoise (Phocoena phocoena) off Oregon and Washington Based on 2021–2022 Aerial Surveys*; NOAA Technical Memorandum NMFS, Southwest Fisheries Science Center: La Jolla, CA, USA, 2023. [CrossRef]
14. Parsons, K.M.; Everett, M.; Dahlheim, M.; Park, L. Water, water everywhere: Environmental DNA can unlock population structure in elusive marine species. *R. Soc. Open Sci.* **2018**, *5*, 180537. [CrossRef] [PubMed]
15. Parsons, K.M.; May, S.A.; Gold, Z.; Dahlheim, M.; Gabriele, C.; Straley, J.M.; Moran, J.R.; Goetz, K.; Zerbini, A.N.; Park, L.; et al. Using eDNA to supplement population genetic analyses for cryptic marine species: Identifying population boundaries for Alaska harbour porpoises. *Mol. Ecol.* **2024**, e17563. [CrossRef] [PubMed]
16. Keener, W.; Webber, M.A.; Szczepaniak, I.D.; Markowitz, T.M.; Orbach, D.N. The sex life of harbor porpoises (*Phocoena phocoena*): Lateralized and aerial behavior. *Aquat. Mamm.* **2018**, *44*, 620–632. [CrossRef]
17. Orbach, D.N.; Brennan, P.L.R.; Hedrick, B.P.; Keener, W.; Webber, M.A.; Mesnick, S.L. Asymmetric and spiraled genitalia coevolve with unique lateralized mating behavior. *Sci. Rep.* **2020**, *10*, 3257. [CrossRef]
18. Webber, M.A.; Keener, W.; Wahlberg, M.; Elliser, C.R.; MacIver, K.; Torres Ortiz, S.; Jakobsen, F.; Hamel, H.; Rieger, A.; Siebert, U.; et al. Sexual Behavior and Anatomy in Porpoises. In *Sex in Cetaceans: Morphology, Behavior, and the Evolution of Sexual Strategies*; Würsig, B., Orbach, D., Eds.; Springer: Cham, Switzerland, 2023; pp. 415–441, ISBN 978-3-031-35650-6. [CrossRef]
19. Nachtshelm, D.A.; Viquerat, S.; Ramirez-Martinez, N.C.; Unger, B.; Siebert, U.; Gilles, A. Small Cetacean in a Human High-Use Area: Trends in Harbor Porpoise Abundance in the North Sea Over Two Decades. *Front. Mar. Sci.* **2021**, *7*, 606609. [CrossRef]
20. IJsseldijk, L.L.; van den Broek, J.; Kik, M.J.L.; Leopold, M.F.; Rebolledo, E.B.; Gröne, A.; Heesterbeek, H. Using marine mammal necropsy data in animal health surveillance: The case of the harbor porpoise in the Southern North Sea. *Front. Mar. Sci.* **2024**, *10*, 1306294. [CrossRef]
21. Benke, H.; Bräger, S.; Dähne, M.; Gallus, A.; Hansen, S.; Honnef, C.G.; Jabbusch, M.; Koblitz, J.C.; Krügel, K.; Liebschner, A.; et al. Baltic Sea harbour porpoise populations: Status and conservation needs derived from recent survey results. *Mar. Ecol. Prog. Ser.* **2014**, *495*, 275–290. [CrossRef]
22. Nabe-Nielsen, J.; Sibly, R.M.; Tougaard, J.; Teilmann, J.; Sveegaard, S. Effects of noise and by-catch on a Danish harbour porpoise population. *Ecol. Mod.* **2014**, *272*, 242–251. [CrossRef]

23. Das, K.; Siebert, U.; Fontaine, M.; Jauniaux, T.; Holsbeek, L.; Bouquegneau, J.-M. Ecological and pathological factors related to trace metal concentrations in harbour porpoises *Phocoena phocoena* from the North Sea and adjacent areas. *Mar. Ecol. Prog. Ser.* **2004**, *281*, 283–295. [CrossRef]
24. Verfuß, U.K.; Honnef, C.G.; Meding, A.; Dähne, M.; Mundry, R.; Benke, H. Geographical and seasonal variation of harbour porpoise (*Phocoena phocoena*) presence in the German Baltic Sea revealed by passive acoustic monitoring. *J. Mar. Biol. Assoc. UK* **2007**, *87*, 165–176. [CrossRef]
25. Sveegaard, S.; Andreasen, H.; Mouritsen, K.N.; Jeppesen, J.P.; Teilmann, J.; Kinze, C.C. Correlation between the seasonal distribution of harbour porpoises and their prey in the Sound, Baltic Sea. *Mar. Biol.* **2012**, *159*, 1029–1037. [CrossRef]
26. Benjamins, S.; Dale, A.; Hastie, G.; Waggitt, J.J.; Lea, M.-A.; Scott, B.; Wilson, B. Confusion reigns? A review of marine megafauna interactions with tidal-stream environments. In *Oceanography and Marine Biology*; CRC Press: Boca Raton, FL, USA, 2015; Volume 53, pp. 1–54. [CrossRef]
27. Benjamins, S.; van Geel, N.; Hastie, G.; Elliott, J.; Wilson, B. Harbour porpoise distribution can vary at small spatiotemporal scales in energetic habitats. *Deep Sea Res. Part II Top. Stud. Oceanogr.* **2016**, *141*, 191–202. [CrossRef]
28. van Beest, F.M.; Teilmann, J.; Dietz, R.; Galatius, A.; Mikkelsen, L.; Stalder, D.; Sveegaard, S.; Nabe-Nielsen, J. Environmental drivers of harbour porpoise fine-scale movements. *Mar. Biol.* **2018**, *165*, 95. [CrossRef]
29. Stalder, D.; van Beest, F.M.; Sveegaard, S.; Dietz, R.; Teilmann, J.; Nabe-Nielsen, J. Influence of environmental variability on harbour porpoise movement. *Mar. Ecol. Prog. Ser.* **2020**, *648*, 207–219. [CrossRef]
30. Dracott, K.; Robinson, C.V.; Brown-Dussault, A.; Birdsall, C.; Barrett-Lennard, L. Porpoises after dark: Seasonal and diel patterns in Pacific harbour porpoise (*Phocoena phocoena* vomerina) aggregations at one of North America's fastest growing ports. *Front. Mar. Sci.* **2022**, *9*, 1010095. [CrossRef]
31. Stedt, J.; Wahlberg, M.; Calström, J.; Nilsson, P.A.; Amundin, M.; Oskolkov, N.; Carlsson, P. Micro-scale spatial preference and temporal cyclicity linked to foraging in harbour porpoises. *Mar. Ecol. Prog. Ser.* **2023**, *708*, 143–161. [CrossRef]
32. Elliser, C.R.; MacIver, K.H.; Green, M. Group characteristics, site fidelity, and photo-identification of harbor porpoises, *Phocoena phocoena*, in Burrows Pass, Fidalgo Island, Washington. *Mar. Mamm. Sci.* **2018**, *34*, 365–384. [CrossRef]
33. Elliser, C.R.; Hall, A. Return of the Salish Sea Harbor Porpoise, *Phocoena phocoena*: Knowledge Gaps, Current Research, and What We Need to Do to Protect Their Future. *Front. Mar. Sci.* **2021**, *8*, 618177. [CrossRef]
34. Morin, P.A.; Forester, B.R.; Forney, K.A.; Crossman, C.A.; Hancock-Hanser, B.L.; Robertson, K.M.; Barrett-Lennard, L.G.; Baird, R.W.; Calambokidis, J.; Gearin, P.; et al. Population structure in a continuously distributed coastal marine species, the harbor porpoise, based on microhaplotypes derived from poor-quality samples. *Mol. Ecol.* **2021**, *30*, 1457–1476. [CrossRef] [PubMed]
35. Williams, R.; Hall, A.; Winship, A. Potential limits to anthropogenic mortality of small cetaceans in coastal waters of British Columbia. *Can. J. Fish. Aquat. Sci.* **2008**, *65*, 1867–1878. [CrossRef]
36. Carlén, I.; Nunny, L.; Simmonds, M.P. Out of sight, out of mind: How conservation is failing European porpoises. *Front. Mar. Sci.* **2021**, *8*, 61748. [CrossRef]
37. Evans, P.G.H. *European Whales, Dolphins and Porpoises: Maine Mammal Conservation in Practice*; Academic Press: London, UK, 2020; pp. 1–306, ISBN 9780128190531.
38. Government of Canada Species at Risk Act: Harbour Porpoise. Available online: https://www.canada.ca/en/environment-climate-change/services/species-risk-public-registry/consultation-documents/harbour-porpoise-2023.html (accessed on 17 January 2025).
39. Washington Department of Fish and Wildlife, Species & Habitats: Harbor Porpoise. Available online: https://wdfw.wa.gov/species-habitats/species/phocoena-phocoena (accessed on 17 January 2025).
40. Scheffer, V.B.; Slipp, J.W. The whales and dolphins of Washington State with the key to the cetaceans of the west coast of North America. *Am. Midl. Nat.* **1948**, *39*, 257–337. [CrossRef]
41. Calambokidis, J.; Steiger, G.H.; Cubbage, J.C. *Marine Mammals in the Southwestern Strait of Juan de Fuca: Natural History and Potential Impacts of Harbor Development in Neah Bay*; Cascadia Research Collective: Olympia, WA, USA, 1987; pp. 1–109.
42. Barlow, J. Harbor porpoise, *Phocoena phocoena*, abundance estimation for California, Oregon, and Washington: I. Ship surveys. *Fish. Bull.* **1988**, *86*, 417–432.
43. Raum-Suryan, K.L.; Harvey, J.T. Distribution and abundance of and habitat use by harbor porpoise, *Phocoena phocoena*, off the northern San Juan Islands, Washington. *Fish. Bull.* **1998**, *96*, 808–822.
44. Gaydos, J.K.; Pearson, S.F. Birds and mammals that depend on the Salish Sea: A compilation. *Northwestern Nat.* **2011**, *92*, 79–94. [CrossRef]
45. Osmek, S.; Calambokidis, J.; Laake, J.; Gearin, P.; Delong, R.; Scordino, J.; Jeffries, S.; Brown, R. *Assessment of the Status of Harbor Porpoise (Phocoena phocoena) in Oregon and Washington Waters*; Department of Commerce, NOAA, National Marine Fisheries Service, Alaska Fisheries Center: Seattle, WA, USA, 1996.

46. Evenson, J.R.; Anderson, D.; Murphie, B.L.; Cyra, T.A.; Calambokidis, J. *Disappearance and Return of Harbor Porpoise to Puget Sound: 20 Year Pattern Revealed from Winter Aerial Surveys*; Technical Report; Washington Department of Fish and Wildlife, Wildlife Program and Cascadia Research Collective: Olympia, WA, USA, 2016.
47. Carretta, J.V.; Oleson, E.; Baker, J.; Weller, D.W.; Lang, A.R.; Forney, K.A.; Muto, M.M.; Hanson, B.; Orr, A.J.; Huber, H.; et al. *U.S. Pacific Draft Marine Mammal Stock Assessments: 2015*; Department of Commerce, NOAA Southwest Fisheries Science Center: San Diego, CA, USA, 2015.
48. Vallejo, G.C.; Grellier, K.; Nelson, E.J.; McGregor, R.M.; Canning, S.J.; Caryl, F.M.; McLean, N. Responses of two marine top predators to an offshore wind farm. *Ecol. Evol.* **2017**, *7*, 8698–8708. [CrossRef]
49. Fenton, H.; Daoust, P.-Y.; Forzán, M.J.; Vanderstichel, R.V.; Ford, J.B.K.; Spaven, L.; Lair, S.; Raverty, S. Causes of mortality of harbor porpoises *Phocoena phocoena* along the Atlantic and Pacific coasts of Canada. *Dis. Aquat. Org.* **2017**, *122*, 171–183. [CrossRef] [PubMed]
50. Calambokidis, J.; Barlow, J. Chlorinated hydrocarbon concentrations and their use for describing population discreteness in harbor porpoises from Washington, Oregon, and California. In *Marine Mammal Strandings in the United States, Proceedings of the Second Marine Mammal Stranding Workshop, Miami, FL, USA, 3–5 December 1987*; Reynolds, J.E., III, Odell, D.K., Eds.; NOAA Technical Report NMFS: Miami, FL, USA, 1991; pp. 101–110.
51. Norman, S.A.; Winfield, Z.C.; Rickman, B.H.; Usenko, S.; Klope, M.; Berta, S.; Dubpernell, S.; Garrett, H.; Adams, M.J.; Lambourn, D.; et al. Persistent Organic Pollutant and Hormone Levels in Harbor Porpoise with B Cell Lymphoma. *Arch. Environ. Contam. Toxicol.* **2017**, *72*, 596–605. [CrossRef] [PubMed]
52. Williams, R.S.; Curnick, D.J.; Brownlow, A.; Barber, J.L.; Barnett, J.; Davison, N.J.; Deaville, R.; Doeschate, M.T.; Perkins, M.; Jepson, P.D., et al. Polychlorinated biphenyls are associated with reduced testes weights in harbour porpoises (*Phocoena phocoena*). *Environ. Int.* **2021**, *150*, 106303. [CrossRef]
53. Ford, J.K.B.; Ellis, G.M.; Barrett-Lennard, L.G.; Morton, A.B.; Palm, R.S.; Balcomb, K.C., III. Dietary specialization in two sympatric populations of killer whales (*Orcinus orca*) in coastal British Columbia and adjacent waters. *Can. J. Zool.* **1998**, *76*, 1456–1471. [CrossRef]
54. Dalheim, M.E.; White, P.A. Ecological aspects of transient killer whales (*Orcinus orca*) as predators in southeastern Alaska. *Wildl. Biol.* **2010**, *16*, 308–322. [CrossRef]
55. Wells, R.S. The role of long-term study in understanding the social structure of a bottlenose dolphin community. In *Dolphin Societies*; Pryor, K., Norris, K.S., Eds.; University of California Press: Berkeley, CA, USA, 1991; pp. 199–255.
56. Connor, R.C.; Smolker, R.A.; Richards, A.F. Two levels of alliance formation among male bottlenose dolphins (*Tursiops* sp.). *Proc. Natl. Acad. Sci. USA* **1992**, *89*, 987–990. [CrossRef]
57. Mann, J.; Connor, R.C.; Barre, L.M.; Heithaus, M.R. Female reproductive success in bottlenose dolphins (*Tursiops* sp.): Life history, habitat, provisioning, and group-size effects. *Behav. Ecol.* **2000**, *11*, 210–219. [CrossRef]
58. Rogers, C.A.; Brunnick, B.J.; Herzing, D.L.; Baldwin, J.D. The social structure of bottlenose dolphins, Tursiops truncatus, in the Bahamas. *Mar. Mamm. Sci.* **2004**, *20*, 688–708. [CrossRef]
59. Kent, E.E.; Mazzoil, M.; McCulloch, S.D.; Defran, R.H. Group characteristics and social affiliation patterns of bottlenose dolphins (*Tursiops truncatus*) in the Indian River Lagoon, Florida. *Fla. Sci.* **2008**, *71*, 149–168.
60. Elliser, C.R.; Herzing, D.L. Community structure and cluster definition of Atlantic spotted dolphins, *Stenella frontalis*, in the Bahamas. *Mar. Mamm. Sci.* **2012**, *28*, e486–e502. [CrossRef]
61. Elliser, C.R.; Volker, C.L.; Herzing, D.L. Integration of a social cluster of Atlantic spotted dolphins (*Stenella frontalis*) after a large immigration event in 2013. *Mar. Mamm. Sci.* **2022**, *39*, 114–130. [CrossRef]
62. Elliser, C.R.; Herzing, D.L. Long-term social structure of a resident community of Atlantic spotted dolphins, *Stenella frontalis*, in the Bahamas 1991–2002. *Mar. Mamm. Sci.* **2014**, *30*, 308–328. [CrossRef]
63. Gaskin, D.E.; Watson, A.P. The harbor porpoise, *Phocoena phocoena*, in Fish Harbour, New Brunswick, Canada: Occupancy, distribution, and movements. *Fish. Bull.* **1985**, *83*, 427–442.
64. Koopman, H.N.; Gaskin, D.E. Individual and geographical variation in pigmentation patterns of the harbour porpoise, *Phocoena phocoena* (L.). *Can. J. Zool.* **1994**, *72*, 135–143. [CrossRef]
65. Flaherty, C.; Stark, S. *Harbor Porpoise (Phocoena phocoena) Assessment, in Washington Sound*; Final Report for Subcontract #80-ABA-3584 to the National Marine Mammal Laboratory; NMFS: Seattle, WA, USA, 1982.
66. Hammond, P.S.; Berggren, P.; Benke, H.; Borchers, D.L.; Collet, A.; Heide-Jørgensen, M.P.; Heimlich, S.; Hiby, A.R.; Leopold, M.F.; Øien, N. Abundance of harbour porpoise and other cetaceans in the North Sea and adjacent waters. *J. Appl. Ecol.* **2002**, *39*, 361–376. [CrossRef]
67. Elliser, C.R.; van der Linde, K.; MacIver, K. Adapting photo-identification methods to study poorly marked cetaceans: A case study for common dolphins and harbor porpoises. *Mamm. Biol.* **2022**, *102*, 811–827. [CrossRef]
68. Mercer, M.C. Observations on distribution and intraspecific variation in pigmentation patterns of odontocete Cetacea in the western North Atlantic. *J. Fish. Res. Board Can.* **1973**, *30*, 1111–1130. [CrossRef]

69. Read, A.J.; Keener, W.; Webber, M.A.; Siebert, U. Harbor porpoise *Phocoena phocoena*. In *Ridgeway and Harrison's Handbook of Marine Mammals, Vol. 1, Coastal Dolphins & Porpoises*, 2nd ed.; Jefferson, T.A., Ed.; Elsevier: San Diego, CA, USA, 2024; pp. 421–486, ISBN 9780443137464.
70. Dunn, H. People's porpoise project 2018–19: Photo-ID methods. *Sea Trust CIC* **2019**, 1–14. Available online: https://www.seatrust.org.uk/wp-content/uploads/2022/11/Peoples-Porpoise-Project-Photo-ID-Methods.pdf (accessed on 1 January 2025).
71. Olie, R.; Zanderink, F. Bruinvissen herkennen: Wendy of Elliptica? *Zoogdier* **2023**, *34*, 12–14. Available online: https://rugvin.nl/wp-content/uploads/2023/12/Zoogdier-december-23-Foto-ID-Bruinvissen.pdf (accessed on 1 January 2025).
72. Hamel, H.; Ortiz, S.T.; Wahlberg, M. Long-term observations reveal short-term mother-calf affiliation in wild harbour porpoises. *Anim. Behav.* **2024**, *219*, 122992. [CrossRef]
73. Shane, S.H. Behavior and ecology of the bottlenose dolphin at Sanibel Island, Florida. In *The Bottlenose Dolphin;* Leatherwood, S., Reeves, R.R., Eds.; Academic Press: San Diego, CA, USA, 1990; pp. 245–265, ISBN 0-12-440280-1.
74. Noren, S.R.; Noren, D.P.; Gaydos, J.K. Living in the fast lane: Rapid development of the locomotor muscle in immature harbor porpoises (*Phocoena phocoena*). *J. Comp. Physiol. B.* **2014**, *184*, 1065–1076. [CrossRef]
75. Adams, J.D.; Speakman, T.; Zolman, E.; Schwacke, L.H. Automating image matching, cataloging, and analysis for photo-identification research. *Aquat. Mamm.* **2006**, *32*, 374–384. [CrossRef]
76. Seabold, S.; Perktold, J. Statsmodels: Econometric and statistical modeling with python. In Proceedings of the SciPly, Austin, TX, USA, 28 June–3 July 2010.
77. Terpilowski, M.A. scikit-posthocs: Pairwise multiple comparison tests in Python. *J. Open Source Softw.* **2019**, *4*, 1169. [CrossRef]
78. Kaps, M.; Lamberson, W.R. *Biostatistics for Animal Science*, 3rd ed.; CABI Publishing: Cambridge, MA, USA, 2017; ISBN 978-1-78639-035-6.
79. Chabanne, D.B.H.; Finn, H.; Bejder, L. Identifying the relevant local population for environmental impact assessments of mobile marine fauna. *Front. Mar. Sci.* **2017**, *4*, 148. [CrossRef]
80. Lockyer, C.; Kinze, C. Status, ecology and life history of harbour porpoise (*Phocoena phocoena*), in Danish waters. *NAMMCO Sci. Publ.* **2003**, *5*, 143–176. [CrossRef]
81. Stern, S.J.; Keener, W.; Szczepaniak, I.D.; Webber, M.A. Return of harbor porpoises (*Phocoena phocoena*) to San Francisco Bay. *Aquat. Mamm.* **2017**, *43*, 691–702. [CrossRef]
82. Anderson, D.; Shuster, L.; Elliser, C.R.; MacIver, K.; Gless, E.J.; Krieger, J.; Hall, A. Harbor Porpoise Aggregations in the Salish Sea. *Oceans* **2023**, *4*, 269–285. [CrossRef]
83. Camphuysen, K.C.J.; Krop, A. Maternal care, calf training and site fidelity in a wild harbour porpoise in the North Sea. *Lutra* **2011**, *54*, 123–126.
84. Zerbini, A.N.; Goetz, K.T.; Forney, K.A.; Boyd, C. Estimating abundance of an elusive cetacean in a complex environment: Harbor porpoises (*Phocoena phocoena*) in inland waters of Southeast Alaska. *Front. Mar. Sci.* **2022**, *9*, 966489. [CrossRef]
85. Zerbini, A.N.; Parsons, K.M.; Goetz, K.T.; Angliss, R.P.; Young, N.C. *Identification of Demographically Independent Populations Within the Currently Designated Southeast Alaska Harbor Porpoise Stock;* NOAA Technical Memorandum NMFS, Alaska Fisheries Science Center: Juneau, AK, USA, 2022. [CrossRef]
86. Johnston, D.W.; Westgate, A.J.; Read, A.J. Effects of fine-scale oceanographic features on the distribution and movements of harbour porpoises *Phocoena phocoena* in the Bay of Fundy. *Mar. Ecol. Prog. Ser.* **2005**, *295*, 279–293. [CrossRef]
87. Nielsen, N.H.; Teilmann, J.; Sveegaard, S.; Hansen, R.G.; Sinding, M.-H.S.; Dietz, R.; Heide-Jørgensen, M.P. Oceanic movements, site fidelity and deep diving in harbour porpoises from Greenland show limited similarities to animals from the North Sea. *Mar. Ecol. Prog. Ser.* **2018**, *597*, 259–272. [CrossRef]
88. Gilles, A.; Adler, S.; Kaschner, K.; Scheidat, M.; Siebert, U. Modelling harbour porpoise seasonal density as a function of the German Bight environment: Implications for management. *Endang. Species Res.* **2011**, *14*, 157–169. [CrossRef]
89. Scheidat, M.; Verdaat, H.; Aarts, G. Using aerial surveys to estimate density and distribution of harbour porpoises in Dutch waters. *J. Sea Res.* **2012**, *69*, 1–7. [CrossRef]
90. Peschko, V.; Ronnenberg, K.; Siebert, U.; Gilles, A. Trends of harbour porpoise (*Phocoena phocoena*) density in the southern North Sea. *Ecol. Indic.* **2016**, *60*, 174–183. [CrossRef]
91. Elliser, C.R.; Hessing, S.; MacIver, K.H.; Webber, M.A.; Keener, W. Harbor porpoises (*Phocoena phocoena* vomerina) catching and handling large fish on the U.S. West Coast. *Aquat. Mamm.* **2020**, *46*, 191–199. [CrossRef]
92. Skov, H.; Thomsen, F. Resolving fine-scale spatio-temporal dynamics in the harbour porpoise *Phocoena phocoena*. *Mar. Ecol. Prog. Ser.* **2008**, *373*, 173–186. [CrossRef]
93. Stedt, J.; Hamel, H.; Ortiz, S.T.; Kristensen, J.H.; Wahlberg, M. Harbour porpoises are flexible predators displaying context-dependent foraging behaviours. *Ecol. Evol.* **2024**, *14*, e70671. [CrossRef]
94. Nehls, G.; Zydelis, R.; Matuschek, R.; Brandt, M.; Diederichs, A.; Hoeschle, C.; Thomsen, F. Impact of high marine traffic on harbor porpoise: Effect on abundance and distribution. In *The Effects of Noise on Aquatic Life;* Popper, A.N., Sisneros, J., Hawkins, A.D., Thomsen, F., Eds.; Springer Nature: Cham. Switzerland, 2024; pp. 1–28, ISBN 9783031104176.

95. Wisniewska, D.M.; Johnson, M.; Teilmann, J.; Siebert, U.; Galatius, A.; Dietz, R.; Madsen, P.T. High rates of vessel noise disrupt foraging in wild harbour porpoises (*Phocoena phocoena*). *Proc. R. Soc. B* **2018**, *285*, 20172314. [CrossRef]
96. Bjørge, A.; Tolley, K.A. Harbor porpoise. In *Encyclopedia of Marine Mammals*, 3rd ed.; Würsig, B., Thewissen, J.G.M., Kovacs, K.M., Eds.; Academic Press: San Diego, CA, USA, 2018; pp. 448–451, ISBN 9780128043271.
97. Møhl-Hansen, U. Investigations on reproduction and growth of the porpoise (*Phocaena phocaena* (L.)) from the Baltic. *Vidensk. Meddr. Dan. Naturh. Foren* **1954**, *116*, 369–396.
98. Sørensen, T.B.; Kinze, C.C. Reproduction and reproductive seasonality in Danish harbour porpoises, *Phocoena phocoena*. *Ophelia* **1994**, *39*, 159–176. [CrossRef]
99. Mann, J.; Smuts, B.B. Natal attraction: Allomaternal care and mother-infant separations in wild bottlenose dolphins. *Anim. Behav.* **1998**, *55*, 1097–1113. [CrossRef]
100. Hill, H.M.; Greer, T.; Solangi, M.; Kuczaj, S.A., II. All mothers are not the same: Maternal styles in bottlenose dolphins (*Tursiops truncatus*). *Int. J. Comp. Psychol.* **2007**, *20*, 35–54. [CrossRef]
101. Dudenski, K.M.; Ribic, C.A.; Hill, H.M.M.; Bolton, T.T. Evidence for maternal style among adult female dolphins when sharing pectoral fin contacts with their calves. *Anim. Behav. Cogn.* **2021**, *8*, 52–68. [CrossRef]
102. Zein, B.; Woelfing, B.; Dähne, M.; Schaffeld, T.; Ludwig, S.; Rye, J.H.; Baltzer, J.; Ruser, A.; Siebert, U. Time and tide: Seasonal, diel and tidal rhythms in Wadden Sea harbour porpoises (*Phocoena phocoena*). *PLoS ONE* **2019**, *14*, e0213348. [CrossRef] [PubMed]
103. Holdman, A.K.; Haxel, J.H.; Klinck, H.; Torres, L.G. Acoustic monitoring reveals the times and tides of harbor porpoise (*Phocoena phocoena*) distribution off central Oregon, U.S.A. *Mar. Mamm. Sci.* **2019**, *35*, 164–186. [CrossRef]
104. Goodwin, L. Diurnal and tidal variations in habitat use of the harbour porpoise (*Phocoena phocoena*) in southwest Britain. *Aquat. Mamm.* **2008**, *34*, 44–53. [CrossRef]
105. Pirotta, E.; Thompson, P.M.; Miller, P.I.; Brookes, K.L.; Cheney, B.; Barton, T.R.; Graham, I.M.; Lusseau, D. Scale-dependent foraging ecology of a marine top predator modelled using passive acoustic data. *Func. Ecol.* **2014**, *28*, 206–217. [CrossRef]
106. Read, A.J.; Hohn, A.A. Life in the fast lane: The life history of harbor porpoises from the Gulf of Maine. *Mar. Mamm. Sci.* **1995**, *11*, 423–440. [CrossRef]
107. Read, A.J. Age at Sexual Maturity and Pregnancy Rates of Harbour Porpoises *Phocoena phocoena* from the Bay of Fundy. *Can. J. Fish. Aquat. Sci.* **1990**, *47*, 561–565. [CrossRef]
108. Calambokidis, J.; Steiger, G.H.; Curtice, C.; Harrison, J.; Van Parijs, S.M.; Halpin, P.N. Biologically important areas for selected cetaceans within U.S. waters—West coast region. In Biologically Important Areas for Cetaceans Within U.S. Water; Van Parijs, S.M., Curtice, C., Ferguson, M.C., Eds. *Aquat. Mamm. (Spec. Issue)* **2015**, *41*, 39–53. [CrossRef]

oceans

MDPI

Article

Investigation of Spatiotemporal Patterns of Harbour Porpoise (*Phocoena phocoena*) Strandings in Swedish Waters for Improved Monitoring and Management

Vigge Ulfsson [1], Hyeyoung Kim [2], Linnea Cervin [3], Anna Roos [3] and Aleksija Neimanis [4,*]

[1] Biology Education Centre, Uppsala University, SE-752 36 Uppsala, Sweden; oakhillvigge@yahoo.se
[2] Department of Epidemiology, Surveillance and Risk Assessment, Swedish Veterinary Agency, SE-751 89 Uppsala, Sweden; hyeyoung.kim@sva.se
[3] Department of Environmental Monitoring and Research, Swedish Museum of Natural History, SE-104 05 Stockholm, Sweden; linnea.cervin@nrm.se (L.C.); anna.roos@nrm.se (A.R.)
[4] Department of Pathology and Wildlife Diseases, Swedish Veterinary Agency, SE-751 89 Uppsala, Sweden
* Correspondence: aleksija.neimane@sva.se

Abstract: Harbour porpoises (*Phocoena phocoena*) are the only cetacean residents found year-round in Swedish waters and they are exposed to numerous natural and anthropogenic threats. Since the in situ monitoring of cetaceans can be difficult, invasive and often expensive, investigation of stranding patterns and examination of stranded animals can be used as a cost-effective source of data to study these elusive animals. The aim of this study was to investigate the spatiotemporal patterns of harbour porpoise stranding reports and the possible underlying causes in Swedish waters over a ten-year period (2014–2023). Additionally, the Swedish stranding network plays a key role in the collection of stranded carcasses for health and disease surveillance, and geographic coverage of the network also was analysed. When making spatial comparisons, the ten-year period was divided into two five-year blocks. Data on 854 stranded harbour porpoises were analysed from the coasts of the Skagerrak, Kattegat, and Baltic Seas. Both significant spatial and temporal patterns could be identified. Strandings peaked in July through September and hotspots occurred along most of the Swedish west coast, with the most frequent hotspots located around Öresund and especially the area around the Kullen peninsula. The spatial patterns of strandings found in this study are consistent with data on porpoise abundance, prey abundance, and gillnet fisheries' efforts. The latter is known to be one of the primary causes of porpoise mortality. Furthermore, the coverage of the Swedish stranding network increased between the two periods, likely reflecting an increased awareness of the carcass-based surveillance program, and gaps requiring network expansion efforts were identified. These results also provide baseline data to enable the continued monitoring of stranding trends, as changes may indicate changes in population distribution, size or mortality rates.

Keywords: harbour porpoise; stranding; *Phocoena phocoena*; cetacean; marine mammal; wildlife monitoring; wildlife health surveillance; spatiotemporal patterns; gillnet fisheries

1. Introduction

Harbour porpoises (*Phocoena phocoena*) are top predators in temperate marine ecosystems and the only permanent cetacean residents in Swedish waters [1]. Globally and in Europe, the harbour porpoise is listed by the International Union for Conservation of Nature (IUCN) as being of Least Concern (LC) [2,3]. There are three genetically and morphometrically separate populations within Swedish waters, including the North Sea population extending into the Skagerrak, the Belt Sea population, and the critically endangered Baltic Sea population [4–6]. Abundance estimates were most recently made by the SCANS-IV surveys and SAMBAH project, where abundances for the North Sea and Belt Sea populations were estimated to include 338,918 (95% CI = 243,063–476,203) and

14,403 (95% CI = 9555–21,769) individuals, respectively [7], while the Baltic Sea population was estimated to include 491 (95% CI = 71–1105) individuals [8]. Estimates for the North Sea population are similar to estimates from previous SCANS surveys. In contrast and of concern, the Belt Sea estimate of 14,403 is considerably lower than earlier estimates of 42,324 (95% CI = 23,368–76,658) in 2016 and 40,475 (95% CI = 25,614–65,041) in 2012 [7,9,10].

Some of the most significant threats to harbour porpoises include environmental contaminants, noise pollution, the loss of habitat, reduced food availability as a result of fishing, and bycatch [11], with bycatch in gillnets continuing to be one of the greatest threats [12–16]. The most significant factors contributing to porpoise bycatch in gillnets are the mesh size, soak time and vessel length [17], with porpoises mainly being bycaught in larger mesh sizes (>120 mm) [18].

Berggren [12] identified gillnets as being responsible for more than 80% of the bycatches of harbour porpoises in Swedish waters. Fisheries with the highest levels of bycatch differed between the three regions assessed by Berggren: Skagerrak, Kattegat and the Baltic Sea. Gillnets set for cod (*Gadus morhua)* accounted for large shares of bycatch in all three areas, with the highest being 72% in the Kattegat Sea. However, the largest share of bycatch observed in the Baltic Sea was from driftnets for salmon (*Salmo salar*). In the Skagerrak Sea, gillnets set for spiny dogfish (*Squalus acanthias*) accounted for most of the bycatch [12]. According to records held by Swedish Museum of Natural History (NRM), harbour porpoises have also been reported as bycatch in Swedish fisheries for lumpfish (*Cyclopterus lumpus*), pollock (*Pollachius pollachius*), Atlantic mackerel (*Scomber scombrus*), turbot (*Scophthalmus maximus*), and langoustine (*Nephrops norvegicus*). However, since Berggren's assessment, there have been noteworthy changes in fisheries activities. Namely, there have been zero quotas on spiny dogfish since 2011 in accordance with council regulation (EU) No 57/2011 [19], but spiny dogfish fisheries have resumed operations in 2023 [20], and targeted fishing on cod has been heavily reduced or prohibited [21]. Yet bycatch still poses a threat to porpoise populations [1,11].

Recent necropsy data on stranded animals examined from Swedish waters show that bycatch, when considered together with probable bycatch, is the most frequent cause of death, supporting previous studies [1]. The carcasses of stranded animals are often in poor states of preservation due to decomposition, handling, scavenging, the delay between stranded animals being reported and collected for necropsy, and the possibility of carcasses drifting for extended periods before stranding. This masks the characteristic signs of bycatch and makes it more difficult to assign a cause of death. Consequently, both the numbers of bycaught individuals and probable bycaught individuals are likely underestimated in the necropsies of stranded porpoises [1,22]. Therefore, an efficient and extensive stranding network is vital to obtaining fresh individuals for necropsies to facilitate the determination of the cause of death and the identification of bycatch and other threats.

Because porpoise populations face numerous threats, harbour porpoises are protected and listed as a species in need of conservation efforts and monitoring in several conventions and directives both in Sweden and internationally [11]. Some examples include the Habitats Directive (92/43/EEC) [23], the Oslo-Paris (OSPAR) Convention [24], the Helsinki Commission (HELCOM) [25], and the Agreement on the Conservation of Small Cetaceans of the Baltic, North East Atlantic, Irish and North Sea (ASCOBANS) [26], for which, the harbour porpoise is named as the flagship species. Despite efforts and conventions, knowledge of the health status of harbour porpoises in Swedish waters is described as inadequate by the Swedish Agency for Marine and Water Management (SwAM) [11]. Efficient and systematic monitoring schemes are essential in augmenting our knowledge of this species.

Monitoring porpoise populations not only helps with assessing the health of the species itself but also provides management with vital information to assess the overall health and stability of marine ecosystems because top predators reflect the health and stability of the environments in which they reside [1,27]. Marine mammals often have long life spans, which, combined with feeding at high trophic levels, leads to the accumulation of environmental contaminants in their tissues [28]. Since harbour porpoises share the coastal environment and food sources with people, they may also function as effective sentinels for human health [1,28]. Furthermore, with the increasing noise pollution and disturbance from commercial shipping and the construction of offshore wind farms, the need to monitor marine mammals is even greater to detect impacts which may even have cascading, negative effects on the rest of the ecosystem [29].

However, monitoring cetaceans in situ, i.e., at sea, can be costly and challenging due to their elusive nature, high mobility, and ability to spend extended periods underwater [1,30,31]. As an alternative to in situ methods, strandings can be used as a cost- and time-effective way of monitoring cetacean populations and they provide excellent opportunities for biological sampling [1,30]. The documentation of stranding reports and the collection of stranded individuals for necropsies over time enables the identification of spatial and temporal changes in stranding patterns, as well as changes in overall population health [1,22]. Several studies in Northern Europe have investigated spatiotemporal trends in cetacean strandings and factors that affect the distribution and cause of strandings, such as life history parameters, population density, fisheries activity, and weather phenomena [22,32,33]. However, to our knowledge, similar studies have not been conducted in Sweden.

The primary aim of this study was to investigate whether or not spatiotemporal patterns in harbour porpoise strandings occur in Sweden, and if they occur, to describe their nature and investigate possible underlying factors. Additionally, this study aimed to investigate the coverage of Sweden's stranding network to identify areas of inadequate coverage where there is a high clustering of stranding reports. Finally, with no similar studies yet conducted in Sweden, this research serves to establish baselines for future monitoring and provides further evidence for the value of using strandings to inform management.

2. Materials and Methods

2.1. Data Collection and Study Area

For this study, stranded harbour porpoises were defined as porpoises that were found onshore, or in rare cases, found dead, drifting at sea with an unknown history or cause of death. Porpoises that were submitted directly from known fishery interactions, i.e., were retrieved from nets set at sea, were excluded, as they did not represent strandings. Reports of stranded harbour porpoises are made by the general public to the Swedish Museum of Natural History (NRM). Since 2021, reports have been submitted to NRM via a web-based formulary (https://marinadaggdjur.nrm.se accessed on 31 October 2023) [34]. Before the website's launch, reports were submitted to NRM via telephone and email and those have retrospectively been added to the reporting database. Reports generally contain the geographic coordinates (latitude and longitude), number of individuals, and observation date. Reports also often include a short comment, and the submission of photographs is encouraged. When they are deemed an animal of interest and resources are available, stranded harbour porpoises are retrieved by the Swedish Veterinary Agency's (SVA) stranding network and submitted for necropsy examination. Examination includes a determination of the cause of death, investigation of other pathologies, surveillance for pathogens and collection and archiving of an extensive suite of samples and data as described in Neimanis et al. [1]. The stranding network includes private citizens, academic institutions, non-governmental organizations, municipalities, counties, the police, the Coast Guard, fishermen, and hunters [35]. The study area encompasses all of Sweden's coastline which runs from 59.0° N, 11.0° E on the west coast to 65.7° N, 24.2° E on the northeast coast.

2.2. Stranding Data, Necropsy Data, and Processing

Data on stranded harbour porpoises from 1 January 2014 to 31 December 2023 were retrieved from NRM and consisted of 1007 reports. Suspected or confirmed double reports (N = 161) and findings from outside of the Swedish territory (N = 4) were excluded from the dataset. Reports containing pictures were quality-controlled to remove other marine fauna that were misidentified as porpoises (N = 1). Furthermore, a few observations were missing coordinates (N = 14). For these, an approximate position was added in cases where the location was clearly described in the comments (N = 13). In some cases, the number of individuals was reported as more than one, while the comment indicated only one finding. In these cases, the numbering of individuals was corrected to one for a conservative estimation (N = 13). Observations where the comment clearly stated that the individual was submitted as bycatch rather than stranded (N = 11) were also omitted from the data set. All analyses were conducted using the individual count rather than the number of reports since some reports include more than one individual.

Necropsy data for the years 2020–2022, provided by the SVA, was acquired from Sweden's dataportal (www.dataportalen.se accessed on 27 November 2023) [36]. Furthermore, necropsy data from 2014–2019 was provided in Neimanis et al. [1] and data from 2023 was directly accessed from the SVA's database, SVALA.

2.3. Seasonality and Cluster Identification

Temporal variations and seasonality in stranding reports were assessed using chi-squared tests to ascertain whether the distribution of yearly and monthly occurrences deviated from a uniform pattern. Linear trends in annual recorded strandings were examined using Pearson correlation statistics to determine whether there was an upward, downward, or stable trend in the recorded strandings. Stranding reports of skeletal remains (N = 16 individuals) were excluded from the temporal analysis since the finding date does not necessarily reflect the stranding date. However, skeletal remains were included in the spatial analyses. Results were visualised using bar plots containing the month and year on the x-axis and the number of stranded individuals on the y-axis.

Spatial analyses utilised administrative-level data for Swedish municipalities. Topographical data for administrative areas were obtained from Lantmäteriet on 17 November 2023 [37]. Only topographical information of administrative regions where stranded porpoises were located was included in the analysis. A grid with 5 $\times$ 5 km cells was created to refine the spatial analysis, employing administrative areas as the spatial framework. Clusters were discerned using the Local Moran's I statistic [38], implemented with the spdep package version 1.2-8 [39]. The Local Moran's statistics were computed with the "greater" alternative and queen contiguity as spatial weights. The alternative "greater" was applied to evaluate only positive autocorrelation, meaning identifying only clusters with low values surrounded by low values, "coldspots", and clusters with high values surrounded by high values, "hotspots". A significance level was calculated for each grid cell where $p < 0.05$ indicates significance. The Local Moran's I statistic was calculated individually each year to assess cluster persistence over time. Additionally, the ratio of animals retrieved for necropsy from each clustered grid cell to the total number of stranded individuals in the corresponding cell was calculated to quantify coverage by the stranding network and pinpoint areas lacking coverage. The difference in coverage over time was calculated using a Wilcoxon rank sum test [40,41].

Maps were created using the packages ggplot2 version 3.4.4 [42], rworldmap version 1.3-8 [43], and ggspatial version 1.1.9 [44]. The 10-year period was split in two. Maps were produced for both 5-year periods (2014–2018 and 2019–2023) to facilitate comparison and because 5-year periods were deemed to be suitable time frames to follow trends for management purposes. Coverage maps were visualised with grids from clusters, but also grids not identified as clusters where individuals had been collected for necropsies. All statistical and spatial analyses were conducted using R version 4.2.2 [45].

3. Results

3.1. Seasonality and Temporal Trends in Stranding Reports

After data processing, the dataset for the temporal analysis contained 814 stranding reports with 838 individuals. Chi-squared tests showed yearly distributions significantly deviating from a uniform pattern ($\chi^2 = 169.35$, d.f. = 9, $p < 0.001$), with the lowest number of stranding reports occurring in 2014 (N = 18) and the highest number of stranding reports occurring in 2022 (N = 147) (Table 1 & Figure 1). Pearson correlation tests showed a significant upward trend in stranded individuals across years (r = 0.853, d.f. = 8, $p = 0.002$). Monthly strandings also differed from a uniform pattern ($\chi^2 = 368.64$, d.f. = 11, $p < 0.001$), with reported strandings peaking in late summer (July and August) and early autumn (September and October) and with minimal reported strandings in the winter months (December, January, and February) (Table 1 & Figure 2).

Table 1. Counts of reported stranded harbour porpoise individuals (*Phocoena phocoena*) across the months and years of 2014–2023. Skeletal remains are included in counts and specified in brackets.

	2014	2015	2016	2017	2018	2019	2020	2021	2022	2023	Total Reported Stranded Individuals
Total	18 (1)	37 (1)	63 (1)	111 (1)	64 (2)	81 (2)	104 (2)	120 (2)	147 (2)	109 (2)	854 (16)
January	0	5	1	4 (1)	2	3	2	3	9	2	31 (1)
February	0	1	5	1	1	2	3	0	11	4	28
March	3	1	1	3	1	11	2	14	9	3	48
April	3	1	1	10	2	2 (1)	5 (1)	5	5	5	39 (2)
May	0	5 (1)	3 (1)	5	1	2	3	14	24	5	62 (2)
June	0	3	1	13	6 (1)	2	4	10 (1)	10 (1)	5	54 (3)
July	3	2	13	11	11	12	22 (1)	16	31	26 (1)	147 (2)
August	5	3	11	25	11	11	31	22	22	21	162
September	2 (1)	3	16	11	16	16	15	18 (1)	10 (1)	20	127 (3)
October	0	4	2	16	8 (1)	16 (1)	7	9	10	13	85 (2)
November	2	7	8	9	2	4	6	8	5	3	54
December	0	2	1	3	3	0	4	1	1	2 (1)	17 (1)

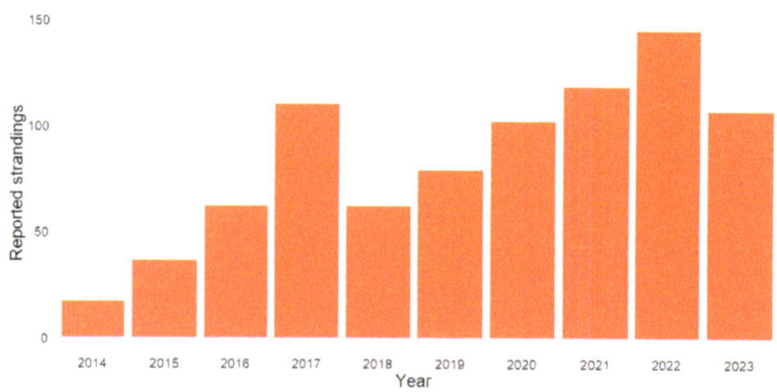

Figure 1. Annual counts of stranded individual harbour porpoises (*Phocoena phocoena*) reported in Sweden for the period of 2014–2023.

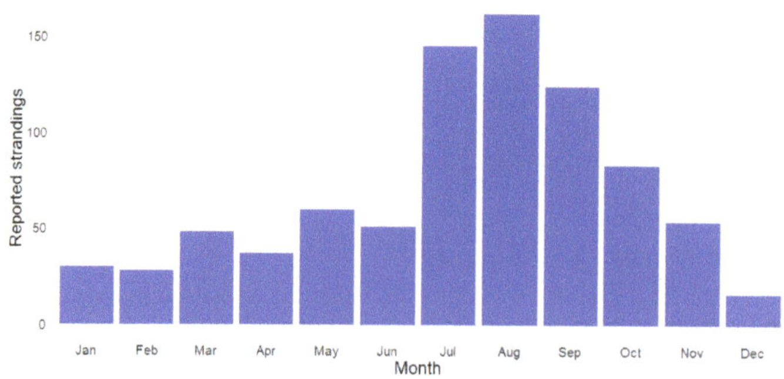

Figure 2. Monthly counts of stranded individual harbour porpoises (*Phocoena phocoena*) reported in Sweden across the period of 2014–2023.

3.2. Distribution of Total Strandings

Data sets for spatial analyses included reports of skeletal remains and contained 829 stranding reports with 854 individuals. Strandings were observed in 187 grids in total (N = 125 in 2014–2018 and N = 156 in 2019–2023), covering essentially the whole Swedish west and south coast. However, almost no strandings were reported on the East Coast, with the exception of a few strandings in the Simrishamn municipality between 2014 and 2018 and single findings in the Ronneby municipality and the western coast of Öland in the Mörbylånga municipality between 2019 and 2023 (Figure 3). Although generally increasing, strandings appear to be reasonably consistent spatially, with the highest stranding numbers occuring in southern Halland along the coastlines of Halmstad and Laholm, and in Scania along Öresund.

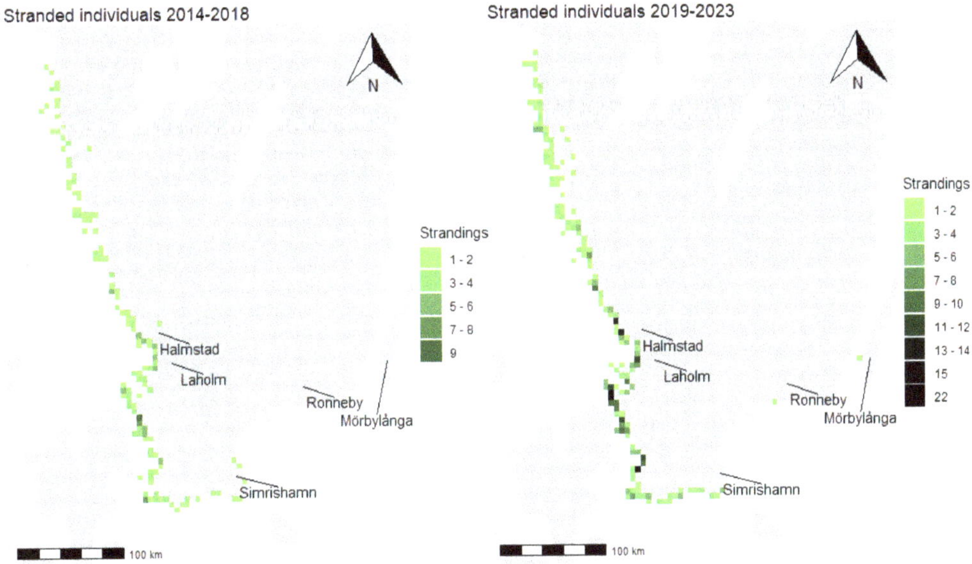

Figure 3. Grid map showing all reports of stranded harbour porpoises (*Phocoena phocoena*) in each grid cell over the two periods of 2014–2018 and 2019–2023. Darker grid cells indicate a higher abundance of stranded individuals.

3.3. Spatial Clusters and Cluster Persistence

Significant clustering was identified for 103 grids for the total period (N = 65 in 2014–2018 and N = 80 in 2019–2023). All identified clusters were grids containing high values surrounded by high values and, therefore, "hotspots". Stranded individuals in clusters (Figure 4) and cluster persistence, i.e., how many years a cluster appears during the period (Figure 5), appear to be very similar, with areas with high strandings being persistent as clusters over time.

Generally, the cluster distribution is similar for both 2014–2018 and 2019–2023, with the cluster appearance being denser in Öresund and southern Halland and gradually becoming more scarce further north, south, and east of these areas. In 2014–2018, no clusters appeared in more than three out of the five years. More persistent clusters (i.e., clusters appearing two to three times) appear around Öresund, especially the northern part along the coastline of the Höganäs municipality known as the Kullen peninsula, but also along the coastlines of the municipalities of Helsingborg, Landskrona, and the coastlines of the municipalities of Halmstad and Laholm in Halland. Clusters occurred more frequently in 2019–2023, with some clusters occurring all five years. Similar to the previous five years, the most persistent clusters are situated along the coasts of Höganäs, the Kullen peninsula,

Helsingborg, and Landskrona, but even more distinctly than previously. Distinct clusters in Halmstad and Laholm also persist. Notably, the largest difference between the two periods is the appearance of persistent clusters along all of Lommabukten, north of Malmö.

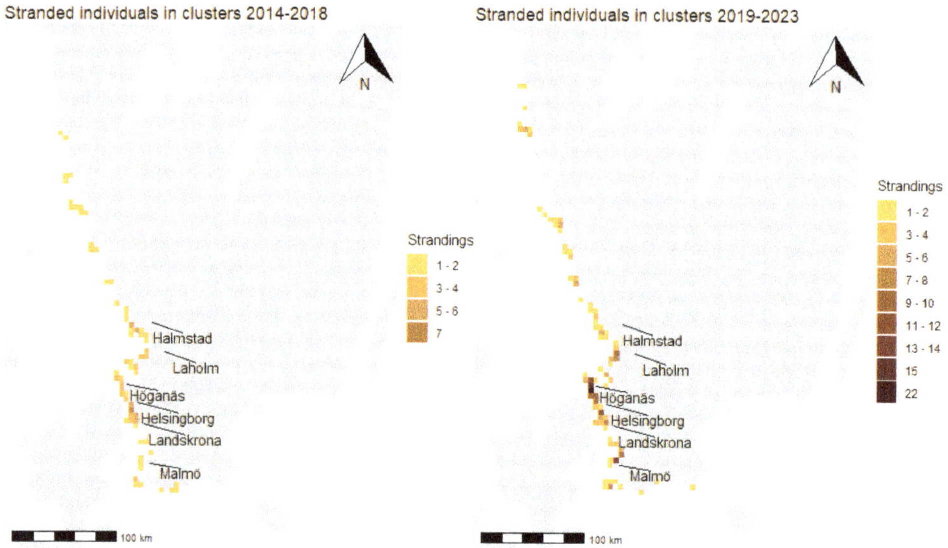

Figure 4. Grid cells with a significant clustering of harbour porpoise (*Phocoena phocoena*) stranding reports (hotspots) and the number of stranded harbour porpoises in each grid for each respective period. Only reports from years where the grid cell was identified as a cluster are included. Darker grid cells indicate higher numbers of strandings.

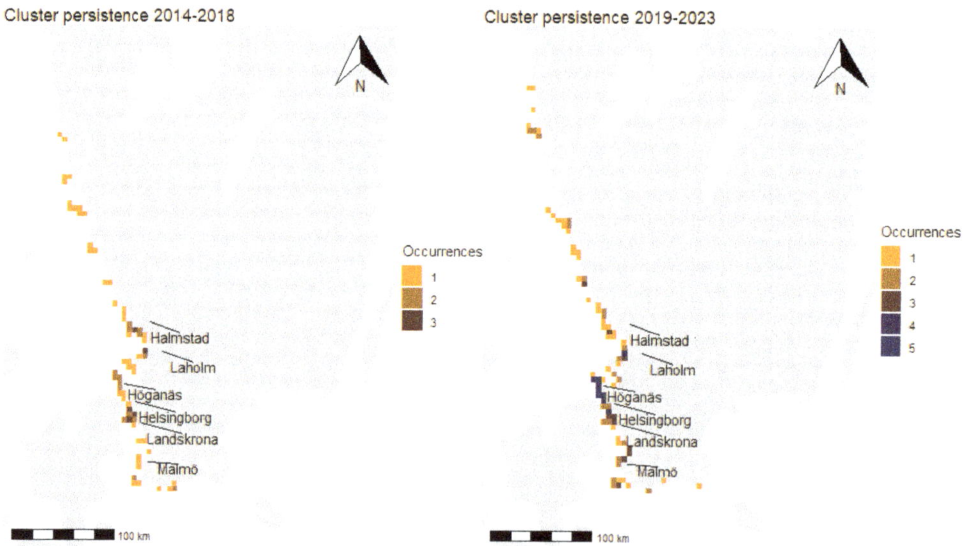

Figure 5. Grid cells with a significant clustering of harbour porpoise (*Phocoena phocoena*) stranding reports (hotspots) and the occurrence of each significant cluster across each respective period. Occurrences refer to the number of years each grid cell has been identified as a significant cluster for each period. Darker grid cells indicate more persistent clusters that occur more often.

3.4. Stranding Network Coverage

Median coverage increased between the two periods of 2014–2018 (Mdn = 0, $\overline{x}$ = 0.0679) and 2019–2023 (Mdn = 0.0839, $\overline{x}$ = 0.175) ($p < 0.001$). During both periods, several individuals located outside of grids identified as clusters were retrieved for necropsy. This includes areas, for both periods, far north on the West Coast and, for the later period, on the South Coast, as well as individuals east of Scania. Most areas in 2014–2018 had nearly zero coverage. Coverage increased in general during 2019–2023 and there were fewer areas with zero coverage. However, some distinct regions with nearly zero coverage remain. These are highlighted with red circles in Figure 6. These regions include the coast of Falkenberg, northern Halmstad, southern Gothenburg, and areas around Lommabukten. While efforts are made to enter all reports of known strandings into the database, two of the grid cells for 2019–2023 had more individuals retrieved for necropsies than there were stranded individuals recorded for that grid (2 in a grid cell of 1 and 3 in a grid cell of 2). These were set to have a coverage of 1.00.

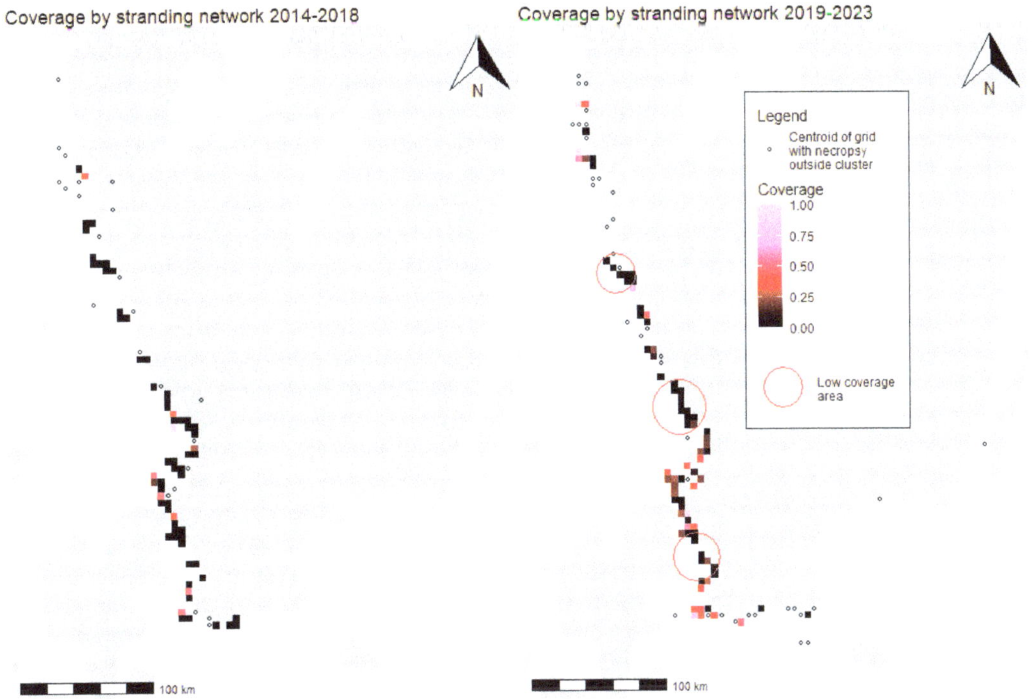

Figure 6. Grid cells with a significant clustering of harbour porpoise (*Phocoena phocoena*) stranding reports (hotspots) and the coverage by the stranding network is represented by the share of stranded individuals in that area retrieved for necropsy. Darker grid cells indicate lower coverage. Small black circles represent the centroids of grid cells not identified as hotspots where porpoises have been recovered for necropsy. Large red circles highlight larger areas with nearly zero coverage for the period of 2019 to 2023.

4. Discussion

Similar to other studies [22,32,46], stranding reports in Sweden have significantly increased over time (r = 0.853, d.f. = 8, $p = 0.002$) (Figure 1). The increasing trend in Sweden is very clear, with the exception of the uptick in stranding reports in 2017. Increased strandings, as suggested by McGovern [32], can be attributed to various factors; these include (1) an actual increase in strandings where population abundance has remained relatively static, (2) an increased porpoise abundance in the area, (3) an increased reporting effort, and (4) currents and weather conditions shifting. Increased strandings as a consequence of a higher porpoise abundance in the area is unlikely, considering that no increase in porpoise abundance has been identified by the SCANS-IV survey (2022) [7] when compared to the SCANS-III survey (2016) [10] in the North Sea or the Kattegat and Belt seas. In fact, the inverse is happening in the Belt Sea, where the estimates from the SCANS-IV survey of 14,403 and the estimates from miniSCANS-II of 17,301 [46] are alarmingly lower than the previous estimates of 42,324 from SCANS-III. Instead, these estimates support the possibility of (1) being true because a population decrease may be caused by a higher mortality and consequently lead to more porpoises stranding, provided the stranding probability of dead animals and the reporting effort remains relatively static. However, a population decrease can also be the result of a population shift away from the survey area or a decreased reproductive rate.

Interestingly, stranding rates spiked in 2017 and stand out when compared to earlier years and the year after, having almost twice as many strandings as 2016, and 2018. According to Gilles et al. [7], a decline in the porpoise abundance in the Belt Sea cannot be excluded, and a further analysis using a Bayesian approach is planned to investigate the status of the Belt Sea population. If these analyses suggest a significant population decline since 2016, it would be interesting to investigate whether or not stranding data regarding the Belt Sea population from Denmark and Germany also indicate stranding peaks in 2017. If so, we hypothesize that this year may have marked a year of unusually high mortality and warrants further investigation of the causes.

Although an actual increase in strandings may have contributed to an increase in reports over time, stranding reports are furthermore influenced by the reporting effort [47] and may have increased as a consequence of raised awareness. In 2021, a new health and disease surveillance program was officially launched, resulting in increased outreach efforts to raise awareness of the need for reporting strandings. This also coincided with the COVID-19 pandemic when people generally spent more time outdoors. Double reports of the same stranded animal have occurred more frequently since the start of 2021, supporting an increased awareness and effort from this time point, and this reinforces the reliability of the data from 2021 onwards. However, double reports were difficult to identify prior to 2020 because digital tools to easily identify double reports were not yet available. This makes it difficult to assess whether double reports actually have increased or whether new tools have facilitated distinguishing double reports. Despite this, the distinct jump in double reports since 2021 and the new digital infrastructure suggest that the effort should be comparable from 2021 and onwards. Under that assumption, strandings appear to be fairly consistent between 2021 and 2023. Whether or not shifting currents or weather conditions have affected strandings over time was beyond the scope of this study and requires further investigation.

Annually, strandings clearly peak in the months of July, August, and September, then gradually decrease to a minimum in December (Figure 2). Summer stranding peaks align with previous studies, both from neighbouring countries, and studies outside of Northern Europe. In German waters, peak strandings were recorded from June to August in the North Sea while they were delayed until July to September in the Baltic Sea [22]. Along the north-eastern coast of the North Sea and in Danish waters, multiple studies consistently also confirm the strong seasonality of strandings in summer [48–50]. Looking outside of Europe, strandings have also been found to peak within July to August in the Sea of Azov [51].

Seasonality in strandings is caused by the variation in the abundance, distribution, or mortality of animals and the non-biological components, oceanographic and effort factors, of the strandings process [33]. Calving season has been discovered to strongly correlate with stranding peaks, with high rates of calves, neonates, and yearlings stranding [22,51]. This is a particularly vulnerable time for neonates and calves that are inexperienced and still depend on their mothers. Mating occurs shortly after calving. Mean conception dates, estimated to be 25 July ± 20.3 days for the Kattegat and Skagerrak seas, and considerably later in the Baltic Sea at 18 August ± 11.8 days [52], strongly align with stranding peaks in those same areas from Siebert et al. [22] and also from this study. Additionally, summer is an energetically demanding period for adult females that are near parturition and/or lactating and a time of independent foraging for yearlings, possibly leading to risky foraging behaviour near gillnets for both groups [51]. Precarious foraging behaviour may not only lead to higher mortality among adult females, but also to their dependent, nursing calves, yielding more strandings in summer [51,53].

Strandings to some extent reflect the relative abundance of harbour porpoises in the surrounding waters and fluctuations in abundance contribute to seasonality in strandings [32,49]. Sveegard et al. [54,55] mapped the abundance of harbour porpoises in Danish waters and along the Swedish west coast using satellite tracking between 1997 and 2007 and acoustic surveys in 2007. They concluded that high-density areas are relatively stable over time and that both satellite tracking and acoustic surveys are powerful tools to monitor porpoise density. Satellite tracking showed porpoise densities to be highest near the Swedish west coast from March to April, while acoustic surveys showed higher abundances from July to August, partially supporting strandings' representativeness of the relative abundance of at-sea populations when compared to the stranding seasonality found in this study.

Lockyer and Kinze [48] suggest the possibility of the reporting effort in summer affecting the pattern of stranding reports, with more people moving around the coastline. While it is true that sightings of live porpoises, just like strandings, often peak in summer [22,48,56], there are also examples of sightings peaking earlier, in February, while the stranding peak remains in summer [50]. Furthermore, studies with a constant effort across years and months in German waters have shown evidence of stranding peaks occurring in summer [22], supporting the probability of these patterns being true increases in strandings, although the contribution of an increased reporting effort in the summer cannot be excluded. Additionally, stranded animals have a higher probability of being observed than sightings of live animals at sea because stranded animals are less mobile, suggesting strandings are a more accurate data source of species occurrence than sightings [32]. However, even though they are less mobile than living specimens, stranded individuals can move due to waves, winds, and currents, occasionally getting completely swept away from the original stranding location.

In accordance with abundance data, all discerned hotspots were located on the west and southern coasts (Figure 5), inhabited by the Belt Sea and North Sea populations [7]. Hotspots increased in both number and frequency in the second period when compared to the first one, aligning with there being more stranding reports in general and an increased reporting effort since 2021. Furthermore, the spatial distribution of hotspots is, in general, similar between the periods, indicating no significant population shifts during the study period. Additionally, when comparing Figures 4 and 5, stranding reports seem to reflect the hotspot occurrence for each respective period, indicating no or very few extreme stranding events have occurred.

Persistent hotspots appear in Öresund, especially in the areas around Kullen. Previous studies, using satellite and acoustic surveys, have pointed out Kullen as a key habitat for harbour porpoises and as the area most important to adult female porpoises in the waters around Scania. Porpoises aggregate around Kullen and central Öresund in the spring, summer, and fall, which overlaps with calving and mating periods [48,57]. Porpoises have also been found to follow specific prey species such as cod, herring, and gobies, with a high occurrence of cod in porpoise stomachs year-round. Cod appears to be an especially important food source in summer when larger cod (20–45 cm), which are known to constitute the majority of consumed cod, accumulate in Öresund [58,59]. Notably, fishing efforts by the Swedish and Danish commercial gillnet fleets are high in Öresund and around Kullen, with studies mapping gillnet fishing efforts in 2018 [18] and for the period of 2010–2020 [17]. The areas with the highest gillnet fishing efforts from these studies are very similar to the most frequently occurring stranding hotspots found in this study, especially when considering larger mesh sizes (≥120 mm), in which porpoises are known to more frequently getcaught. We demonstrate a clear association between harbour porpoise abundance, gillnet fisheries' effort, prey abundance, and stranded porpoises. However, to what extent each factor influences strandings is unclear.

Hotspot occurrence relies on stranded porpoises being reported. Naturally, the geography of the areas in which strandings occur can influence the extent of detection and, consequently, the reporting of stranded animals [47]. For example, the coastal landscape of Halland and Scania is, in general, flat, with numerous sandy beaches which are easily accessible. On the contrary, the coastal landscape around Gothenburg and further north consists of a rocky, sometimes inaccessible archipelago, often with steep cliffs. Therefore, the geography on the Swedish west coast might additionally influence the detection rate and reporting rate and would hypothetically influence rates on the northern half of the Swedish west coast negatively. Furthermore, the extent of oceanographic conditions' and weather conditions' effect on stranding frequency and location requires further research.

The increase in stranding network coverage is likely influenced by the establishment of an official health and disease surveillance program for harbour porpoises in Sweden in 2021. Starting in 2016, a limited number (10–15) of post-mortem examinations of stranded and bycaught porpoises were funded under the directive of the SwAM as part of Sweden's environmental monitoring [1]. Since the new health and disease surveillance program was launched in 2021, over 40 porpoises have been examined via necropsy per year [36]. Targeted outreach efforts have been made since 2021 to increase awareness of the need for reporting strandings. Furthermore, the stranding network has expanded since the establishment of the surveillance program to facilitate the recovery of stranded porpoises and is thereby contributing to more extensive coverage.

Many necropsied individuals were recovered from areas outside of clusters. This can be explained by the nature of the health and disease surveillance program. The recovery of carcasses is largely influenced by the geographical location of the contacts in the stranding network and the accessibility of the locations of stranded porpoises. Additionally, some porpoises are prioritized for retrieval for necropsy. Examples may include individuals from threatened populations, stranding anomalies in time and space, and recently deceased individuals, which yield post-mortem investigations of higher quality [60]. Targeted efforts to recover porpoises from the Baltic Sea Proper can be ascribed to the conservation status of the critically endangered Baltic subpopulation, giving it high surveillance priority. The Baltic subpopulation at times overlaps with the Belt Sea population around southeastern Scania, which helps explain the high recovery efforts in those areas despite a lack of clustering.

5. Conclusions

There has been an increasing trend in the stranding reports of harbour porpoises over the last 10 years in Sweden. The underlying causes cannot be determined with certainty, but the increasing trend can likely, in part, be attributed to the increased reporting efforts from the increased awareness due to outreach efforts and the launch of a new health and disease surveillance program in 2021. Porpoise abundance remains seemingly static for the North Sea population, while the Belt Sea population experienced a possible decrease which could help explain the increase in reported strandings. However, whether the increase in stranding reports marks an actual increase in morality requires further investigation.

Stranding patterns showed a significant seasonality, similar to other studies on harbour porpoise strandings in the same or surrounding areas [22,33,48,49], with most stranding reports occurring from July to September. Notably, the stranding reports' peak coincides with calving season for porpoises in the waters surrounding Sweden. Clustering exclusively occurs on the west and southern coast of Sweden, reflecting the relative porpoise abundance in the surrounding waters and remains relatively similar over the study period. Furthermore, Öresund, and especially the area around the Kullen peninsula, a suspected calving ground, displays a frequent clustering of porpoise strandings. The same area shows high abundances of porpoises, high prey abundance, and high fishing efforts with gillnets, providing support for the connection between porpoise abundance, prey abundance, gillnet fisheries' effort, and strandings. Further research is required to investigate the relative contributions of each factor to help inform appropriate management actions including limiting access to vulnerable areas at certain times. Public awareness campaigns are already in place to minimize the disturbance of porpoises in high-density areas such as around the Kullen peninsula and to encourage the reporting of stranded animals. Such initiatives could be expanded to other hotspots. The influence of weather and oceanographic variables such as currents on strandings also needs further investigation.

We also conclude that coverage has increased from the first period of 2014–2018 to the second period of 2019–2023. However, some larger areas with significant clustering still lack coverage, including the coast of Falkenberg, southern Gothenburg, northern Halmstad, and certain areas around Lommabukten. Extending the stranding network through targeted outreach efforts to cover these gaps will facilitate the retrieval of stranded animals and will result in post mortem examinations of higher quality to maximize insights into the health status of the population, possible threats, and causes of mortality to inform management. Furthermore, outreach efforts appear to have a positive effect on reporting rates and the quality of data has increased in recent years. Further improvements on the reporting forms, such as accurate carcass measurement tools and detailed pictures, would facilitate determining the age class, degree of decomposition, and identifying bycatch marks, yielding important information for the management of Swedish harbour porpoise populations and potentially identify areas of ecological importance, such as calving grounds, and areas of conflict between wildlife and human activity.

Supplementary Materials: The following supporting information can be downloaded at: https://www.mdpi.com/article/10.3390/oceans5020010/s1, Datafile S1: Dataset of harbour porpoise (*Phocoena phocoena*) strandings in Sweden 2014–2023.

Author Contributions: Conceptualization, A.N. and A.R.; methodology, all authors; validation, V.U. and L.C.; formal analysis, V.U. and H.K.; data curation, L.C. and A.R.; writing—original draft preparation, V.U.; writing—review and editing, H.K., L.C., A.R. and A.N.; visualization, V.U. and H.K.; supervision, H.K., L.C., A.R. and A.N.; project administration, A.N.; funding acquisition, A.N. and A.R. All authors have read and agreed to the published version of the manuscript.

Funding: This research was partly funded by the Swedish Agency for Marine and Water Management (SwAM) contract numbers 66-21 and 67-21 as part of Sweden's environmental monitoring.

Institutional Review Board Statement: Not applicable as no live animals were used in this study.

Informed Consent Statement: Not applicable.

Data Availability Statement: All data presented in this study are available in the Supplementary Materials.

Acknowledgments: The authors thank Ulrika Larsson Pettersson (Swedish Veterinary Agency) and Annika Strömberg, Jessica Åsbrink and Pia Östenson (Swedish Museum of Natural History) for their assistance with the entry of retrospective stranding reports into the database. We also are indebted to all of the people who have reported stranded porpoises over the years.

Conflicts of Interest: The authors declare no conflicts of interest. The funders had no role in the design of the study; in the collection, analyses, or interpretation of the data; in the writing of the manuscript; or in the decision to publish the results.

References

1. Neimanis, A.; Stavenow, J.; Ågren, E.O.; Wikström-Lassa, E.; Roos, A.M. Causes of Death and Pathological Findings in Stranded Harbour Porpoises (*Phocoena phocoena*) from Swedish Waters. *Animals* **2022**, *12*, 369. [CrossRef] [PubMed]
2. Braulik, G.T.; Minton, G.; Amano, M.; Bjørge, A. *Phocoena phocoena* (amended version of 2020 assessment). *IUCN Red List Threat. Species* **2023**, *1*, e.T17027A247632759. [CrossRef]

3. Sharpe, M.; Berggren, P. *Phocoena phocoena* (Europe assessment). *IUCN Red List Threat. Species* **2023**, *1*, e.T17027A219010660. Available online: https://www.iucnredlist.org/species/17027/219010660 (accessed on 16 January 2024).
4. Wiemann, A.; Andersen, L.W.; Berggren, P.; Siebert, U.; Benke, H.; Teilmann, H.; Lockyer, C.; Pawliczka, I.; Skóra, K.; Roos, A.; et al. Mitochondrial Control Region and Microsatellite Analyses on Harbour Porpoise (*Phocoena phocoena*) Unravel Population Differentiation in the Baltic Sea and Adjacent Water. *Conserv. Genet.* **2010**, *11*, 195–211. [CrossRef]
5. Galatius, A.; Kinze, C.; Teilmann, J. Population Structure of Harbour Porpoises in the Batlic Region: Evidence of Separation Based on Geometric Morphometric Comparisons. *J. Mar. Biol. Assoc. U. K.* **2012**, *92*, 1669–1676. [CrossRef]
6. Carlström, J.; Carlén, I.; Dähne, M.; Hammond, P.S.; Koschinski, S.; Owen, K.; Sveegaard, S.; Tiedemann, R. *Phocoena phocoena* (Baltic Sea subpopulation). *IUCN Red List Threat. Species* **2023**, *1*, e.T17031A50370773. Available online: https://www.iucnredlist.org/species/17031/50370773 (accessed on 16 January 2024).
7. Gilles, A.; Authier, M.; Ramirez-Martinez, N.C.; Araújo, H.; Blanchard, A.; Carlström, J.; Eira, C.; Dorémus, G.; Fernández-Maldonado, C.; Geelhoed, S.C.V.; et al. Estimates of Cetacean Abundance in European Atlantic Waters in Summer 2022 from the SCANS-IV Aerial and Shipboard Surveys. 2023; p. 64. Available online: https://www.tiho-hannover.de/itaw/scans-iv-survey (accessed on 16 January 2024).
8. Amundin, M.; Carlström, J.; Thomas, L.; Carlén, I.; Koblitz, J.; Teilmann, J.; Tougaard, J.; Tregenza, N.; Wennerberg, D.; Loisa, O.; et al. Estimating the abundance of the critically endangered Baltic Proper harbour porpoise (*Phocoena phocoena*) population using passive acoustic monitoring. *Ecol. Evol.* **2022**, *12*, e8854. [CrossRef] [PubMed]
9. Viquerat, S.; Herr, H.; Gilles, A.; Peschko, V.; Siebert, U.; Sveegaard, S.; Teilmann, J. Abundance of harbour porpoises (*Phocoena phocoena*) in the western Baltic, Belt Seas and Kattegat. *Mar. Biol.* **2014**, *161*, 745–754. [CrossRef]
10. Hammond, P.S.; Lacey, C.; Gilles, A.; Viquerat, S.; Börjesson, P.; Herr, H.; Macelod, K.; Ridoux, V.; Santos, M.B.; Teilmann, J.; et al. Estimates of Cetacean Abundance in European Atlantic Waters in Summer 2016 from the SCANS-III Aerial and Shipboard Surveys. 2021. Available online: https://scans3.wp.st-andrews.ac.uk/files/2021/06/SCANS-III_design-based_estimates_final_report_revised_June_2021.pdf (accessed on 16 January 2024).
11. Swedish Agency for Marine and Water Management. Åtgärdsprogram för Tumlare. 2021: 11. Available online: https://www.havochvatten.se/download/18.764c2cb917a6c5c85b03c106/1625648670494/rapport-2021-atgardsprogram-for-tumlare-v2.pdf (accessed on 29 November 2023).
12. Berggren, P. Bycatches of the harbour porpoise (*Phocoena phocoena*) in the Swedish Skagerrak, Kattegat and Baltic Seas 1973–1993. *Rep. Int. Whale. Comm.* **1994**, *15*, 211–215.
13. Vinther, M. Bycatches of harbour porpoises (*Phocoena phocoena* L.) in Danish set-net fisheries. *J. Cetacean Res. Manag.* **1999**, *1*, 123–135. [CrossRef]
14. Berggren, P.; Wade, P.R.; Carlström, J.; Read, A.J. Potential limits to anthropogenic mortality for harbour porpoises in the Baltic region. *Biol. Conserv.* **2002**, *103*, 313–322. [CrossRef]
15. Vinther, M.; Larsen, F. Updated estimates of harbour porpoise (*Phocoena phocoena*) bycatch in the Danish North Sea bottom-set gillnet fishery. *J. Cetacean Res. Manag.* **2004**, *6*, 12–24. [CrossRef]
16. HELCOM. *HELCOM Red List of Baltic Sea Species in Danger of Becoming Extinct*; Baltic Sea Environment Proceedings No. 140; Helsinki Commission: Helsinki, Finland, 2013.
17. Kindt-Larsen, L.; Glemarec, G.; Berg, C.W.; Königson, S.; Kroner, A.; Søgaard, M.; Lusseau, D. Knowing the fishery to know the bycatch: Bias-corrected estimates of harbour porpoise bycatch in gillnet fisheries. *Proc. R. Soc. B* **2023**, *290*, 20222570. [CrossRef]
18. Glemarec, G.; Königson, S.; Kindt-Larsen, L. Bycatch in Baltic Sea Commercial Fisheries: High-Risk Areas and Evaluation of Measures to Reduce Bycatch. HELCOM ACTION. 2021. Available online: https://helcom.fi/wp-content/uploads/2021/11/Bycatch-in-Baltic-Sea-commercial-fisheries.pdf (accessed on 26 January 2024).
19. Regulation 57/2011. Council Regulation (EU) No 57/2011 of 18 January 2011 Fixing for 2011 the Fishing Opportunities for Certain Fish Stocks and Groups of Fish Stocks, Applicable in EU Waters and, for EU Vessels, in Certain Non-EU Waters. Available online: http://data.europa.eu/eli/reg/2011/57/2011-08-04 (accessed on 17 January 2024).
20. Swedish Agency of Marine and Water Management. Begränsat Fritidsfiske Efter Pigghaj Möjligt. Available online: https://www.havochvatten.se/arkiv/aktuellt/2023-06-02-begransat-fritidsfiske-efter-pigghaj-mojligt.html (accessed on 7 December 2023).
21. Bryhn, A.C.; Bergek, S.; Bergström, U.; Casini, M.; Dahlgren, E.; Ek, C.; Hjelm, J.; Königson, S.; Ljungberg, P.; Lundström, K.; et al. Which factors can affect the productivity of cod stocks in the Baltic Sea, Kattegat and Skagerrak? *Ocean Coast Manag.* **2022**, *223*, 106154. [CrossRef]
22. Siebert, U.; Gilles, A.; Lucke, K.; Ludwig, M.; Benke, H.; Kock, K.; Scheidat, M. A decade of harbour porpoise occurrence in German waters–Analyses of aerial surveys, incidental sightings and strandings. *J. Sea Res.* **2006**, *56*, 65–80. [CrossRef]
23. Directive 92/43/EEC. Council Directive (EU) 92/43/EEC of 21 May 1992 on the Conservation of Natural Habitats and of Wild Fauna and Flora. Available online: http://data.europa.eu/eli/dir/1992/43/oj (accessed on 17 January 2024).
24. OSPAR Commission. *Convention for the Protection of the Marine Environment of the North-East Atlantic*; OSPAR Commission: London, UK, 1992.
25. HELCOM. *Convention on the Protection of the Marine Environment of the Baltic Sea Area*; Helsinki Commission: Helsinki, Finland, 1992.
26. ASCOBANS. *Agreement on the Conservation of Small Cetaceans of the Baltic, North East Atlantic, Irish and North Seas*; ASCOBANS: Bonn, Germany, 1991.

27. Moore, S.E. Marine Mammals as Ecosystem Sentinels. *J. Mammal.* **2008**, *89*, 534–540. [CrossRef]
28. Bossart, G.D. Marine Mammals as Sentinel Species for Oceans and Human Health. *Vet. Pathol.* **2011**, *48*, 676–690. [CrossRef] [PubMed]
29. Kiszka, J.; Heithaus, M.; Wirsing, A. Behavioural drivers of the ecological roles and importance of marine mammals. *Mar. Ecol. Prog.* **2015**, *523*, 267–281. [CrossRef]
30. Peltier, H.; Dabin, W.; Daniel, P.; Van Canneyt, O.; Dorémus, G.; Huon, M.; Ridoux, V. The significance of stranding data as indicators of cetacean populations at sea: Modelling the drift of cetacean carcasses. *Ecol. Indic.* **2012**, *18*, 278–290. [CrossRef]
31. Wund, S.; Méheust, E.; Dars, C.; Dabin, W.; Demaret, F.; Guichard, B.; Jauniaux, T.; Labrut, S.; Spitz, J.; Van Canneyt, O.; et al. Strengthening the health surveillance of marine mammals in the waters of metropolitan France by monitoring strandings. *Front. Mar. Sci.* **2023**, *10*, 1116819. [CrossRef]
32. McGovern, B.; Gulloch, R.M.; O'Connell, M.; Berrow, S. Temporal and spatial trends in stranding records of cetaceans on the Irish coast, 2002–2014. *J. Mar. Biol. Assoc. U. K.* **2018**, *98*, 977–989. [CrossRef]
33. IJsseldijk, L.L.; ten Doeschate, M.T.I.; Brownlow, A.; Davison, N.J.; Deaville, R.; Galatius, A.; Gilles, A.; Haelters, J.; Jepson, P.D.; Keijl, G.O.; et al. Spatiotemporal mortality and demographic trends in a small cetacean: Strandings to inform conservation management. *Biol. Conserv.* **2020**, *249*, 108733. [CrossRef]
34. Rapportera Tumlare Och Sälar. Available online: https://marinadaggdjur.nrm.se/ (accessed on 31 October 2023).
35. Vår Övervakning av Marina Däggdjur. Available online: https://www.sva.se/vilda-djur/marina-daggdjur/var-overvakning-av-marina-daggdjur/ (accessed on 7 November 2023).
36. Open Data, Examinations of Cetaceans in Swedish Waters, 2020–2022. Available online: https://www.dataportal.se/en/datasets/59_2060/open-data-examinations-of-cetaceans-in-swedish-waters-2020-2022#ref=?p=1&q=marine%20mammal&s=2&t=20&f=&rt=dataset$esterms_IndependentDataService$esterms_ServedByDataService&c=false (accessed on 27 November 2023).
37. Topografi 250 Nedladdning, Vector. Available online: https://geotorget.lantmateriet.se/geodataprodukter/topografi-250-nedladdning-vektor (accessed on 17 November 2023).
38. Anselin, L. Local Indicators of Spatial Association–LISA. *Geogr. Anal.* **1995**, *27*, 93–115. [CrossRef]
39. Bivand, R.; Wong, D. Comparing implementations of global and local indicators of spatial association. *TEST* **2018**, *27*, 718–748. [CrossRef]
40. Wilcoxon, F. Individual Comparisons by Ranking Methods. *Biometrics* **1945**, *1*, 80–83. [CrossRef]
41. Mann, H.B.; Whitney, D.R. On a Test of Whether one of Two Random Variables is Stochastically Larger than the Other. *Ann. Math. Stat.* **1947**, *18*, 50–60. [CrossRef]
42. Wickham, H. *ggplot2: Elegant Graphics for Data Analysis*; Springer: New York, NY, USA, 2016.
43. South, A. rworldmap: A new R package for Mapping Global Data. *R J.* **2011**, *3*, 35–43. [CrossRef]
44. Dunnington, D. ggspatial: Spatial Data Framework for ggplot2. 2023. Available online: https://paleolimbot.github.io/ggspatial/ (accessed on 15 January 2024).
45. R Core Team. *R: A Language and Environment for Statistical Computing*; R Foundation for Statistical Computing: Vienna, Austria, 2023. Available online: https://www.R-project.org/ (accessed on 15 January 2024).
46. Unger, B.; Nachtsheim, D.; Martínez, N.R.; Siebert, U.; Kyhn, L.A.; Balle, J.D.; Teilmann, J.; Calrström, J.; Owen, K.; Gilles, A. *MiniSCANS-II: Aerial Survey for Harboir Porpoises in the Western Baltic Sea, Belt Sea, the Sound and Kattegat in 2020*; Danish Environmental Protection Agency: Odense, Denmark; German Federal Agency for Nature: Bonn, Germany; Conservation and Swedish Agency for Marine and Water Management: Göteborg, Sweden, 2021; p. 28.
47. Authier, M.; Peltier, H.; Dorémus, G.; Dabin, W.; Van Canneyt, O.; Ridoux, V. How much are strandings records affected by variation in reporting rates? A case study of small delphinids in the Bay of Biscay. *Biodivers. Conserv.* **2014**, *23*, 2591–2612. [CrossRef]
48. Lockyer, C.; Kinze, C. Status, ecology and life history of harbour porpoise (*Phocoena phocoena*), in Danish waters. *NAMMCO Sci. Publ.* **2003**, *5*, 143–175. [CrossRef]
49. Peltier, H.; Baagøe, H.J.; Camphuysen, K.C.J.; Czeck, R.; Dabin, W.; Daniel, P.; Deaville, R.; Haelters, J.; Jauniaux, T.; Jensen, L.F.; et al. The stranding anomaly as population indicator: The case of harbour porpoise *Phocoena phocoena* in North-Western Europe. *PLoS ONE* **2013**, *8*, e62180. [CrossRef]
50. IJsseldijk, L.L.; Camphuysen, K.C.J.; Keijl, G.O.; Troost, G.; Aarts, G. Predicting harbour porpoise strandings based on near-shore sightings indicates elevated temporal mobility rates. *Front. Mar. Sci.* **2021**, *8*, 668038. [CrossRef]
51. Vishnyakova, K.; Gol'din, P. Seasonality of strandings and bycatch of harbour porpoises in the Sea of Azov: The effects of fisheries, weather conditions, and life history. *ICES Mar. Sci.* **2014**, *72*, 981–991. [CrossRef]
52. Börjesson, P.; Read, A.J. Variation in timing of conception between populations of the harbour porpoise. *J. Mammal.* **2003**, *84*, 948–955. [CrossRef]
53. Woodley, T.H.; Read, A.J. Potential rates of increase of a harbour porpoise (*Phocoena phocoena*) population subjected to incidental mortality in commercial fisheries. *Can. J. Fish. Aquat. Sci.* **1991**, *48*, 2429–2435. [CrossRef]
54. Sveegaard, S.; Teilmann, J.; Tougaard, J.; Dietz, R.; Mouritsen, K.N.; Desportes, G.; Siebert, U. High-density areas for harbor porpoises (*Phocoena phocoena*) identified by satellite tracking. *Mar. Mamm. Sci.* **2011**, *27*, 230–246. [CrossRef]
55. Sveegaard, S.; Teilmann, J.; Berggren, P.; Mouritsen, K.; Gillespie, D.; Tougaard, J. Acoustic surveys confirm the high-density areas of harbour porpoises found by satellite tracking. *ICES J. Mar. Sci.* **2011**, *68*, 929–939. [CrossRef]

56. Berrow, S.D.; Whooley, P.; O'Connell, M.; Wall, D. *Irish Cetacean Review (2000–2009)*; Irish Whale and Dolphin Group: Kilrush, Ireland, 2016; p. 60.
57. Teilmann, J.; Dietz, R.; Sveegaard, S. *The Use of Marine Waters of Skåne nu Harbour Porpoises in Time and Space*; Technical Report No. 236; Aarhus University, DCE—Danish Centre for Environment and Energy: Roskilde, Denmark, 2022; p. 76.
58. Sveegaard, S. Spatial and Temporal Distribution of Harbour Porpoises in Relation to Their Prey. Ph.D. Thesis, Aarhus University, Aarhus, Denmark, 2011.
59. Sveegaard, S.; Andreasen, H.; Mouritsen, K.; Jeppesen, J.P.; Teilmann, J.; Kinze, C. Correlation between the seasonal distribution of harbour porpoises and their prey in the Sound, Baltic Sea. *Mar. Biol.* **2012**, *159*, 1029–1037. [CrossRef]
60. ASCOBANS/ACCOBAMS. Best Practice on Cetacean Post Mortem Investigation and Tissue Sampling. 2019. Available online: https://www.ascobans.org/en/document/best-practice-cetacean-post-mortem-investigation-and-tissue-sampling (accessed on 26 January 2024).

oceans

Article

Baleen–Plastic Interactions Reveal High Risk to All Filter-Feeding Whales from Clogging, Ingestion, and Entanglement

Alexander J. Werth [1,*], Shirel R. Kahane-Rapport [2], Jean Potvin [3], Jeremy A. Goldbogen [4] and Matthew S. Savoca [4]

1 Department of Biology, Hampden-Sydney College, Hampden-Sydney, VA 23943, USA
2 Department of Biological Sciences, California State University, Fullerton, CA 92831, USA
3 Department of Physics, WATER Institute, Saint Louis University, St. Louis, MO 63103, USA
4 Department of Oceans, Hopkins Marine Station, Stanford University, Pacific Grove, CA 93950, USA
* Correspondence: awerth@hsc.edu

Abstract: Baleen whales are ecosystem sentinels of microplastic pollution. Research indicates that they likely ingest millions of anthropogenic microparticles per day when feeding. Their immense prey consumption and filter-feeding behavior put them at risk. However, the role of baleen, the oral filtering structure of mysticete whales, in this process has not been adequately addressed. Using actual baleen tissue from four whale species (fin, humpback, minke, and North Atlantic right) in flow tank experiments, we tested the capture rate of plastics of varying size, shape, and polymer type, as well as chemical residues leached by degraded plastics, all of which accumulated in the baleen filter. Expanded polystyrene foam was the most readily captured type of plastic, followed by fragments, fibers, nurdles, and spherical microbeads. Nurdle and microbead pellets were captured most readily by right whale baleen, and fragments were captured by humpback baleen. Although not all differences between polymer types were statistically significant, buoyant polymers were most often trapped by baleen. Plastics were captured by baleen sections from all regions of a full baleen rack, but were more readily captured by baleen from dorsal and posterior regions. Baleen–plastic interactions underlie various risks to whales, including filter clogging and damage, which may impede feeding. We posit that plastics pose a higher risk to some whale species due to a combination of factors, including filter porosity, diet, habitat and geographic distribution, and foraging ecology and behavior. Certain whale species in specific marine regions are of the greatest concern due to plastic abundance. It is not feasible to remove all plastic from the sea; most of what is there will continue to break into ever-smaller pieces. We suggest that higher priorities be accorded to lessening humans' dependence on plastics, restricting entry points of plastics into the ocean, and developing biodegradable alternatives.

Keywords: Cetacea; mysticete; baleen whales; plastic; pollution; filtration

1. Introduction

In many ways, synthetic plastic polymers represent one of the greatest success stories of modern human history. Relative to other materials such as paper, wood, and glass, plastics offer five major benefits: they are exceptionally durable, lightweight, moldable (hence the name "plastic"), inexpensive, and easy to produce. Unfortunately, these clear advantages have proven disadvantageous in the environmental landscape of the twenty-first century [1]. The two traits that most distinguish plastics—and a large part of their appeal—are their persistence and ubiquity. Plastics break into smaller pieces via chemical processes including extended UV radiation, interactions with seawater, and mechanical forces such as waves, tides, and currents [1]. Recent research indicates that recycling efforts merely generate additional plastic pieces of smaller size, facilitating their spread [2]. Plastic pollution research has suffered from a lack of harmonization of terms, until recently [3]. Experts now classify pieces by size, ranging from very large mega- (>1 m) to increasingly smaller macro- (<1 m), meso- (<2.5 cm), microplastics, (<5 mm), and nanoplastics smaller than 1 μm [3]. Plastics of all sizes are further characterized as primary or secondary, depending on whether they exist in their originally manufactured form or if they represent broken or otherwise degraded pieces smaller than their original form.

Although our dawning awareness of plastics' downsides developed slowly, it has culminated in a current flood of studies underscoring these risks [4–10]. Plastic pollution has now been found throughout the lithosphere, hydrosphere, atmosphere, cryosphere, and biosphere [11–15]. Although estimates vary, over 170 trillion pieces now litter the ocean [16], forming a global "plastic smog" with its own microbial ecosystem or "plastisphere" [14,17]. Another recent study [18] determined that the annual input of new plastics into the ocean is, at 500 kilotons, slightly less than previously estimated, but nonetheless concluded that the residence time of plastics is greater than formerly thought, such that plastics persist longer and, if not removed, have more time to break and spread. Further studies indicate that plastics wreak immunological, endocrine, and other physiological damage [19–21]. Marine microbes hitch rides on plastic debris [22] and plastics alter gut microbiota and weaken immune systems, leaving humans and animals vulnerable to disease [23,24].

Plastics are particularly insidious in aquatic environments because they often float, and thus, most types of organisms readily encounter them [18,25]. Plastics accumulate along current fronts and form shelters that attract organisms [26,27]. They are easily ingested by primary consumers that obtain nutrients via indiscriminate filter feeding [27,28]. In addition, several studies [29–37] report that aquatic animals may preferentially seek to ingest plastics due to the appealingly "tasty" chemicals that plastics release in water, including various oligomers and monomers that include micronutrients or mimic organic compounds. In this way, plastics easily enter trophic webs, binding all inhabitants of an ecosystem.

Large filter-feeding organisms are at greater risk of plastic ingestion for multiple reasons. Due to the simple yet invariant rules of thermodynamics, larger organisms must consume more food. Although biodilution occurs in ascending trophic pyramids, filter-feeding whales consume colossal amounts of food but do so low on the trophic pyramid [31], placing them at a particularly unique risk. Further, there is a high risk of filter feeders ingesting small plastics not only directly from the environment (i.e., in ingested seawater) but also indirectly via trophic (dietary) transfer—that is, within bodies of ingested organisms [27,38–41]. Even if some particles pass undigested through the guts of organisms that ingest plastics directly and via trophic transfer, there is a higher likelihood that various internal tissues and organs will retain and assimilate pollutants [15].

As the largest animals on Earth, whales (Mammalia: Cetacea) are at an exceptionally high risk of ingesting plastics [11,28,41–43]. A flurry of recent studies has documented the presence of various-sized plastics within digestive tracts or incorporated into tissues of whales [44]. This involves several species of toothed whales (Odontoceti), including sperm [45–48], pilot, beluga [49], and beaked whales [50–53], as well as dolphins and porpoises [54–56]. Plastics have also been found in baleen whales (Mysticeti), including right [57], gray [28], fin [58–62], Bryde's [27], sei [27], minke [63], and humpback whales [64]. As a consequence of their need for large quantities of food, mysticetes forage mainly where prey is abundant. Cetaceans are often observed within the Great Pacific Garbage Patch [65], but they likely forage little in oligotrophic gyres where plastics accumulate substantially, and such accumulations are likely unimportant for baleen–plastic interaction and ingestion. More concerning are heavily polluted locations where mysticetes are known to forage regularly, such as off the southern and eastern coasts of Asia and in the Mediterranean Sea [5,66].

Because mysticetes are filter feeders that engulf and process huge volumes of prey- (and plastic-) laden seawater each day when feeding, they are likely to capture and ingest plastics both directly from seawater and indirectly from bodies of their prey items due to the scale and efficiency of their filtration [40]. Recent studies have estimated the amount of plastic fragments ingested by baleen whales: as much as ten million pieces per day for a blue whale [40]. Numerous recent empirical studies have supported these levels of plastic ingestion in separate ecosystems and baleen whale species [27,28,40]. However, the oral baleen filter itself has not yet received sufficient attention for its role in removing and accumulating environmental plastics from seawater within the whale's mouth. Indeed, this project originated indirectly via unforeseen observations from laboratory flow tank experiments using buoyant plastic particles to study baleen's functional and biomechanical properties [67]. It was discovered that plastic particles, intended simply to indicate flow directions and serve as a stand-in for tiny particulate prey, were captured at unexpectedly high rates by the baleen filter [68]. Therefore, this study was undertaken to determine and more precisely characterize the risks posed to baleen whales from their standard mode of filter feeding. Specifically, we aimed to determine how well whales' oral baleen filter captures plastic pieces of varying size and type with experiments deploying actual baleen tissue in a circulating flow tank (flume). Further, we sought to investigate factors affecting capture, including parameters associated with different plastics (such as size, density, and buoyancy) and with whales themselves (such as swim speed during foraging and forces associated with ingestion, intraoral filtration, and water expulsion). A relationship between removal of plastic via baleen entanglement/clogging and plastic accumulation at the oropharynx/ingestion is also proposed. Our greater aim was to use results of this comparative experimental study to determine more accurately the specific risks posed to whales by plastic filtration and ingestion, as well as to better understand ways to prevent or mitigate these risks.

Although published studies regarding interactions between cetaceans and marine plastic pollution have not specifically focused on the role of the keratinous oral baleen filter in collecting plastics, previous accounts have indicated or suggested at least seven potentially harmful consequences (Figure 1). These include (1) direct and indirect (i.e., trophic) capture of micro- and macroscopic plastic pieces by the baleen filter, after which any plastic retained by the filter can (2) remain within baleen and clog the filter, temporarily or permanently impeding its future function, or alternatively fall off the baleen and be swept or expelled outside the mouth, or (3) be channeled into the digestive tract. Plastic can pass through the gut untouched, (4) cause intestinal impactions (or gut "plasticosis" [63,69,70], or (5) be assimilated or otherwise incorporated into muscle or other body tissues, particularly if the plastic pieces are of very small size [20,44,71]. Further, larger pieces of plastic debris such as ropes, nets, buoys, or other debris (often from discarded fishing gear [72,73]) can (6) entangle whale bodies or body parts, including flippers and flukes, hindering whales' locomotion, respiration, feeding, and other activities, or (7) lodge within baleen racks, not so much clogging as damaging the filter by bending plates and opening gaps between them, which obviously affects filtration adversely. Our study specifically focused on basic baleen–plastic interactions. Additional projects are needed to better assess risks of micro- and nanoplastic assimilation and sequestration within tissues [44], as well as larger-scale filtration dynamics of impaired baleen racks due to gear entanglement.

Finally, we sought to investigate whether particular species or populations of whales are at a higher risk from plastic pollution, including not only the oral capture of plastics but also larger-scale feeding-related risks such as entanglement and filter clogging. We investigated baleen from the skim-feeding North Atlantic right whale (Balaenidae) and three species of lunge-feeding rorquals (Balaenopteridae). In summary, we found that plastic pollution poses serious risks to all cetacean species. An array of diverse behavioral, ecological, and morphological factors, including but not limited to varying foraging strategies, geographic distributions and habitats, and filter porosities, result in differing and often elevated risks to each whale species.

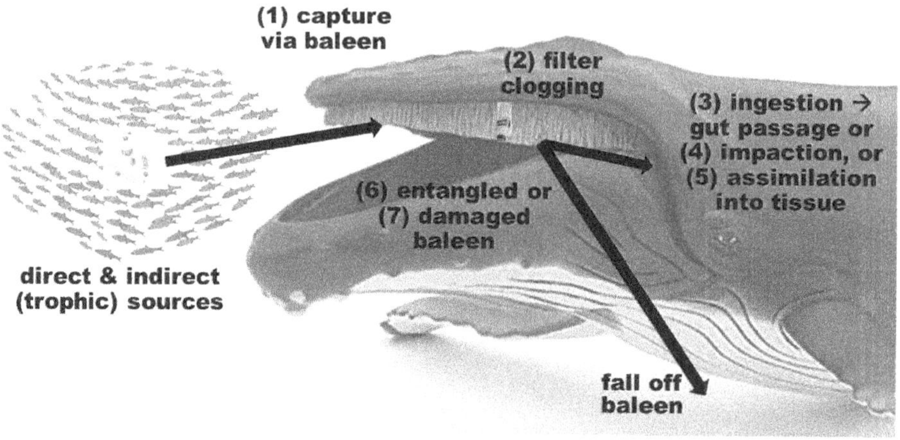

Figure 1. Plastic debris can enter a whale's mouth directly (via primary/original or secondary/broken pieces) and indirectly via trophic transfer, with multiple feeding-related pathways and seven deleterious outcomes: (1) capture in the baleen filter, after which plastic can (2) remain in and clog the filter or fall out of the filter and move out of the mouth. Plastic caught and then dislodged from the filter can also be (3) swallowed, leading either to passage through the gut, (4) impaction of plastic within the gut, or (5) assimilation of absorbed plastic within whale tissues. Larger plastic pieces can also (6) entangle any part of the body, harming foraging and filtration, or (7) temporarily or permanently damage the filter by breaking or bending baleen plates, creating gaps in the filter.

2. Material and Methods

2.1. Biological Specimens

We used baleen plates from humpback (*Megaptera novaeangliae*), fin (*Balaenoptera physalus*), common or Northern minke (*Balaenoptera acutorostrata*), and North Atlantic right (*Eubalaena glacialis*) whales. The collection was carried out by the NOAA/NMFS Northeast Marine Mammal Stranding and Disentanglement Network (VA, MA) or the NOAA/NMFS Southeast Marine Mammal Stranding Network (NC, FL) under U.S. Permits 17350 and 18786. All baleen specimens were obtained from adult animals that died naturally and were procured by certified stranding networks in accordance with MMPA and other applicable statutes. All samples came from specimens that were stranded along the Atlantic coast of USA, except for 11.8 kg of humpback whale (specimen #RMNH.MAM.5506) baleen, which was obtained from the Naturalis Biodiversity Center, Leiden, The Netherlands, on 3 January 2018, from a whale that was stranded along the North Sea coast of the Netherlands in 2012, following CITES permitting (transaction #2082682 to the Smithsonian National Museum of Natural History, Washington, DC, USA). Baleen samples were frozen for shipment and storage, then thawed at room temperature and kept submerged in flowing water for at least seven days prior to flow tank testing.

2.2. Flow Tank Testing

Baleen plates were cut into 18 cm long × 10 cm wide sections and assembled into "mini-racks" of eight pieces, each separated by a 1 cm intra-plate gap (from a plastic or cardboard strip), simulating the normal arrangement of baleen in vivo [74]. Where possible, mini-racks were created by cutting sections from adjacent plates; for the North Atlantic right whale baleen, four sections were cut from each of two adjacent plates in a rack to create the eight-piece mini-rack (Figure 2). For a later set of experiments, plate sections were specifically cut from plates in different regions of the whale mouth (i.e., plate sections came from different locations along a full baleen rack), as described later. For a few final trials, larger humpback whale baleen mini-rack units were used (Figure 3C,D), including

24 plate sections as expanded versions of the mini-rack units. Note that only major baleen plates (not medial minor plates [75]) were cut to produce the sections comprising each baleen mini-rack.

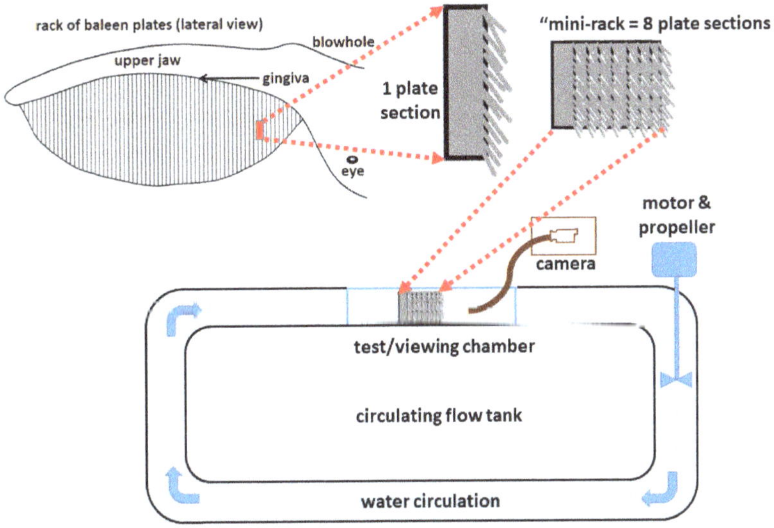

Figure 2. Schematic diagram of experimental setup (as described in text), which involved the cutting of 10 × 18 cm sections of keratinous baleen, combined to form eight-section "mini-racks" of baleen tissue. Mini-racks were then mounted in the test chamber of a flume through which water circulated and into which plastic debris was introduced. Flow trials, videorecorded via underwater camera, were analyzed to study plastic capture and other baleen–plastic interactions.

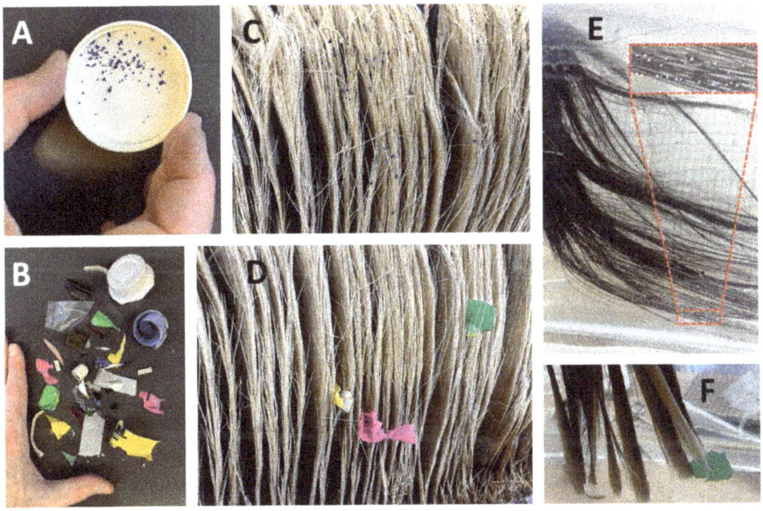

Figure 3. Plastic debris of multiple sizes, shapes, and types were used in flow tank trials, including (**A**) tiny microbeads and (**B**) assorted microplastic fragments of varying shapes and polymer types. (**C**,**D**) Plastic microbeads and macrofragments, respectively, captured by humpback whale baleen (note: with >8 plates, as in mini-racks used in comparative trials). (**E**) Screenshot from a video of right whale baleen mini-rack capturing microbeads, with enlarged inset in dashed red box; (**F**) screenshot from a video of humpback whale baleen mini-rack capturing microplastic fragments.

For all species, each baleen plate section was approximately 0.3 cm thick, and each piece included one edge of the lingual or medial margin of a baleen plate featuring freely eroded fringes (Figure 2), also known as baleen bristles or hairs (Figure 3E). The same mini-rack was used for each trial of baleen from a particular species. Each mini-rack was secured by clamping it to a metal rod, then submerged below the water surface in a 90 L circulating flow tank (flume). The flume was made of PVC sections in a vertical loop, modeled on a design by S. Vogel [67], with a transparent Plexiglas top in which a flat viewing window was installed through which the ruled grid behind the test chamber could be seen (Figures 2 and 3E). The working section of the tank's test chamber had a length of 70 cm and a cross-sectional area of 900 cm^2, with 1–2% blockage from baleen samples when present. Water flow speed through the tank varied from 0.15–1.4 m/s. Varying flow speeds were achieved by (a) selecting different motor speeds, (b) using metal propellers of differing diameter, and (c) adjusting a rheostat to alter the motor's input voltage. Before and after experimental trials, flow speed was checked and calibrated with a digital flow meter (model MFP51, Geopacks, Hatherleigh, Devon, UK). Most trials were performed with flow speeds commensurate with intraoral currents generated in vivo during swimming in the range of 0.15–0.9 m/s, which is consistent with published swimming speeds of balaenid and balaenopterid whales during skim- or lunge-feeding at the surface or deeper in the water column [76–80]. Baffles at each end of the test section promoted laminar flow. After preliminary experiments testing the effects of water temperature on plastic capture rate (and finding no effect), all flow experiments were performed in freshwater or seawater at 19 °C. Baleen was submerged before flow commenced, with plates arranged in the tank (Figure 2) perpendicular to flow (i.e., a 90° angle of attack), simulating the direction of flow in a whale mouth in vivo [67].

2.3. Experimental Design and Statistical Power Analysis

Given the breadth and complexity of our experimental setup (testing baleen–plastic interactions from baleen of four whale species in multiple ways, including five different shapes, three sizes, and six types of plastic debris in varying flow speeds, etc.), we used a linear mixed-effects statistical model to analyze correlation and variance–covariance structures. We conducted a basic power analysis prior to data collection to ensure that our null hypothesis testing would be valid and robust. Using WebPower 4.5.2 (an online R package), we determined the minimum sample size needed to avoid Type II (false-negative) errors due to insufficient statistical power: for basic functional (debris capture) testing, N = 5 replicates (for each plastic variable; N = 450 total) with 0.80 power (p), representing greater than 80% confidence for significance at the 0.05 level. This was the same (N = 450) for each flow speed used in the videorecorded flume testing (30 s trials). For the residue analysis, we used a minimum N = 12 samples per species, with N = 3 flume trial replicates per each flow speed (total N = 216). Data from each component of the study were analyzed statistically (using SigmaPlot 13.0) via parametric paired t-tests to compare different treatments from the same species, including linear regression via Pearson's tests to study relationships between variables, and ANOVA (with post-hoc Bonferroni correction) to compare results of plastic interactions in different species. All data are shown with error bars = standard deviation.

2.4. Video Analysis

Flow sequences (recorded at or edited to 30 s in duration) were videorecorded laterally and anteriorly underwater from within the flume (Figure 2) and from outside the flume via a planar viewing window on the side of the test chamber, by a digital endoscope (VideoFlex SD, Umarex Laserliner, Arnsberg, Germany) with an illuminated 17 mm camera head recording JPEG images and AVI video (30 frames per second; 5/25/50 cm focal distances). Anterior views were used to set up the experimental protocol, but for continuity and uniformity, only lateral views were used in the final analysis. The recorded field of view covered the entire baleen mini-rack (Figure 3E,F). Video was viewed frame-by-frame via

GoPro Studio v. 2.5.7, with landmarks digitized via Tracker v. 4.92. Kinematic analysis focused on the number of plastic pieces captured (defined as adhering to baleen for $\geq$2 s) relative to water flow speed. Initially, flow tank runs were scored simply by summing the total number of plastic pieces captured within the entire field of view over the 30 s of the trial; this count provided a clear, easy, and effective method and a useful basis of comparison. However, this method depended on the density of plastic debris encountering baleen. Even if each trial began with standardized density, pieces did not easily recirculate within the flume at the same rate. Therefore, trials were rescored not by simple counting but instead as a percentage of the total number of visible plastic pieces passing through the videorecorded field of view within the 30 s trial (=capture rate for N particles or NCR).

For each set of experimental variables, trials were conducted $\geq$5 times (total N = 450 for each flow speed, except for trials on plastic residues, with N = 3 or total N = 216), with paired *t*-tests/Pearson's tests and ANOVA testing of data from replicates, as outlined in the previous section.

As mentioned above, the cutoff for debris "capture" in video analysis was set at 2 s. This duration was determined after initial analysis because many particles (including pellets, fibers, fragments, etc.) were observed to briefly contact baleen tissue during their flow through and around the mini-racks, but in a majority of cases (>50%), particles "bounce" off the baleen almost immediately or adhere to it very briefly (~1 s) before flowing away. Particles that touch elements of the baleen filter for >1 s generally remain trapped there for long durations (>2 s), although they too can be washed away, particularly in the case of smaller (microplastic) particles. Hence, the 2 s cutoff was used to define capture.

Determining actual plastic ingestion into the gut requires accounting for possibilities of particles either moving past one baleen plate (Figure 1) toward the next downstream plate or directly to the oropharynx or else passing through the filter to exit into the ocean due to smaller particle size relative to fringe porosity and intra-baleen plate gap width. Estimating such a number from capture percentages (as defined above) was done as follows. Using particle counts, one starts by considering the total number (N_{tot}) of particles soon to arrive at the location of a baleen plate (Figure 1), a number also equal to the sum of the number of particles passing through baleen while following the crossflow between plates [67,81] ($N_{through}$), those bypassing the plate altogether (N_{bypass}), and the plastic entangling with fringes ("capture"; $N_{entangle}$). This sum can be further expressed in terms of the capture rate measurement described above (NCR), yielding $N_{entangle}/N_{bypass} = NCR$, and through fringe–baleen crossflow [82], estimated via the ratio $N_{through}/N_{total} = \varphi(\rho_{particles}/\rho_{water})C_{CFF}$. Here, parameters φ, $\rho_{particles}$, ρ_{water}, C_{CFF} correspond to fringe-plate section porosity [83], mass density of the plastic particulates, mass density of seawater, and water flow fraction passing through the baleen filter [81,82], respectively. Inserting this into the particle count expressed as a fraction of the total (N_{total}) yields the following equation, after solving for the bypass ratio, N_{bypass}/N_{total}:

$$\frac{N_{bypass}}{N_{total}} = \left(\frac{1}{1+NCR}\right)\left(1-\varphi\frac{\rho_{particles}}{\rho_{waters}}C_{CFF}\right) \quad (1)$$

Note that where plastic particulates are larger than the largest fringe gaps, the value of the crossflow coefficient (C_{CFF}) effectively becomes zero.

2.5. Plastic Pollutants Used

Baleen tissues of different whale species were exposed in each series of comparative trials to one of several kinds of plastics, which varied according to (a) shape, (b) size, and (c) chemical type. A fourth series of preliminary trials focused on (d) chemical residues leached from plastics, as explained below. Plastic debris varied in buoyancy: some types/sizes were highly buoyant, especially at low flow speeds, whereas other pieces of neutral buoyancy flowed through the middle of the tank's water column. Some pieces sank to the bottom as flow stopped when trials concluded.

The density of plastic pieces (i.e., number of fragments per volume of water) used in trials varied according to the shape, size, and type of plastic. Densities ranged from approximately 15,000 spheres per m^3 for latex microbeads (average diameter 710 um) to 10–50 fragments per m^3 for larger macro- and mesoplastics. Although there is tremendous variance in the concentration of plastic debris in natural marine environments, these concentrations are generally high. However, this study focused less on generating realistic ingestion estimates and more on developing a mechanistic understanding of how baleen captures synthetic polymer particles. Note that not all combinations or permutations of plastic shape, size, etc., were possible or attempted for this study. For example, trials that compared plastic capture by whale species used solely microbead spheres and macro- and mesoplastic fragments; trials that compared different plastic polymer types used only fragments of micro-, macro-, and mesoplastics. A drain at the bottom of the tank allowed water and any plastic to be removed from the tank, especially between trials using different types or sizes of plastic debris. Each flow trial involved only each type of plastic (i.e., fibers or fragments), not combined plastics of multiple shapes or sizes, etc., with the notable exception that microbeads were used in some trials of larger plastics because the microbeads were useful in visualizing flow. In these cases where microbeads were used, even though the focus was on another type of plastic, a smaller density of beads was used; approximately 10% of the tested density where the focus was on microbead capture (i.e., ~1000–2000 beads/m^3 instead of 15,000 beads/m^3).

Microplastic (<5 mm) pieces were allowed to recirculate through the circular flume, but larger pieces and all sections of plastic cord were contained by screens installed at each end of the test section. This was done to prevent any plastic debris from entangling the propeller or otherwise interfering with flow, but it meant that debris remained in the test section and recirculated to a lesser degree than smaller pieces. Video analysis indicated that approximately 30% of plastic pieces were stuck to the downstream screen, especially at high flow speeds ($\geq$75 cm/s). Nonetheless, most pieces were buoyant and easily recirculated, allowing many chances for circulating plastic to be trapped by the baleen filter.

2.5.1. Plastic Shapes

Plastics used in the flow trials included five different forms: (1) fragments of irregular size, (2) linear fibers, and (3–5) three kinds of spherical pellets: tiny microbeads, expanded polysytene (hereafter called "foam") spheres, and nurdles used as components for injection molding or other manufacturing. Some trials also used 10 cm long sections of plastic cord rope, made of polypropylene or nylon.

2.5.2. Plastic Sizes

Additionally, trials included three general sizes of plastics, especially for the fragments: (1) microplastics less than 5 mm in size/diameter, (2) macroplastics from 5 mm to 2.5 cm, and (3) mesoplastics larger than 2.5 cm. Spherical pellets ranged from transparent, white, or blue microplastic latex beads averaging 710 µm in diameter (Sargent–Welch #50024, 0696-00-K, Rochester, NY, USA) to red or white macroplastic nurdles ranging from 5 to 8 mm in diameter, made of acrylonitrile butadiene styrene (ABS) and obtained in 2005 from LEGOLAND California (Carlsbad, CA, USA). The 10 cm long sections of plastic rope were of varying diameter: 3 mm, 8 mm, and 14 mm. No large megaplastics (i.e., >1 m) or microscopic nanoplastics (<1 µm) were used in the flow tank experiments for this study.

2.5.3. Plastic Resin/Polymer Types

Limited trials of micro-, macro-, and mesoplastic fragments focused on different chemical types of plastics, with types based on the universal international recycling scheme. Thus, trials were conducted with type #1 polyethylene terephthalate (PET), #2 high density polyethylene (HDPE), #3 polyvinyl chloride (PVC), #4 low density polyethylene (LDPE), #5 polypropylene (PP), and #6 polystyrene (PS). Together, these polymers make up ~95% of all plastic production and range from positively to negatively buoyant in seawater [84]. Plastics of category/type #7 (all other plastics) were not specifically used. All plastics used for this experiment involved "found" objects collected as outdoor and indoor environmental debris in central Virginia, USA. Fragments for each of the six recycling types 1–6 were obtained from debris as follows: type #1 (PET) from clear water and other bottles; #2 (HDPE) from milk/detergent/soda jugs, toys, mats, and water bottles; #3 (PVC) from wire coatings, packaging, toys, and bags; #4 (LDPE) from shopping bags, sacks, and wrappers; #5 (PP) from bottles/caps, containers, yogurt cups, and plastic tape; #6 (PS) from egg cartons, foam trays, and fast food "clamshell" containers. All plastics of these types were identified by the recycling code on the plastic source from which the fragments were created by tearing or snapping by hand or by cutting with scissors or an X-Acto razor knife.

2.5.4. Plastic Residues

Finally, some trials used chemicals leached from plastics after sitting in glass beakers of heated (40–45 °C) water for >10 days. Although these chemical residues from mechanically broken/degraded plastics were not tested for final identification, we believe, based on published literature regarding similar plastics and their degradation [62,85], that these residues were primarily styrene monomers and dimers and polystyrene oligomers, although some samples may have included bisphenol A (BPA) as well. Small amounts (10–20 mL) of these residues were poured into the moving water of the flow tank so that their interaction with baleen tissue could be recorded and studied. Unlike the trials with different plastic debris pieces of varying shape, size, and type, in which the capture rate was analyzed by video analysis, the capture of plastic residues could not be effectively observed via video. Thus, the baleen mini-racks were visually examined, both within the water-filled tank and especially in the air after removal from the flow tank, once the flow had ceased. Photography could then document the presence of residues on flat portions of baleen plates [68]. Although chemical residues were also observed on baleen fringes, it was ultimately deemed too difficult and subjective to determine their presence there. Attempts were made to judge amounts of residue on baleen plates as high, low, or intermediate, but this also proved difficult, so any chemical residue was simply scored as present or absent. As with trials using plastic pieces, all tank water was drained, flushed, and replaced following flow trials investigating plastic residues.

3. Results

3.1. Comparative Species Results

Micro- and macroplastic pieces, including spherical pellets (microbeads averaging 710 μm in diameter and mesoplastic nurdles ranging from 5 to 8 mm in diameter) and irregularly shaped mesoplastic fragments (5 mm to 2.5 cm in diameter), were readily captured by all whale baleen specimens used in the flow tank experiments (Figure 3). For all shapes of plastic, all baleen samples (all species) captured, through three trials, an average of 29.53 (±4.33)% of pieces videorecorded flowing through the baleen mini-rack. For microbeads, the average capture rate (all species) was 17.0 (±4.0)%, for macroplastic nurdle pellets, it was 23.67 (±4.51)%, for fibers it was 27.67 (±6.03)%, for foam pieces it was 42.67 (±2.08)%, and for irregularly shaped microplastic fragments, it was 35.67 (±5.03)%.

Capture rates varied by species but were most widespread for microbeads and fragments, with more fragments captured than beads (Figure 4). Beads had a capture rate of 15 (±4.33)% in fin whale baleen, 17 (±4.1)% in humpback whale baleen, 13.67 (±4.0)% in minke whale baleen, and 20.33 (±1.53)% in right whale baleen. For fragments, the capture rates were 31.67 (±5.13)% in fin, 35.67 (±5.0)% in humpback, 24.67 (±6.0)% in minke, and 25.0 (±2.65)% in right whale baleen. Thus, although the results did not demonstrate statistical significance, right whale baleen captured more beads than baleen from any of the three rorqual species tested (p-value = 0.33; r = 0.69), whereas humpback baleen captured the most fragments (p-value = 0.48) (Figure 4).

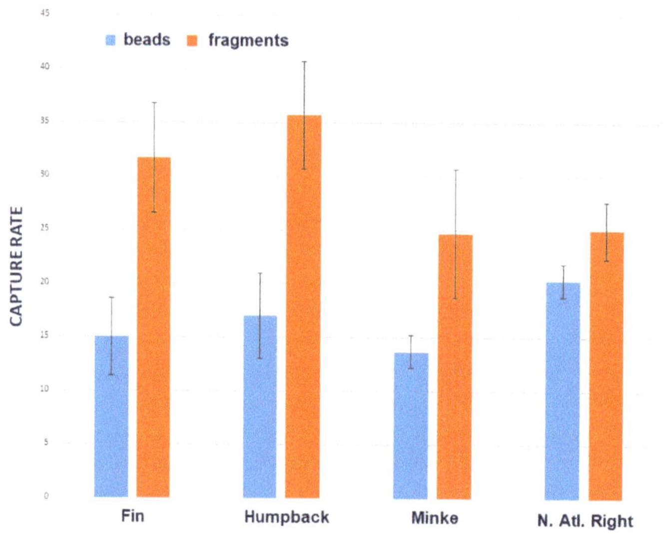

Figure 4. Capture rate results for bead and fragment shape trials (all flow speeds averaged) with baleen from four whale species. Right whale baleen captured more beads, whereas humpback baleen captured the most fragments. Error bars = standard deviation.

3.2. *Effect of Fragment Size*

There was an unsurprisingly strong positive correlation between the sizes of plastic pieces used in the trials and their capture rate (r = 0.89; *p*-value = 0.03; Figure 5). Larger pieces were significantly more easily captured; they also were retained within the baleen filter for longer durations than smaller pieces (r = 0.71; *p*-value = 0.08), which, even if captured, were more likely to be quickly swept free from baleen at all flow speeds.

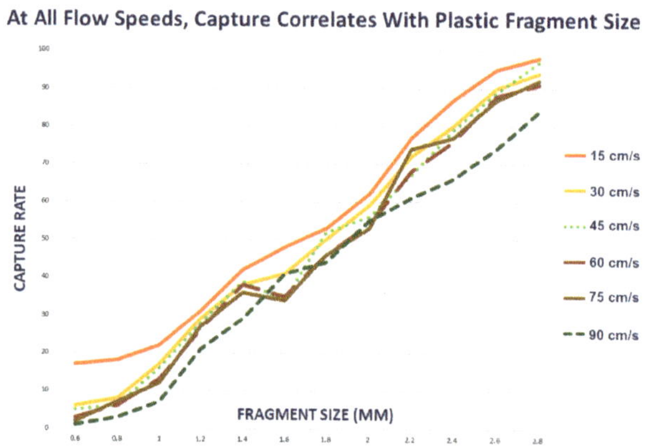

Figure 5. Results from flow trials (only humpback baleen data for plastic macrofragments displayed here) show no statistically significant differences with varying flow, but larger plastic fragments were captured at all flow speeds.

3.3. *Effect of Plastic Fragment Shape*

Trials with different kinds of plastic debris showed that different micro- and macroplastic pieces are captured and retained differently within the baleen filter (Figure 6). After experimenting with these shapes and concluding that humpback whale baleen achieved the highest overall capture rates for all shapes except microbeads (caught most readily by right whale baleen), more replicates of trials (5× for each shape) were conducted with humpback baleen. Capture rate results for humpback baleen were as follows (Figure 6): 17 (±4.1)% for microbeads, 23.67 (±4.51)% for nurdles, 27.67 (±6.03)% for fibers, 43.67 (±2.08)% for foam, and 35.67 (±5.03)% for fragments. Foam pieces were captured at the highest rate, more than twice that of microbeads; relative to other plastic pieces, foam was the only shape to demonstrate (barely) any significantly different rate of capture (p-value = 0.047). Irregularly shaped microplastic fragments were the second most captured pieces. The overall capture rate for all types/shapes was 29.53 (±4.33)%.

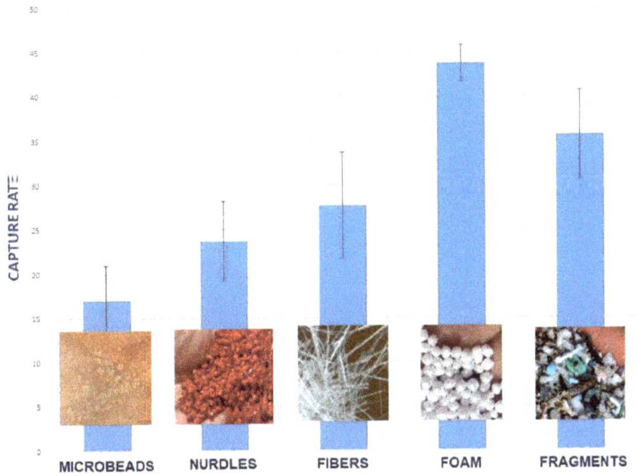

Figure 6. Results from flow trials (only humpback baleen data at 75 cm/s flow speed displayed here) showed differences among five shapes/types of plastic debris. Error bars = standard deviation.

3.4. *Effect of Flow Parameters*

We sought to test the impact of varying flow regimes on the capture rate of plastic debris in baleen of all tested species. To do this, we initially adjusted parameters such as angle of attack (simulating different pathways of flow into, through, and out of the mouth), water temperature (simulating different seasons, habitats, or geographic regions), and flow speed (simulating different speeds of whale swimming during foraging). We found no statistically significant or noteworthy effects of these variables (on baleen of any species; p-value = 0.32), except for a slightly negative correlation between capture and flow speed for all species (r = −0.19; Figure 5). In faster flows, baleen fringes captured fewer plastic particles, or if pieces did contact baleen briefly, they were not in contact long enough (≥2 s) to be counted as "captured" because they were quickly swept away.

3.5. *Effect of Plastic Polymer Type*

Trials examining interactions of humpback whale baleen with different chemical types of plastic polymer or resin, following the universal recycling code, showed some differences, although not all were statistically significant (Figure 7). The capture rates by plastic recycling code were as follows: Type #1 (PET) 18 ± 4%, #2 (HDPE) 24 ± 5%, #3 (PVC) 14 ± 3.5%, #4 (LDPE) 33 ± 6%, #5 (PP) 21 ± 5%, #6 (PS) 38 ± 8%. Polystyrene and low-density polyethylene were captured at significantly higher rates (r = 0.82; p-value = 0.04), while polyethylene terephthalate and polyvinyl chloride had the lowest rates (p-value = 0.26).

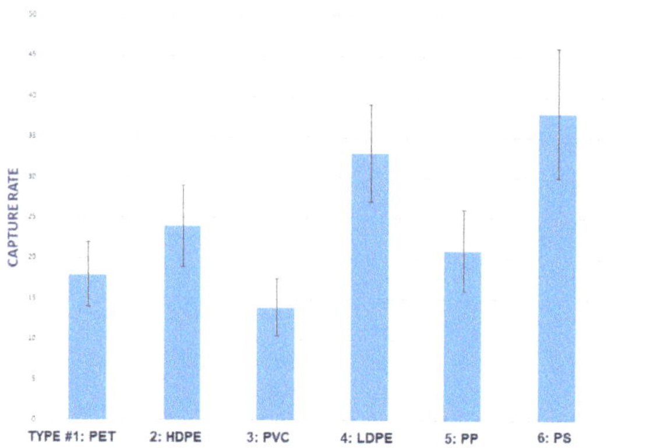

Figure 7. Results from flow trials (only humpback baleen data at 75 cm/s flow speed displayed here) showed differences in the six types of plastic polymers tested, arranged by universal recycling code numbers. Error bars = standard deviation.

3.6. Differential Capture within Regions of Baleen Rack

An attempt was made to determine whether baleen taken from different regions of a complete baleen rack varied in its ability to capture plastic, given that baleen tissue may differ slightly between regions, particularly in terms of the length, diameter, and density of fringe "hairs" [67,82,83]. For three of the four whale species tested (fin, humpback, and right), mini-racks were compared using baleen from nine distinct spatial locations within the mouth (Figure 8): dorsal anterior (DA), mid-anterior (MA), ventral anterior (VA), dorsal central (DC), mid-central (MC), ventral central (VC), dorsal posterior (DP), mid-posterior (MP), and ventral posterior (VP). Results of fin and humpback whale baleen did not show significant differences between regions. Right whale baleen did show two clear trends, with slightly more plastic captured toward the dorsal (upper) and posterior region of the rack, and a slightly higher overall capture rate for the dorsal central region (Figure 8), but again, these were not statistically significant (p-value = 0.31).

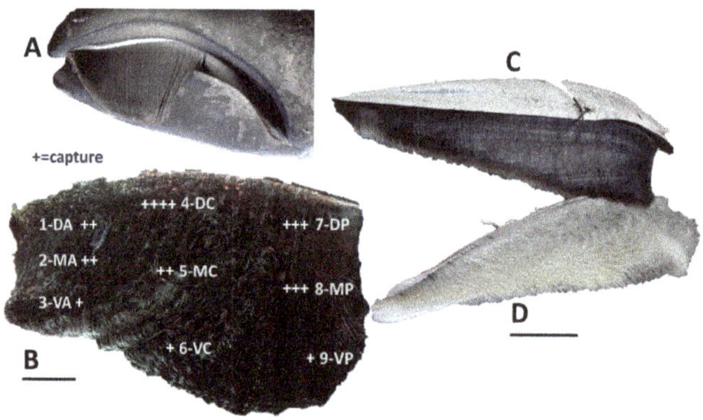

Figure 8. Plastic pieces were captured differently in tests using baleen sections from nine locations along a full baleen rack (all scale bars = 1 m): dorsal, medial, and ventral regions of anterior, central, and posterior plates (e.g., DA = dorsal anterior region). (**A**) Lateral side of a bowhead whale rack (similar to that of a right whale) and (**B**) medial side of a right whale rack; (**C**,**D**) lateral and medial sides of a fin whale rack. In (**B**), + signs indicate the relative degree of capture of plastic debris by both right and humpback whale baleen: lowest in the ventral regions, higher in mid-baleen, and highest in dorsal regions.

3.7. Capture of Plastic Residues

Trials to determine whether and how well baleen captured chemical residues leached from degraded and heated plastics confirmed that residues were captured by baleen from each of the four tested whale species. Residue capture was scored as the overall number of baleen surfaces (eight plates with two flat sides each, for 16 total sides) with adhering chemical residue as confirmed by observation in water or air following a 30-flow trial. Residue adherence ranged from 9–27 ± 3.5%, and there was a slight but insignificant trend toward higher capture at higher flow speeds (Figure 9), seen in all species ($r = 0.27$; p-value = 0.31). The residues appeared on baleen tissue as spots that were lightly or darkly discolored or iridescent (Figure 9). No species proved obviously better at capturing or adhering plastic residue. It initially appeared that humpback and right whale baleen had more residue, but upon closer review, it was deemed that this was likely an artifact from the baleen of each species initially having different colors and variably shiny or matte surfaces depending on the pigmentation of the flattened outer keratin layer [68].

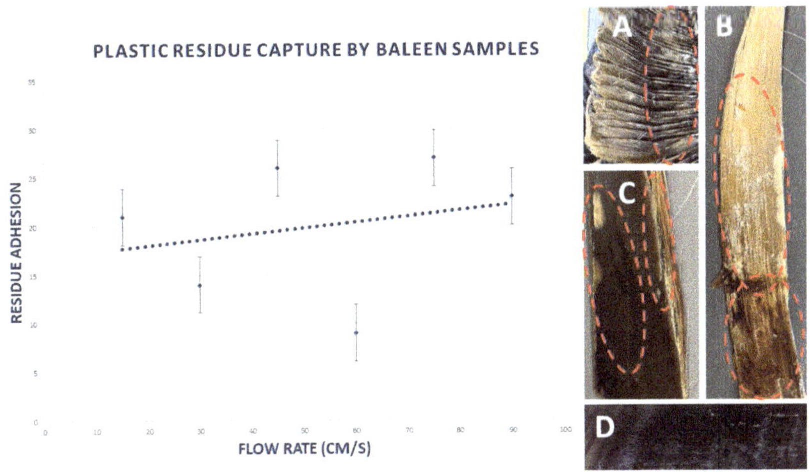

Figure 9. Capture and adhesion of leached plastic residues on baleen did not vary by flow speed (or region within a full rack). Photos (**A–C**) show sections of humpback baleen, with discolored (light and dark) adhering residues indicated by dotted ovals. Photo (**D**) shows a humpback whale baleen sample with normal wear but no chemical residues (e.g., control image). Error bars = standard deviation.

3.8. Filter Bypass and Ingestion Rate Estimation

Plastic particulates not flowing cross-wise through baleen [81] or tangling within fringes bypass the filter and accumulate near the oropharynx for ingestion (Figure 1). Estimating such mass can be done via Equation (1), after inferring the value of N_{total} and fringe porosity at each baleen plate (Figure 8), a step that was not possible with the circulating flow setup used here. Estimating filter bypass fraction (N_{bypass}/N_{total}) at each plate can be achieved after considering likely values for porosity (φ~10–20%) [83], crossflow mass fraction (C_{CFF}~0.3–0.5 without fringes [81] and <0.1 with fringes [82]), and the plastic-to-water mass ratio ($\rho_{particles}/\rho_{water}$; kg/m^3 of plastic in the sample over kg/m^3 of sample's water.) Plastics encountered at densities of 10 or fewer particles per cubic meter [86] per Equation (1) would yield N_{bypass}/N_{total}~$(1 + \text{NCR})^{-1}$~0.74–0.86 at each station per Figure 4. Compounding particle capture effects of entanglement ("capture") while moving posteriorly past three plates (Figure 8) would yield~$0.8^3 = 0.51$ of the mass arriving at the oropharynx (one of the scenarios considered in [40]). Obviously, entanglement increases clogging and reduces the total particles being ingested, but it also interferes with the crossflow prey–water separation process.

4. Discussion

4.1. Comparative Species Results

Analysis of flume results revealed that baleen of all tested mysticete species readily captured plastic pieces of all sizes, shapes, and types. This demonstrates that baleen's ability to collect plastic pollution is a potentially major concern for fin, humpback, minke, and right whales, and, given the spread of results for all tested species, including error bars (Figure 4), very likely for all baleen whale species. Previous studies have suggested that the capture of plastic is a problem for filter feeders [31], but previous estimates of mysticete plastic ingestion [40] have focused on ingestion via trophic transfer. Our ongoing research confirms that baleen itself plays a prominent role in collecting plastic pollution, leading to direct ingestion (Figure 1).

In our study, right whale baleen captured slightly more small spherical pieces (microplastic beads) than baleen of other species, and humpback whale baleen captured slightly more macro- (5–25 mm) and mesoplastic (>25 mm) plastic fragments than baleen of other tested species, but these differences were not significant (Figure 4). Flow regimes also had no significant effects on plastic capture by baleen (Figure 5). However, recall that "capture" was defined as plastic pieces clinging to baleen for at least two seconds, and that in faster flows, pieces were more likely to be swept free from baleen. In life, such pieces might be briefly (<2 s) filtered from water and rapidly swept into the mouth, where they might pass unharmed through the gut (Figure 1), or, alternatively, where they might become impacted in the digestive tract or even absorbed (particularly in the case of small micro- and nanoplastics) and assimilated into bodily tissues [44]. Thus, our arbitrary definition of "capture" (with a two-second cutoff), although helpful for methodological purposes, might not be realistic for the real-world effects of plastic pollution in vivo. The issue of temporary capture of plastic debris via baleen leading to ingestion versus permanent capture leading to reduced functionality of the filter (or other impairment of foraging ability) is a focus of ongoing research. Nonetheless, our main conclusion is that all baleen is effective at capturing plastic, placing all mysticetes at risk.

4.2. Analysis of Diverse Plastic Pollutants

Baleen captured larger pieces of plastic debris more readily (Figure 5), although the above-mentioned caveat that smaller pieces might still be filtered from seawater into a whale's mouth and then quickly released into the mouth means that plastics of all sizes probably pose risks to whales. However, our study suggests that plastics of all sizes pose different threats to baleen whales (Figure 10). Mesoplastics and smaller pieces pose a higher risk of being swallowed and potentially assimilated into bodily tissues. This is true whether tiny plastics are encountered directly or indirectly via trophic transfer (i.e., within bodies of ingested prey items). Larger macro- and mesoplastics are also at risk of being swallowed, where they are more likely than small debris items to become impacted within the digestive system (i.e., stomach chambers and sections of the small and large intestines). Pieces larger than mesoplastics would be too large to be swallowed given the narrow diameter of the esophageal opening [87]. However, meso- and macroplastics appear to pose the highest risk of filter clogging (Figure 10). Megaplastics (>1 m), the largest pieces of plastic debris, also (like smaller pieces) pose risks of clogging the filter and esophagus, but in addition, they carry further risks of body entanglement and temporary or permanent damage to individual baleen plates or an entire rack of baleen. All mysticetes are obligate filter feeders. However, megaplastics are unlikely to be found in schools of aggregating bulk prey. Even if plastic pieces do not enter a whale's gut, they can, if captured by the baleen filter, still create major feeding-related problems (Figure 1) if they clog or damage the filter or if they impair motions of a whale's jaws, head, or other body parts (such as the flippers, flukes, and tail stock) essential to locomotion or other feeding-related movements. Like clogging, entanglement and related deformation bending of baleen can create gaps within the filter, altering flow regimes and interfering with normal filtration and prey collection processes.

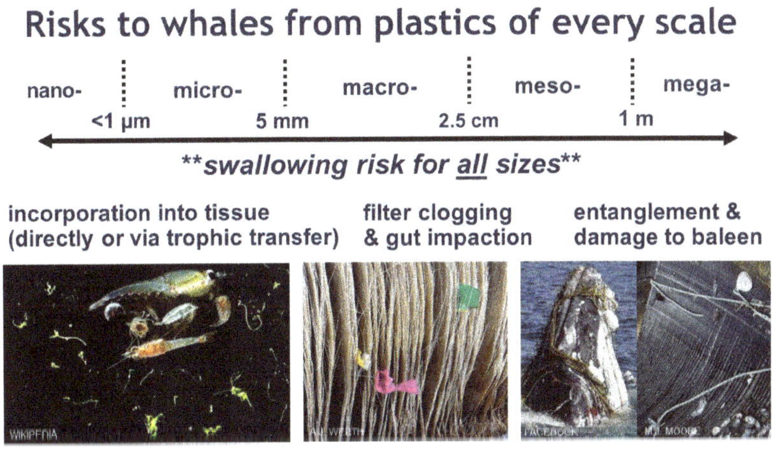

Figure 10. Baleen whales are vulnerable to plastic pollution of all sizes. At the lower end of the scale, nano- and microplastics are more likely to be swallowed, possibly leading to incorporation within body tissue. Macro- and mesoplastics are more likely to clog the baleen filter or, once swallowed, lodge within the gut. Megaplastics are more likely to entangle a whale or body part, including a full baleen rack, leading to serious transient or permanent effects on filtration.

With regard to different shapes of plastics, we found that pieces of foam, irregularly shaped fragments, fibers, nurdles, and microbeads were all readily captured (in that descending order of frequency; Figure 6). Plastic resins of all types were captured easily (by baleen of all tested species), although polystyrenes and polyethylenes (both high- and low-density PE) were captured most frequently (Figure 7). This might be due to differential buoyancy: PS and PE tend to float, whereas PET and PVC (captured by baleen at lower rates) tend to sink. It is worth noting, with regard to potential baleen–plastic interactions, that PEs are the most commonly found plastic within the water column [88], and quite specifically where mysticetes commonly feed [40]. PEs and PP are known to drift long distances and are the most commonly found plastic at the surface [89]. Still, most consumer products are a blend of different plastic polymer types, along with numerous dyes and other chemical additives [90].

Our study also indicated that baleen obtained from all regions within a full rack, and thus from all regions of the mouth, captures plastic pieces (Figure 8). The precise directions, degrees, and forces of flow within the mouth during baleen bulk filtration are not entirely known; this remains an active area of investigation. However, it is crucial that no matter where within a whale's mouth plastic is encountered—no matter where in the baleen filter it makes contact—there is a good chance the plastic will be indiscriminately filtered from water and may cause adverse impacts. Plastics are collected by baleen's medial fringes (on the internal or lingual side of a rack) but also in spaces between plates. Clogging, entanglement, and damaged (e.g., bent) plates can occur throughout the entire baleen filter.

Whales can remove small, accumulated plastics from their baleen filter by shaking or nodding the head [74]. However, this would most likely cause released plastic debris to be swallowed, thereby trading one pathology for another. Further, larger pieces of clogging or entangled plastics cannot be removed from baleen by head shaking.

Balaenid (bowhead and right) whales are continuous skim feeders that use long (3+ m), finely-fringed baleen plates to separate and retain tiny (1–10 mm) copepods wherever they accumulate in dense patches, whether at the surface, near the bottom, or at mid-water depth [74]. The fine filter of these skimming whales [91,92] and long duration of filtration activity [92] put them at high risk of acquiring and ingesting tiny microbeads or other abundant plastic debris (Figure 4). Balaenids' slow-speed, long-duration foraging behavior, often at or near the surface [67], also results in frequent entanglement with fishing gear or other ropes (Figure 10), which can quickly or slowly kill a whale or create lasting damage to its oral filter. In contrast, balaenopterid or rorqual (groove-throated) whales, including the fin, humpback, and minke species included in this study, normally feed on larger schooling fish or invertebrates, which are engulfed within a single mouthful via rapid gulping lunges at all depths [74]. Rorquals include the largest, most active whales, putting them at high risk of direct or indirect (trophic) ingestion of all kinds of plastic debris [40]. Although baleen of gray whales, *Eschrichtius robustus*, was not included in this experimental study, it has been documented [28] that gray whales are at high risk due to the fact that nearly all of their feeding involves suction-based ingestion of shallow-water coastal sediments, which are typically rich with nutritious benthic invertebrates but also where much plastic debris accumulates from riverine and coastal runoff.

Although biological risks attributable to chemical residues from the production and degradation of plastics are not yet clearly known [62,85], these residues are also captured or adhere to the alpha-keratin surfaces of baleen tissue (Figure 9). We suggest that these as of yet unknown risks might be substantial if, like plastic pieces, such chemical residues are swallowed, incorporated into whale tissues, ultimately clogging or impairing the baleen filter's effectiveness, or otherwise affecting a whale's behavior and foraging efficiency.

4.3. Risk Factors of Plastics for Various Cetacean Species

Kahane-Rapport et al. [40] calculated that whales consume up to ten million pieces of plastic per day—potentially trillions over the course of a lifetime. The risk of plastic ingestion for diverse whale species varies according to multiple parameters: ecological (e.g., feeding method, depth, and behavior), biogeographic (habitat and distribution), oceanographic and bathymetric (e.g., currents and gyres in marine circulation patterns), and other abiotic and biological factors, such as size and lifespan, not to mention differences in the baleen filter itself (number and dimensions of plates and eroded fringes, etc.).

Because odontocetes (toothed whales, dolphins, and porpoises) target individual food items, they can preferentially seek and ingest buoyant plastics that visually, acoustically, or chemically resemble prey (for example, plastic bags resembling salps or jellyfish), or that are simply located in regions where cetaceans habitually feed. Necropsies of stranded sperm, pilot, and beaked whales (and other odontocetes) have revealed numerous diverse plastics found lodged within the stomach chambers or intestines, including nets, ropes, and dozens of plastic bags. Plastic ingestion may be the primary cause of death or a major contributing factor. Once a cetacean ingests one such plastic item, it may come to regard other such items as food, starting a habit of ingesting plastics [9–12,45–54]. Further, whales that are sick or weakened may ingest plastics because they are less able to capture live prey, creating a vicious cycle of plastic ingestion [45]. The bulk filtration of mysticetes creates different risks than the precisely targeted feeding of odontocetes seeking individual items. The high volume of water filtered during foraging, whether continuous (as in balaenid skim feeding) or intermittent (as in lunge feeding of rorquals or suction ingestion of gray whales), poses another major risk factor for baleen whales relative to toothed whales. The experimental study reported here outlined slightly different yet overall largely similar risks to varying whale species due to their particular type of baleen tissue.

Plastic, like prey, is not randomly distributed in the ocean. Unfortunately, the same oceanographic forces that aggregate the prey species of whales also aggregate plastic pollutants, often in the same locations. Because foraging is energetically expensive, whales and other marine organisms seek to maximize caloric intake while minimizing energy expenditure when locating and acquiring food. For this reason, they seek densely concentrated aggregations of zooplankton or schooling fish, which often form at current fronts or other interfaces created by currents and gyres. These frequently attract larger organisms such as forage fish, which, in turn, attract whales. Further, many organisms have been shown to be actively attracted to chemicals within or leached by plastics [33–35], some of which appear to be similar to dimethyl sulfide (DMS). This common organosulfur compound produces the characteristic "seafood odor", which is apparently a key attractant for many foraging marine species, including whales [36,37]. For these reasons, marine plastic debris accumulates, both freely within seawater and eventually inside bodies of small organisms that whales feed on, by means and in locations that attract whales and promote ingestion by whales. In short, the factors that enable plastics to accumulate non-randomly in marine habitats are the very same factors that play key roles in whale feeding. Whether whales encounter plastics via trophic transfer from schooling krill, copepods, or forage fish they eat, or from freely floating or submerged plastic particles that are indiscriminately filtered by the intraoral baleen tissue, all cetaceans have a high probability of locating and ingesting plastic. Oral "play" with plastics, as documented in many cetacean species [93], further increases risks of ingestion and entanglement.

Nonetheless, recent studies have reported that the vast majority (>90%) of plastic ingested by baleen whales is due to trophic transfer [27,40], thus bypassing the role of baleen capture in this process. Consequently, we believe that focusing on regions of overlap between mysticete feeding grounds and extreme levels of water-bound plastics shows where our findings could be most relevant and applicable to inform risk assessments. We consider highly polluted regions first, then identify species that frequent these heavily contaminated waters.

In the open ocean, the highest densities of plastic occur in oligotrophic subtropical gyres [94,95]. Although mysticetes traverse these regions on their seasonal migrations, they are unlikely to do much foraging while there. However, there are sedentary populations in regions with exceptionally high plastic pollution that may be at threat. A prime example involves resident fin whales in the Mediterranean Sea, a region that has among the highest plastic debris densities anywhere on Earth [96]. As a result, this fin whale population has already been suggested as a bioindicator of plastic pollution in the Mediterranean Sea [97]. In Southeast Asian waters, Eden's whales (*B. edeni*) and the little-known Omura's whale (*B. omurai*) face threats because this region has the highest rates of plastic discharge to the nearshore marine environment anywhere on the planet [98]. Indian Ocean blue whales (*B. musculus indica*) spend their entire lives in the tropical and subtropical Indian Ocean [99], which in addition to having an emerging plastic gyre [100,101], also has a rapidly expanding human population on all sides of the basin. In other words, plastic pollution–wildlife interactions in this region are likely to worsen. Fin and common minke whales that inhabit the Yellow and East China Seas are also at risk for the same reasons [56]. The highly endangered Rice's whale (*B. ricei*) in the Gulf of Mexico should also be monitored as well; the holotype specimen was killed by an ingested plastic fragment that perforated its intestine [102].

4.4. Reducing and Mitigating Threats to Baleen-Filtering Whales from Plastic Pollution

Threats to whales and other marine life from plastic pollution, both primary and secondary (i.e., broken or degraded pieces from a larger original source), have been increasingly publicized. This includes plastic ingestion directly from the water column and trophic transfer. Many solutions have been proposed, but two simple solutions are ultimately most likely to reduce the threat. First, more empirical research is needed about baleen–plastic interactions and effects, from filter clogging to swallowing and incorporation within bodily tissues. Second, plastic flow into the sea should be addressed by focusing less on recycling and cleanup, and more on halting new plastic production, use, and environmental entry. Many small- and large-scale efforts are underway to ascertain the scope of the marine plastic pollution problem and develop effective strategies to mitigate it. However, as many scientists and environmentalists have pointed out, the best strategy is to prevent new plastics from reaching the sea via rivers and wastewater treatment centers, perhaps by using lightweight degradable "sponges" [103] or bioinspired filters based on whale baleen [67].

In what is surely one of history's most heartbreaking ironies, prior to the twentieth century, the role of modern plastics was held by baleen [104]. Baleen's exceptional tensile strength and pliant flexibility made it an ideal material for corset stays, buggy whips, skirt hoops, umbrella ribs, and mechanical or kitchen implements [105]. Baleen's value during the era of industrial whaling peaked in 1853, with over 5.6 million pounds, mostly from North Atlantic right whales, sold in U.S. ports for almost $2,000,000 [64]. Centuries if not millennia earlier, native subsistence hunters fashioned baleen into baskets, armor, weapons, implements, and varied pieces of artwork [105]. Like later whalers, indigenous peoples appreciated baleen's plastic-like properties, namely its tough pliability, resistance to degradation, and ease of shaping into various useful forms.

Plastics currently comprise 83% of ocean trash [106], most of it in the form of microplastics (beads/cosmetics, clothing fibers, nurdle resin pellets, and secondary fragments), macroplastics (mainly single-use food wrappers, bags, bottles and caps, utensils, straws and stirrers, and cigarette filters), and megaplastics (fishing lines, nets, floats, and traps). There is also dawning awareness of the immense scope of microplastic debris (in air and on land, but ultimately migrating to marine habitats) from regular use and frictional erosion of vehicle tires [107]. In the near future, society's planned plastic production is estimated to outpace all planned mitigation efforts to reduce its flux in the environment. Therefore, marine plastic pollution will continue to worsen [108]. Recycling is insufficient to tackle this challenge; reducing and reusing plastics will have a far greater impact [109]. Although science and technology can alleviate the problem by fueling the development and use of photo- and biodegradable polymers or other new materials [110–112], these advances will not address the overwhelming amounts of plastic already in the environment, nor in the bodies of whales and other sea life. Scientists and activists have long speculated that the oral baleen filter, which served Mysticeti so favorably as a key innovation in their evolutionary radiation, could now present a major drawback by aggregating and ingesting anthropogenic pollutants, and thus serve as an evolutionary trap in the Anthropocene. To ensure whales' survival, we must better understand precisely how their baleen functions as we simultaneously work to eliminate plastics from their environment.

Author Contributions: Project conceptualization, A.J.W., S.R.K.-R., J.P., J.A.G. and M.S.S.; flow experiments, A.J.W.; data collection and curation, A.J.W.; data analysis, A.J.W., S.R.K.-R., J.P. and M.S.S.; draft writing and figure preparation, A.J.W.; draft editing and supervision, A.J.W., S.R.K.-R., J.P., J.A.G. and M.S.S. All authors have read and agreed to the published version of the manuscript.

Funding: MSS received funding from the National Science Foundation (PRFB 1906332) and the National Geographic Society (EC-53352R-18).

Institutional Review Board Statement: Ethical review and approval were waived for this study due to the fact that no live animals were used. No animals were harmed for this study. All baleen specimens were obtained from adult animals that died naturally and were procured by certified stranding networks in accordance with MMPA and other applicable statutes.

Data Availability Statement: The data presented in this study are available upon request from the corresponding author. The data are not publicly available because they remain in use by the authors.

Acknowledgments: We are grateful to the many biologists and staff supervisors who facilitated the collection and transfer of the baleen tissues, and for approving associated permit requests. Irvin Robertson expertly constructed the flow tank, following a general design by Steve Vogel. We thank many scientists for the helpful discussions on whales, filtration, and plastic pollution, especially Dave Cade, Peter van de Graaf, Peter Madsen, Shemar Blakeney, Adrian Cothren, Didier Buston, Bertrand Denis, Olivier Adam, Coralie Barber, Katie Register, Joy Reidenberg, Todd Sformo, Michael Moore, and Craig George.

Conflicts of Interest: The authors declare no conflicts of interest.

References

1. Gallo, F.; Fossi, C.; Weber, R.; Santillo, D.; Sousa, J.; Ingram, I.; Nadal, A.; Romano, D. Marine litter plastics and microplastics and their toxic chemicals components: The need for urgent preventive measures. *Environ. Sci. Eur.* **2018**, *30*, 13. [CrossRef]
2. Brown, E.; MacDonald, A.; Allen, S.; Allen, D. The potential for a plastic recycling facility to release microplastic pollution and possible filtration remediation effectiveness. *J. Hazard. Mater. Adv.* **2023**, *10*, 100309. [CrossRef]

3. GESAMP (IMO/FAO/UNESCO-IOC/UNIDO/WMO/IAEA/UN/UNEP/UNDP/ISA Joint Group of Experts on the Scientific Aspects of Marine Environmental Protection). Guidelines for the monitoring and assessment of plastic litter and microplastics in the ocean. In *GESAMP Reports & Studies Series*; Kershaw, P.J., Turra, A., Galgani, F., Eds.; GESAMP: London, UK, 2019; Volume 99, 130p, ISSN 1020-4873.
4. Bucci, K.; Tulio, M.; Rochman, C.M. What is known and unknown about the effects of plastic pollution: A meta-analysis and systematic review. *Ecol. Appl.* **2020**, *30*, e02044. [CrossRef]
5. Gregorietti, M.; Atzori, F.; Carosso, L.; Frau, F.; Pellegrino, G.; Sarà, G.; Arcangeli, A. Cetacean presence and distribution in the central Mediterranean Sea and potential risks deriving from plastic pollution. *Mar. Pollut. Bull.* **2021**, *173*, 112943. [CrossRef]
6. Koelmans, A.A.; Redondo-Hasselerharm, P.E.; Nor, N.H.M.; de Ruijter, V.M.; Mintenig, S.M.; Kooi, M. Risk assessment of microplastic particles. *Nat. Rev. Mater.* **2022**, *7*, 138–152. [CrossRef]
7. Law, K.L. Plastics in the marine environment. *Annu. Rev. Mar. Sci.* **2017**, *9*, 205–229. [CrossRef] [PubMed]
8. MacLeod, M.; Arp, H.P.H.; Tekman, M.B.; Jahnke, A. The global threat from plastic pollution. *Science* **2021**, *373*, 61–65. [CrossRef]
9. Panti, C.; Baini, M.; Lusher, A.; Hernandez-Milan, G.; Bravo Rebolledo, E.L.; Unger, B.; Syberg, K.; Simmonds, M.K.; Fossi, M.C. Marine litter: One of the major threats for marine mammals. *Environ. Pollut.* **2019**, *247*, 72–79. [CrossRef]
10. Santos, R.G.; Machovsky-Capuska, G.E.; Andrades, R. Plastic ingestion as an evolutionary trap: Toward a holistic understanding. *Science* **2021**, *373*, 56–60. [CrossRef] [PubMed]
11. Fossi, M.C.; Baini, M.; Simmonds, M.P. Cetaceans as ocean health indicators of marine litter impact at global scale. *Front. Environ. Sci.* **2020**, *8*, 586627. [CrossRef]
12. Scacco, U.; Mancini, E.; Marcucci, F.; Tiralongo, F. Microplastics in the deep: Comparing dietary and plastic ingestion data between two Mediterranean bathyal opportunistic feeder species, *Galeus melastomus*, Rafinesque 1810 and *Coelorinchus caelorhincus*, Risso 1810, through stomach content analysis. *J. Mar. Sci. Eng.* **2022**, *10*, 624. [CrossRef]
13. Jacob, H.; Besson, M.; Swarzenski, P.W.; Lecchini, D.; Metian, M. Effects of virgin micro- and nano-plastics on fish: Trends, meta-analysis and perspectives. *Environ. Sci. Technol.* **2020**, *54*, 4733–4745. [CrossRef] [PubMed]
14. Bergmann, M.; Allen, S.; Krumpen, T.; Allen, D. High levels of microplastics in the Arctic sea ice alga *Melosira arctica*, a vector to ice-associated and benthic food webs. *Environ. Sci. Technol.* **2023**, *57*, 6799–6807. [CrossRef] [PubMed]
15. Gopinath, P.M.; Twayana, K.S.; Ravanan, P.; Thomas, J.; Mukherjee, A.; Jenkins, D.F.; Chandrasekharan, N. Prospects on the nano-plastic particles internalization and induction of cellular response in human keratinocytes. *Part. Fibre Toxicol.* **2021**, *18*, 35. [CrossRef] [PubMed]
16. Eriksen, M.; Cowger, W.; Erdle, L.M.; Coffin, S.; Villarrubia-Gómez, P.; Moore, C.J.; Carpenter, E.J.; Day, R.H.; Thiel, M.; Wilcox, C. A growing plastic smog, now estimated to be over 170 trillion plastic particles afloat in the world's oceans—Urgent solutions required. *PLoS ONE* **2023**, *18*, e0281596. [CrossRef] [PubMed]
17. Zettler, E.R.; Mincer, T.J.; Amaral-Zettler, L.A. Life in the "plastisphere: Microbial communities on plastic marine debris. *Environ. Sci. Technol.* **2013**, *47*, 7137–7146. [CrossRef] [PubMed]
18. Kaandorp, M.L.A.; Lobelle, D.; Kehl, C.; Dijkstra, H.A.; van Sebille, E. Global mass of buoyant marine plastics dominated by large long-lived debris. *Nat. Geosci.* **2023**, *16*, 689–694. [CrossRef]
19. Lu, Y.; Zhang, Y.; Deng, Y.; Jiang, W.; Zhao, Y.; Geng, J.; Ding, L.; Ren, H. Uptake and accumulation of polystyrene microplastics in zebrafish (*Danio rerio*) and toxic effects in liver. *Environ. Sci. Technol.* **2016**, *50*, 4054–4060. [CrossRef]
20. Barboza, L.G.A.; Lopes, C.; Oliveira, P.; Bessa, F.; Otero, V.; Henriques, B.; Raimundo, J.; Caetano, M.; Vale, C.; Guilhermino, L. Microplastics in wild fish from North East Atlantic Ocean and its potential for causing neurotoxic effects, lipid oxidative damage, and human health risks associated with ingestion exposure. *Sci. Total Environ.* **2020**, *717*, 134625. [CrossRef]
21. Ramsperger, A.F.R.M.; Narayana, V.K.B.; Gross, W.; Mohanraj, J.; Thalakkat, M.; Greiner, A.; Schmalz, H.; Kress, H.; Laforsch, C. Environmental exposure enhances the internalization of microplastic particles into cells. *Sci. Adv.* **2020**, *6*, eabd1211. [CrossRef]
22. Lear, L.; Padfield, D.; Dowsett, T.; Jones, M.; Kay, S.; Hayward, A.; Vos, M. Bacterial colonisation dynamics of household plastics in a coastal environment. *Sci. Total Environ.* **2022**, *838*, 156199. [CrossRef]
23. Focardi, A.; Moore, L.R.; Raina, J.B.; Seymour, J.R.; Paulsen, I.T.; Tetu, S.G. Plastic leachates impair picophytoplankton and dramatically reshape the marine microbiome. *Microbiome* **2022**, *10*, 179. [CrossRef]
24. Fackelmann, G.; Pham, C.K.; Rodríguez, Y.; Mallory, M.L.; Provencher, J.F.; Baak, J.E.; Sommer, S. Current levels of microplastic pollution impact wild seabird gut microbiomes. *Nat. Ecol. Evol.* **2023**, *7*, 698–706. [CrossRef] [PubMed]
25. Ryan, P.G. Ingestion of plastics by marine organisms. In *Hazardous Chemicals Associated with Plastics in the Marine Environment, The Handbook of Environmental Chemistry*; Takada, H., Karapanagioti, H., Eds.; Springer: Cham, Switzerland, 2016; Volume 78, pp. 235–266. [CrossRef]
26. Gove, J.M.; Whitney, J.L.; McManus, M.A.; Williams, G.J. Prey-size plastics are invading larval fish nurseries. *Proc. Natl. Acad. Sci. USA* **2019**, *116*, 24143–24149. [CrossRef] [PubMed]
27. Zantis, L.J.; Bosker, T.; Lawler, F.; Nelms, S.E.; O'Rorke, R.; Constantine, R.; Sewell, M.; Carroll, E.L. Assessing microplastic exposure of large marine filter feeders. *Sci. Total Environ.* **2022**, *818*, 121815. [CrossRef] [PubMed]
28. Torres, L.G.; Brander, S.M.; Parker, J.I.; Bloom, E.M.; Norman, R.; Van Brocklin, J.E.; Lasdin, K.S.; Hildebrand, L. Zoop to poop: Assessment of microparticle loads in gray whale zooplankton prey and fecal matter reveal high daily consumption rates. *Front. Mar. Sci.* **2023**, *10*, 1201078. [CrossRef]

29. Allen, A.S.; Seymour, A.C.; Rittschof, D. Chemoreception drives plastic consumption in a hard coral. *Mar. Pollut. Bull.* **2017**, *124*, 198–205. [CrossRef] [PubMed]
30. Savoca, M.S.; Tyson, C.W.; McGill, M.; Slager, C.J. Odours from marine plastic debris induce food search behaviours in a forage fish. *Proc. Roy. Soc. B* **2017**, *284*, 20171000. [CrossRef] [PubMed]
31. Germanov, E.S.; Marshall, A.D.; Bejder, L.; Fossi, M.C.; Lonergan, N.R. Microplastics: No small problem for filter-feeding megafauna. *Trends Ecol. Evol.* **2018**, *33*, 227–232. [CrossRef]
32. Axworthy, J.B.; Padilla-Gamiño, J.L. Microplastics ingestion and heterotrophy in thermally stressed corals. *Sci. Rep.* **2019**, *9*, 18193. [CrossRef] [PubMed]
33. Pfaller, J.B.; Goforth, K.M.; Gil, M.A.; Savoca, M.S.; Lohmann, K.J. Odors from marine plastic debris elicit foraging behavior in sea turtles. *Curr. Biol.* **2020**, *30*, R213–R214. [CrossRef] [PubMed]
34. Procter, J.; Hopkins, F.E.; Fileman, E.S.; Lindeque, P.K. Smells good enough to eat: Dimethyl sulfide (DMS) enhances copepod ingestion of microplastics. *Mar. Pollut. Bull.* **2019**, *138*, 1–6. [CrossRef] [PubMed]
35. Savoca, M.S.; Wihlfeil, M.E.; Ebeler, S.E.; Nevitt, G.A. Marine plastic debris emits a keystone info chemical for olfactory foraging seabirds. *Sci. Adv.* **2016**, *2*, e1600395. [CrossRef] [PubMed]
36. Savoca, M.S.; Nevitt, G. Evidence that dimethyl sulfide facilitates a tritrophic mutualism between marine primary producers and top predators. *Proc. Natl. Acad. Sci. USA* **2014**, *111*, 4157–4161. [CrossRef] [PubMed]
37. Bouchard, B.; Barnagaud, J.-Y.; Poupard, M.; Glotin, H.; Gauffier, P.; Torres Ortiz, S.; Lisney, T.J.; Campagna, S.; Rasmussen, M.; Célérier, A. Behavioural responses of humpback whales to food-related chemical stimuli. *PLoS ONE* **2019**, *14*, e0212515. [CrossRef]
38. Burkhardt-Holm, P.; N'Guyen, A. Ingestion of microplastics by fish and other prey organisms of cetaceans, exemplified for two large baleen whale species. *Mar. Pollut. Bull.* **2019**, *144*, 224–234. [CrossRef] [PubMed]
39. Alava, J.J. Modeling the bioaccumulation and biomagnification potential of microplastics in a cetacean foodweb of the Northeastern Pacific: A prospective tool to assess the risk exposure to plastic particles. *Front. Mar. Sci.* **2020**, *7*, 566101. [CrossRef]
40. Kahane-Rapport, S.R.; Czapanskiy, M.F.; Fahlbusch, J.A.; Friedlaender, A.S.; Calambokidis, J.; Hazen, E.L.; Goldbogen, J.A.; Savoca, M.S. Field measurements reveal exposure risk to microplastic ingestion by filter-feeding megafauna. *Nat. Commun.* **2022**, *13*, 6327. [CrossRef]
41. Zantis, L.J.; Carroll, E.L.; Nelms, S.E.; Bosker, T. Marine mammals and microplastics: A systematic review and call for standardization. *Environ. Pollut.* **2021**, *269*, 116142. [CrossRef]
42. Eisfeld-Pierantonio, S.M.; Pierantonio, N.; Simmonds, M.P. The impact of marine debris on cetaceans with consideration of plastics generated by the COVID-19 pandemic. *Environ. Pollut.* **2022**, *300*, 118967. [CrossRef]
43. Kruse, K.; Knickmeier, K.; Brennecke, D.; Unger, B.; Siebert, U. Plastic Debris and Its Impacts on Marine Mammals. In *Marine Mammals*; Brennecke, D., Knickmeier, K., Pawliczka, I., Siebert, U., Wahlberg, M., Eds.; Springer: Cham, Switzerland, 2023; pp. 49–62. [CrossRef]
44. Merrill, G.B.; Hermabessiere, L.; Rochman, C.M.; Nowacek, D.P. Microplastics in marine mammal blubber, melon, & other tissues: Evidence of translocation. *Environ. Pollut.* **2023**, *335*, 122252. [CrossRef]
45. Tarpley, R.J.; Marwitz, S. Plastic ingestion by cetaceans along the Texas coast: Two case reports. *Aquat. Mamm.* **1993**, *19*, 93–98.
46. de Stephanis, R.; Giménez, J.; Carpinelli, E.; Gutierrez-Exposito, C.; Cañadas, A. As main meal for sperm whales: Plastics debris. *Mar. Pollut. Bull.* **2013**, *69*, 206–214. [CrossRef]
47. Brentano, R.; Petry, M.V. Marine debris ingestion and human impacts on the pygmy sperm whale (*Kogia breviceps*) in southern Brazil. *Mar. Pollut. Bull.* **2020**, *150*, e110595. [CrossRef] [PubMed]
48. Tonay, A.M.; Öztürk, A.A.; Salman, A.; Dede, A.; Danyer, I.A.; Danyer, E.; Öztürk, B. Stranding records of sperm whale (*Physeter macrocephalus*) on the Turkish coast in 2019-2020 with a note on the opportunistic sampling of stomach content. *J. Black Sea/Mediterr. Environ.* **2021**, *27*, 281–293.
49. Moore, R.C.; Loseto, L.; Noel, M.; Etemadifar, A.; Brewster, J.D.; MacPhee, S.; Bendell, L.; Ross, P.S. Microplastics in beluga whales (*Delphinapterus leucas*) from the Eastern Beaufort Sea. *Mar. Pollut. Bull.* **2020**, *150*, 110723. [CrossRef] [PubMed]
50. Secchi, E.R.; Zarzur, S. Plastic debris ingested by a Blainville's beaked whale, *Mesoplodon densirostris*, washed ashore in Brazil. *Aquat. Mamm.* **1999**, *25*, 21–24.
51. Lusher, A.L.; Hernandez-Milian, G.; O'Brien, J.; Berrow, S.; O'Connor, I.; Officer, R. Microplastic and microplastic ingestion by a deep diving, oceanic cetacean: The True's beaked whale *Mesoplodon mirus*. *Environ. Pollut.* **2015**, *199*, 185–191. [CrossRef] [PubMed]
52. Abreo, N.A.S.; Macusi, E.D.; Blatchley, D.D.; Cuenca-Ocay, G.C. First evidence of plastic ingestion by the rare Deraniyagala's beaked whale. *Int. J. Ecol. Cons.* **2016**, *19*, 16–36.
53. Kelly, N.E.; Feyre, L.; Gavel, H.; Trela, O.; Ledwell, W.; Breeze, H.; Marotte, E.C.; McConney, L.; Whitehead, H. Long term trends in floating plastic pollution within a marine protected area identifies threats for endangered northern bottlenose whales. *Environ. Res.* **2023**, *227*, 115686. [CrossRef]
54. Alexiadou, P.; Foskolos, I.; Frantzis, A. Ingestion of macroplastics by odontocetes of the Greek Seas, Eastern Mediterranean: Often deadly! *Mar. Pollut. Bull.* **2019**, *146*, 67–75. [CrossRef] [PubMed]
55. López-Martínez, S.; Giménez-Luque, E.; Molina-Pardo, J.L.; Manzano-Medina, S.; Arribas-Arias, H.; Gavara, R.; Morales-Caselles, C.; Rivas, M.L. Plastic ingestion by two cetacean groups: Ziphiidae and Delphinidae. *Environ. Pollut.* **2023**, *333*, 121932. [CrossRef]

56. Sá, S.; Torres-Pereira, A.; Ferreira, M.; Monteiro, S.S.; Fradoca, R.; Sequeira, M.; Vingada, J.; Eira, C. Microplastics in Cetaceans Stranded on the Portuguese Coast. *Animals* **2023**, *13*, 3263. [CrossRef]
57. Alzugaray, L.; Di Martino, M.; Beltramino, L.; Rowntree, V.J.; Sironi, M.; Uhart, M.M. Anthropogenic debris in the digestive tract of a southern right whale (*Eubalaena australis*) stranded in Golfo Nuevo, Argentina. *Mar. Pollut. Bull.* **2020**, *161A*, 111738. [CrossRef]
58. Fossi, M.C.; Panti, C.; Guerranti, C.; Copola, D.; Gianetti, M.; Marsili, L.; Minutoli, R. Are baleen whales exposed to the threat of microplastics? A case study of the Mediterranean fin whale (*Balaenoptera physalus*). *Mar. Pollut. Bull.* **2012**, *64*, 2374–2379. [CrossRef] [PubMed]
59. Fossi, M.C.; Marsili, L.; Baini, M.; Gianetti, M.; Coppola, D.; Guerranti, C.; Caliani, I.; Minutoli, R.; Lauriano, G.; Finoia, M.G.; et al. Fin whales and microplastics: The Mediterranean Sea and the Sea of Cortez scenarios. *Environ. Pollut.* **2016**, *209*, 68–78. [CrossRef] [PubMed]
60. Fossi, M.C.; Baini, M.; Panti, C.; Baulch, S. Impacts of marine litter on cetaceans: A focus on plastic pollution. In *Marine Mammal Ecotoxicology*; Fossi, M.C., Panti, C., Eds.; Academic Press: San Diego, CA, USA, 2018; pp. 147–184. [CrossRef]
61. Im, J.; Joo, S.; Lee, Y.; Kim, B.Y.; Kim, T. First record of plastic debris ingestion by a fin whale (*Balaenoptera physalus*) in the sea off East Asia. *Mar. Pollut. Bull.* **2020**, *159*, 111514. [CrossRef]
62. Garcia-Garin, O.; Sahyoun, W.; Net, S.; Vighi, M.; Aguila, A.; Ouddane, B.; Vikingsoon, G.A.; Chosson, V.; Borrell, A. Intrapopulation and temporal differences of phthalate concentrations in North Atlantic fin whales (*Balaenoptera physalus*). *Chemosphere* **2022**, *300*, e134453. [CrossRef]
63. Jauniaux, T.; Haelters, J.; Degraer, S.; Coignoul, F. Fatal plastic impaction in a minke whale (*Balaenoptera acutorostrata*). In Proceedings of the 28th Annual Conference of European Cetacean Society: Marine Mammals as Sentinels of Changing Environment, Liege, Belgique, 5–9 April 2014. poster 168.
64. Besseling, E.; Foekema, E.M.; Van Franeker, J.A.; Leopold, M.F.; Kühn, S.; Bravo Rebolledo, E.L.; Heße, E.; Mielke, L.; Ijzer, J.; Kamminga, P.; et al. Microplastic in a macro filter feeder: Humpback whale *Megaptera novaeangliae*. *Mar. Pollut. Bull.* **2015**, *95*, 248–252. [CrossRef]
65. Gibbs, S.E.; Kent, S.; Slat, B.; Morales, D.; Fouda, L.; Reisser, J. Cetacean sightings within the Great Pacific Garbage Patch. *Mar. Biodivers.* **2019**, *49*, 2021–2027. [CrossRef]
66. Guerrini, F.; Lorenzo, M.; Renato, C. Modeling plastics exposure for the marine biota: Risk maps for fin whales in the Pelagos Sanctuary (North-Western Mediterranean). *Front. Mar. Sci.* **2019**, *6*, 299. Available online: https://www.frontiersin.org/articles/10.3389/fmars.2019.00299 (accessed on 24 May 2023). [CrossRef]
67. Werth, A.J. Flow-dependent porosity of baleen from the bowhead whale (*Balaena mysticetus*). *J. Exp. Biol.* **2013**, *216*, 1152–1159. [CrossRef] [PubMed]
68. Werth, A.J.; Blakeney, S.M.; Cothren, A.I. Oil adsorption does not structurally or functionally alter whale baleen. *Roy. Soc. Open Sci.* **2019**, *6*, e182194. [CrossRef] [PubMed]
69. Charlton-Howard, H.S.; Bond, A.L.; Rivers-Auty, J.; Lavers, J.L. 'Plasticosis': Characterising macro- and microplastic-associated fibrosis in seabird tissues. *J. Hazard. Mater.* **2023**, *450*, 131090. [CrossRef] [PubMed]
70. Rivers-Auty, J.; Bond, A.L.; Grant, M.L.; Lavers, J.L. The one-two punch of plastic exposure: Macro- and micro-plastics induce multi-organ damage in seabirds. *J. Hazard. Mater.* **2023**, *442*, 130117. [CrossRef] [PubMed]
71. Yu, W.; Ye, T.; Zhang, Y.; Xu, R.; Lei, Y.; Chen, Z.; Yang, Z.; Zhang, Y.; Song, J.; Yue, X.; et al. Global estimates of daily ambient fine particulate matter concentrations and unequal spatiotemporal distribution of population exposure: A machine learning modelling study. *Lancet Planet. Health* **2023**, *7*, e209–e218. [CrossRef] [PubMed]
72. Jerbi, H.; Pérez, W.; Tellechea, J.S. Ingestion of floating net debris in sperm whales and larynx strangulation with gillnet parts in bottlenose dolphins on the coast of Tunisia. *Aquat. Mamm.* **2021**, *47*, 239–244. [CrossRef]
73. Mghili, B.; Keznine, M.; Analla, M.; Aksissou, M. The impacts of abandoned, discarded and lost fishing gear on marine biodiversity in Morocco. *Ocean. Coast. Manag.* **2023**, *239*, e106593. [CrossRef]
74. Werth, A.J. Marine Mammals. In *Feeding: Form, Function, and Evolution in Tetrapod Vertebrates*; Schwenk, K., Ed.; Academic Press: New York, NY, USA, 2000; pp. 475–514.
75. Williamson, G.R. Counting and measuring baleen and ventral grooves of whales. *Sci. Rep. Whales Res. Inst.* **1973**, *24*, 279–292.
76. Carroll, G.M.; George, J.C.; Lowry, L.F.; Coyle, K.O. Bowhead whale (*Balaena mysticetus*) feeding near Point Barrow, Alaska, during the 1985 spring migrations. *Arctic* **1987**, *40*, 105–110. [CrossRef]
77. Nowacek, D.P.; Johnson, M.; Tyack, P.; Shorter, K.A.; McLellan, W.A.; Pabst, D.A. Buoyant balaenids: The ups and downs of buoyancy in right whales. *Proc. R. Soc. Lond.* **2001**, *268*, 1811–1816. [CrossRef]
78. Goldbogen, J.A.; Calambokidis, J.; Croll, D.A.; Harvey, J.T.; Newton, K.M.; Oleson, E.M.; Schorr, G.; Shadwick, R.E. Foraging behavior of humpback whales: Kinematic and respiratory patterns suggest a high cost for a lunge. *J. Exp. Biol.* **2008**, *211*, 3712–3719. [CrossRef]
79. Simon, M.J.; Johnson, M.; Tyack, P.; Madsen, P.T. Behaviour and kinematics of continuous ram filtration in bowhead whales (*Balaena mysticetus*). *Proc. R. Soc. B* **2009**, *276*, 3819–3828. [CrossRef] [PubMed]
80. Simon, M.; Johnson, M.; Madsen, P.T. Keeping momentum with a mouthful of water: Behavior and kinematics of humpback whale lunge feeding. *J. Exp. Biol.* **2012**, *215*, 3786–3798. [CrossRef] [PubMed]

81. Werth, A.J.; Potvin, J. Baleen hydrodynamics and morphology of cross-flow filtration in balaenid whale suspension feeding. *PLoS ONE* **2016**, *11*, e0150106. [CrossRef]
82. Zhu, Y.; Hu, D.; Yang, G. Theoretical analysis of the hydrodynamic filtering system in the balaenid whales suspension feeding. *Bioinspiration Biomim.* **2020**, *16*, e26006. [CrossRef]
83. Vandenberg, M.L.; Cohen, K.E.; Rubin, R.D.; Goldbogen, J.A.; Summer, A.P.; Paig-tran, E.M.; Kahane-Rapport, S.R. Formation of a fringe: A look inside baleen morphology using a multimodal visual approach. *J. Morphol.* **2023**, *284*, e21574. [CrossRef]
84. Andrady, A.L. Microplastics in the marine environment. *Mar. Pollut. Bull.* **2011**, *62*, 1596–1605. [CrossRef]
85. Saido, K.; Kodera, Y.; Kimukai, H.; Sato, H.; Okabe, A.; Koizumi, K.; Takatama, K.; Kwon, B.G.; Chung, S.Y.; Nishimura, M.; et al. Waste polystyrene degradation in the world oceans: Newly identified sources of contamination. *Res. Sq.* 2020; *preprint v3*. [CrossRef]
86. Sridhar, A.; Kanna, D.; Kapoor, A.; Prabhakar, S. Extraction and detection methods of microplastics in food and marine systems: A critical review. *Chemosphere* **2022**, *286*, e131653. [CrossRef]
87. Werth, A.J.; Rita Espada, D.; Lysiak, N.S.; George, J.C.; Sformo, T.L. Baleen turnover and gut transit in mysticete whales and its environmental implications. *Polar Biol.* **2020**, *43*, 707–723. [CrossRef]
88. Choy, C.A.; Robison, B.H.; Gagne, T.O.; Erwin, B.; Firl, E.; Halden, R.U.; Hamilton, J.A.; Katija, K.; Lisin, S.E.; Rolsky, C.; et al. The vertical distribution and biological transport of marine microplastics across the epipelagic and mesopelagic water column. *Sci. Rep.* **2019**, *9*, 7843. [CrossRef]
89. Kashiwabara, L.M.; Kahane-Rapport, S.R.; King, C.; DeVogelaere, M.; Goldbogen, J.A.; Savoca, M.S. Microplastics and microfibers in surface waters of Monterey Bay National Marine Sanctuary, California. *Mar. Pollut. Bull.* **2021**, *165*, 112148. [CrossRef]
90. Rochman, C.M.; Brookson, C.; Bikker, J.; Djuric, N.; Earn, A.; Bucci, K.; Athey, S.; Huntington, A.; McIlwraith, H.; Munno, K.; et al. Rethinking microplastics as a diverse contaminant suite. *Environ. Toxicol. Chem.* **2019**, *38*, 702–711. [CrossRef] [PubMed]
91. Werth, A.J. Models of hydrodynamic flow in the bowhead whale filter feeding apparatus. *J. Exp. Biol.* **2004**, *207*, 3569–3580. [CrossRef]
92. van der Hoop, J.M.; Nousek-McGregor, A.E.; Nowacek, D.P.; Parks, S.E.; Tyack, P.; Madsen, P.T. Foraging rates of ram-filtering North Atlantic right whales. *Funct. Ecol.* **2019**, *33*, 1290–1306. [CrossRef]
93. Rodríguez, Y.; Silva, M.A.; Pham, C.K.; Duncan, E.M. Cetaceans playing with single-use plastics (SUPs): A widespread interaction with likely severe impacts. *Mar. Pollut. Bull.* **2023**, *194*, 115428. [CrossRef]
94. Cózar, A.; Echevarria, F.; González-Gordillo, I.; Duarte, C.M. Plastic debris in the open ocean. *Proc. Natl. Acad. Sci. USA* **2014**, *111*, 10239–10244. [CrossRef]
95. van Sebille, E.; Wilcox, C.; Lebreton, L.; Maximenko, N.; Hardesy, B.D.; van Franekere, J.A.; Eriksen, M.; Siegel, D.; Galgani, F.; Lavender Law, K. A global inventory of small floating plastic debris. *Environ. Res. Lett.* **2015**, *10*, 124006. [CrossRef]
96. Cózar, A.; Sanz-Martin, M.; Marti, E.; González-Gordillo, I.; Ubeda, B.; Gálvez, J.A.; Irigoien, X.; Duarte, C.M. Plastic accumulation in the Mediterranean Sea. *PLoS ONE* **2015**, *10*, e121762. [CrossRef]
97. Fossi, M.C.; Copploa, D.; Baini, M.; Giannetti, M.; Guerranti, C.; Marsli, L.; Panti, C.; de Sabata, E.; Clò, S. Large filter feeding marine organisms as indicators of microplastic in the pelagic environment: The case studies of the Mediterranean basking shark (*Cetorhinus maximus*) and fin whale (*Balaenoptera physalus*). *Mar. Environ. Res.* **2014**, *100*, 17–24. [CrossRef]
98. Jambeck, J.R.; Geyer, R.; Wilcox, C.; Siegler, T.R.; Perryman, M.; Andrady, A.; Narayan, R.; Lavender, K. Plastic waste inputs from land into the ocean. *Science* **2015**, *347*, 768–771. [CrossRef]
99. de Vos, A.; Brownell, R.L.; Tershy, B.; Croll, D. Anthropogenic threats and conservation needs of blue whales, *Balaenoptera musculus indica*, around Sri Lanka. *J. Mar. Sci.* **2016**, *8420846*, 8420846. [CrossRef]
100. Connan, M.; Perold, V.; Dilley, B.J.; Barbraud, C.; Cherel, Y.; Ryan, P.G. The Indian Ocean 'garbage patch': Empirical evidence from floating macro-litter. *Mar. Pollut. Bull.* **2021**, *169*, 112559. [CrossRef]
101. Pattiaratchi, C.; van der Mheen, M.; Schlundt, C.; Narayanaswamy, B.E.; Sura, A.; Hajbane, S.; White, R.; Kumar, N.; Fernandes, M.; Wijeratne, S. Plastics in the Indian Ocean: Sources, transport, distribution, and impacts. *Ocean Sci.* **2022**, *18*, 1–28. [CrossRef]
102. Rosel, P.E.; Wilcox, L.A.; Yamada, T.K.; Mullin, K.D. A new species of baleen whale (*Balaenoptera*) from the Gulf of Mexico, with a review of its geographic distribution. *Mar. Mammal Sci.* **2021**, *37*, 577–610. [CrossRef]
103. Fu, J.; Liu, N.; Peng, Y.; Wang, G.; Wang, X.; Wang, Q.; Lv, M.; Chen, L. An ultra-light sustainable sponge for elimination of microplastics and nanoplastics. *J. Hazard. Mater.* **2023**, *456*, 131685. [CrossRef]
104. Dubner, S. The first great American industry. *Freakonomics* **2023**, *549*, 1–10. Available online: https://freakonomics.com/podcast/the-first-great-american-industry/ (accessed on 24 May 2023).
105. Stevenson, C.H. *Whalebone: Its Production and Utilization*; US Government Printing Office: Washington, DC, USA, 1907; Volume 626, pp. 1–12.
106. Virginia Marine Debris Reduction Plan, 2021–2025. U.S. NOAA Office of Response and Restoration Marine Debris Program. 2021. Available online: https://marinedebris.noaa.gov/report/virginia-marine-debris-reduction-plan (accessed on 12 June 2023).
107. Tan, Z.; Berry, A.; Charalambides, M.; Mijic, A.; Pearse, W.; Porter, A.; Ryan, M.P.; Shorten, R.N.; Stettler, M.E.J.; Tetley, T.D.; et al. *Tyre Wear Particles Are Toxic for Us and the Environment*; Imperial College London Report; Imperial College London: London, UK, 2023. [CrossRef]

108. Borrelle, S.B.; Ringma, J.; Law, K.L.; Monnahan, C.C.; Lebreton, L.; McGivern, A.; Murphy, E.; Jambeck, J.; Leonard, G.H.; Hilleary, M.A.; et al. Predicted growth in plastic waste exceeds efforts to mitigate plastic production. *Science* **2020**, *369*, 1515–1518. [CrossRef] [PubMed]
109. Barnett, M.J.; Hancock, P.I.; Klotz, L.E.; Attari, S.Z. Recycling bias and reduction neglect. *Nat. Sustain.* **2023**, *6*, 1418–1425. [CrossRef]
110. Fernández-d'Arlas, B. Tough and functional cross-linked bioplastics from sheep wool keratin. *Sci. Rep.* **2019**, *9*, 14810. [CrossRef] [PubMed]
111. Tian, L.; van Putten, R.-J.; Gruter, G.J.M. Plastic pollution: The role of (bio)degradable plastics and other solutions. In *Biodegradable Polymers in the Circular Plastics Economy*; Dusselier, M., Lange, J.-P., Eds.; Wiley: Hoboken, NJ, USA, 2022; pp. 59–82.
112. Perera, P.; Herath, H.; Paranagama, P.A.; Wijesinghe, P.; Attanayake, R.N. Wood decay fungi show enhanced biodeterioration of low density polyethylene in the absence of wood in culture media. *PLoS ONE* **2023**, *18*, e0288133. [CrossRef] [PubMed]

oceans

MDPI

Article

Preliminary Studies on Site Fidelity, Residence Index, and Population Size of Irrawaddy Dolphins in West Penang, Malaysia

Nurul Filzati Ali and Leela Rajamani *

Centre for Marine and Coastal Studies, Universiti Sains Malaysia, 11800 USM, Penang, Malaysia; nurulfilzati@yahoo.com
* Correspondence: leelarajamani@usm.my

Abstract: The Irrawaddy dolphin is found in the coastal and estuarine areas of West Penang, Malaysia. Studies were conducted to estimate the site fidelity, residence index, and population size of Irrawaddy dolphins in West Penang. Photo-identification studies were conducted using boat surveys from 2019 to 2021. Thirty-nine marked Irrawaddy dolphins were identified, with thirty-six newly identified individuals and three individuals observed in 2013. Resightings of four individuals indicated that they were found north of Sungai Burung and Sungai Pinang in West Penang. The majority of individuals had low sighting rates, ranging from 2.6 to 7.7%, with three individuals having medium sighting rates, the highest being 15.4%. The residence index was 0.01 for all 36 individuals, and the highest value of 0.36 was recorded for one individual. Using open population models and closed models, the population size was determined to be 64 or and 52, respectively. The results suggest that although there is a population present, it is probably open, as the residence index is low. The population size appeared to be stable from 2013 to 2021. This information will inform conservation managers of the best way forward for the conservation of Irrawaddy dolphins in Penang.

Keywords: Irrawaddy dolphin; Penang; population size; site fidelity; residence index

1. Introduction

The Irrawaddy dolphin (*Orcaella brevirostris*) is an inshore cetacean species found in small patchy populations in coastal, estuarine, and riverine habitats [1]. In Malaysia they have been found in Penang [2,3], Cowie Bay, Sandakan and Kinabatangan (Sabah) [4,5], Kuching Bay, Sarawak [6], and Matang, Perak [7]. Irrawaddy dolphin population size have generally been estimated to be small, e.g., in Cowie Bay, Sabah, in 2010, with a population of 28 individuals (95% confidence limits (CL) = 28–34) [4]; Kuching Bay, Sarawak, with 149 individuals (95% CL = 151–360 with CV = 22.5%) [6]; and Penang Island (32 to 52 individuals) [3], as have other populations in its range. Irrawaddy dolphins are classified as endangered (EN) under the IUCN Red List of Threatened Species for coastal species but are critically endangered for river species and in the Malampaya Sound subpopulation in the Philippines [1,8]. The patchy and fragmented distribution of Irrawaddy dolphins in both coastal waters and rivers renders them particularly vulnerable to threats from human activities concentrated in the same areas [1]. An increase in the number of mortalities among Irrawaddy dolphins has largely been due to threats such as gillnets, which cause incidental mortality in small-scale fisheries [1,9,10]. Habitat loss—particularly from dams in riverine populations and degradation from declining or altered freshwater flows affecting estuarine populations—is a looming conservation threat with the potential to extirpate subpopulations and further fragment the already patchy distribution of the species [1].

The west coast of Penang, Malaysia, is known to host at least three species of inshore cetaceans, namely, the Irrawaddy dolphin (*Orcaella brevirostris*), the Indo-Pacific finless porpoise (*Neophocaena phocaenoides*), and the Indo-Pacific humpback dolphin (*Sousa chinensis*) [2]. These three species have been observed since the beginning of 2013 when formal boat surveys for cetaceans were initiated. Notably, the Irrawaddy dolphin has been predominantly observed. Using photo-identification techniques, a catalog for the Irrawaddy dolphin was created with 30 individuals that were identified regardless of their right dorsal fins or left dorsal fins [11]. Since then, an initial population for Irrawaddy dolphins using mark–recapture techniques is estimated to include 31 to 52 individuals for both the closed and open population models [3]. Further investigations need to be conducted to analyze the current population status of Irrawaddy dolphins in Penang. In this paper, we will discuss the site fidelity and residence index and provide a new estimate of the population size based on photo-identification studies conducted from 2019 to 2021. This information is important, as it provides us with information on how often the Irrawaddy dolphin is present in the western part of Sungai Pinang, the degree of residency, and population size, which will enable us to determine what kind of conservation measures are needed for this species in this region.

2. Materials and Methods

2.1. Study Area

Penang Island is approximately 293 km^2 in area and located on the western seaboard of Peninsular Malaysia in the Northern Malacca Strait (within latitudes 5°12′ N to 5°30′ N and longitudes 100°09′ E to 100° 26′ E). The exact study area was on the western coastline of Penang Island. It comprises 14 km of mangrove coastline with a shallow, mesotidal estuarine habitat where freshwater comes from six different rivers, including the two largest rivers, Sungai Pinang and Sungai Burung [12,13].

2.2. Boat-Based Surveys and Photo-Identification Studies

Boat-based surveys using transects were used to conduct photo-identification studies and population size estimation, as well as to determine the residency index and site fidelity.

2.2.1. Data Collection

Surveys were conducted from a small fiberglass boat (7.6 m size) between 07:30 am to 2:00 pm (weather permitting) for 5 days every month. The study area covered approximately 17.3 km × 6.0 km (103.8 km^2) and included water depths (0.5 m > depth > 29.5 m). Two survey routes were used to ensure complete coverage of the study site and avoid potential bias, one parallel to the shore and the other at a 45° angle in a zigzag pattern.

When both the parallel line and zigzag routes were completed, we considered one survey to be completed (Figure 1). During the survey, two observers, with the aid of binoculars, usually stood at the front of the boat, alternating shifts every 20 min and checking for the presence of animals. At the start of each survey, effort status and environmental conditions were recorded, such as sea state, wave height, visibility, and glare. These conditions were recorded every 20 min or when dolphins or porpoises were encountered. In addition, the position of the vessel was constantly logged, using a GPS Garmin Montana 360 (Garmin Inc., Olathe, Kansas, KS, USA). Surveys were conducted at sea conditions of Beaufort $\leq$ 3.

A dolphin sighting is when the occurrence of a dolphin happens during the search effort and ends when 15 min have elapsed since the last surface of the cetacean [3].

2.2.2. Photo-Identification and Population Size Estimation

Photos were taken with a Nikon D3200 DSLR camera with a 70–300 mm lens, and the images were graded according to their quality, with only those of sufficient quality used to identify individual dolphins [14]. While the boat traveled alongside the dolphins, attempts were made to photograph the left and right sides of the dorsal fins of individual dolphins.

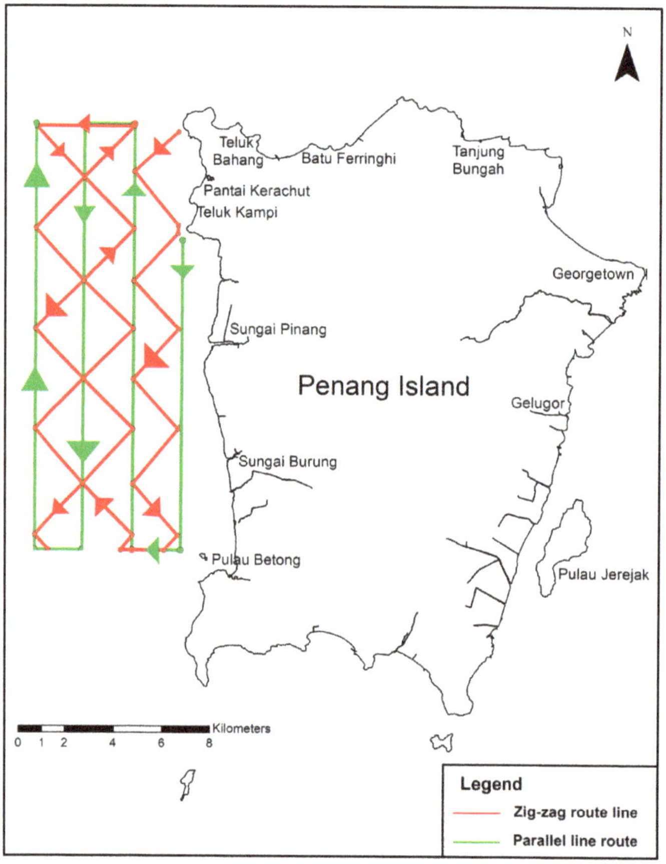

Figure 1. Two types of survey routes where the boat traveled, consisting of a parallel line route and a zigzag route.

Photographs that lacked sufficient markers to be classified as unique were also collected. There were three sets of photo IDs: left dorsal fins (LDFs), right dorsal fins (RDFs), and regardless of the side of the fin (OBP) (see Supplementary Materials).

The dolphin photos were sorted into three folders: Grade 1 (the image of the dolphin or the dorsal fin is less than 10% of the photo's total height; the image is out of focus or blurry); Grade 2 (the photo is in focus, but it does not contain the dorsal fin or the dorsal fin is not properly placed to identify the marks); and Grade 3 (the photo is focused, the dorsal fin is parallel, and the mark is easily recognizable) using the Quality Grading Criteria for Hong Kong HKZMB Photo-Identification Catalog (Lindsay Porter, Southeast Asia Marine Mammal Research, Hong Kong, personal communication). From the photos that we collected and sorted, each dolphin individual was classified based on the fin shape, lesion, scars, and cuts (long-lasting marks). Each unique individual was assigned an ID code with 'OBP' (standing for *Orcaella brevirostris* Penang) as the prefix, followed by the number (e.g., OBP001). Based on the catalog previously created by Rodriguez-Vargas [11], we named the existing individuals, and we provided new names for the new dolphin individuals we discovered.

Two types of population estimates were calculated: closed population and open population. Closed population models were used because it was assumed that there was little change from birth, death, immigration, or emigration during this study period [15].

While open population models consider variation in population size during the study period when assumptions of closure are not met, it can be useful to compare estimates from both open and closed models [16].

Four different models for closed populations were used to determine the best-fit model for population size estimation, which were Mb (behavior-dependent), Mt (time-dependent), Mo (constant), and M(h) (heterogeneity) [17]. The models with the lowest Akaike Information Criterion (AIC) were used to determine the best model for each data set.

The POPAN formulation for Jolly–Seber was used to compare four different models for open populations, where p = capture probabilities, ϕ (phi) = survival probabilities, and PENT = the probability of the entrance of other individuals in the parameterization as a function of time [18].

The four different models are:

- p (t), ϕ (t), PENT (t) = Capture and survival probabilities are time-dependent.
- p (.), ϕ (t), PENT (t) = Capture probabilities are constant, and survival probabilities are time-dependent.
- p (t), ϕ (.), PENT (t) = Capture probabilities are time-dependent, and survival probabilities are constant.
- p (.), ϕ (.), PENT (t) = Capture and survival probabilities are constant.

Since MARK only estimates the number of marked animals, a correction factor that includes the unmarked animals was used to calculate the total population size [6]. The number of identifiable fins per sighting was summed for all sightings and divided by the total sum of both identified and unidentified fins per sighting for all sightings [6]. The resulting mean proportion, p, was used as a correction factor for population estimate estimates (N) that only used the sighting histories of distinctive individuals [19]. The 95% confidence interval, CI, was computed using the following formulas:

$$95\%\ \text{lower interval} = N_{corrected} \times p, \text{ and} \quad (1)$$

$$95\%\ \text{upper interval} = \frac{N_{corrected}}{p} \quad (2)$$

2.2.3. Site Fidelity

Information from photo-identification studies during boat-based surveys was analyzed to provide information on site fidelity and residence index [20–22]. The site fidelity patterns of individually identified dolphins were determined based on their resight rate and presence across seasons [22,23].

Sighting rates were classified into three categories according to the proportion of the number of sightings of an identified dolphin to the total number of surveys, which was, in this case, 39 surveys. Sighting rates less than 10% were classified as low sighting rates (LSR), 10–30% as moderate sighting rates (MSR), and more than 30% as high sighting rates (HSR). Dolphins were considered residents if they had moderate to high sighting frequencies throughout the year (i.e., northeast monsoon and southwest monsoon) [22]. Dolphins identified during the same season in consecutive years but not during intervening seasons were defined as seasonal residents [23]. Occasional visitors were dolphins that had low sighting rates but were present in all seasons [22]. Transients were dolphins that had low sighting rates and were only observed in one season [22].

The residence index (RI) was calculated to quantify the occurrences of each individual of *Orcaella brevirostris* [20] by determining the total number of sightings of the individual with the number of months it was seen. The formula is

$$RI = S \times M/100,$$

where RI = residence index, S = total number of sightings of an individual, and M = total number of months in which this particular individual was seen.

3. Results

A total of 39 surveys in 87 days were completed covering 3745 km, with a daily average of 39.43 km + 1.54 SE (see Figure 2, Table A1). The total survey time was 280.4 h, with 7.5 h dedicated to photo identification. Off-effort sightings took up a total of 41 min. Fifty-two Irrawaddy dolphins were sighted during the survey (Figure 3).

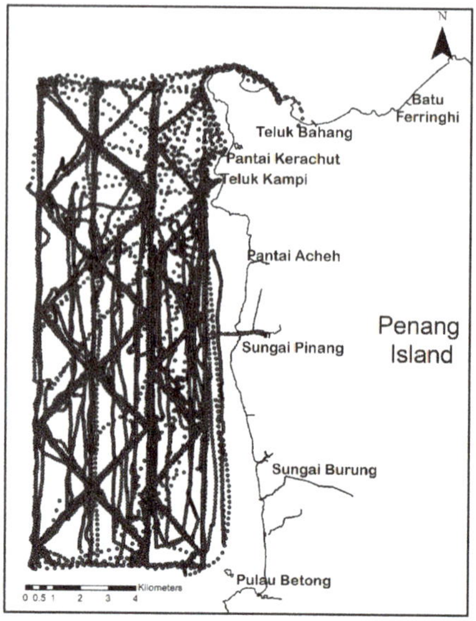

Figure 2. Actual survey path followed by the boat during all the surveys in West Penang.

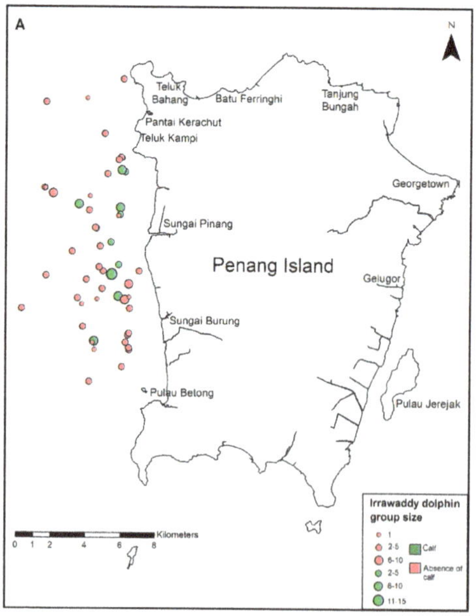

Figure 3. Irrawaddy dolphin encounters in the western coastal waters of Penang Island (February 2019 to April 2021) during on-effort surveys [13] (Permission was granted by the publisher, the Lee Kong Chian Natural History Museum.).

3.1. Photo-Identification Studies and Discovery of New Animals

We obtained a total of 11,056 photos of Irrawaddy dolphins. From these photos, 30.7% (3398 photos) contained images without dolphins, 66.5% (7347 photos) contained Grade 1 photos, 1.56% (173 photos) contained Grade 2 photos, and 1.25% (138 photos) contained Grade 3 photos.

Based on the results of the discovery curve, it appears that most of the individuals were discovered in February (14 animals), April (7), and May (5) of 2019, amounting to a total of 26 individuals (Figure 4). Fewer individuals were discovered in July (4), September (3), and November (2), amounting to a total of nine individuals discovered in the second half of 2019 (Figure 4). No new individuals were found in June, August, or December. In the year 2020, two new individuals were discovered in March, and one new individual was discovered in November. Most of the dolphin individuals sighted in February, March, July, and November 2020 were the resighted individuals from the year 2019. No dolphins were found in January, June, September, or October 2020. In 2021, one new individual was discovered in March. No new individual dolphins were found in January, February, or April.

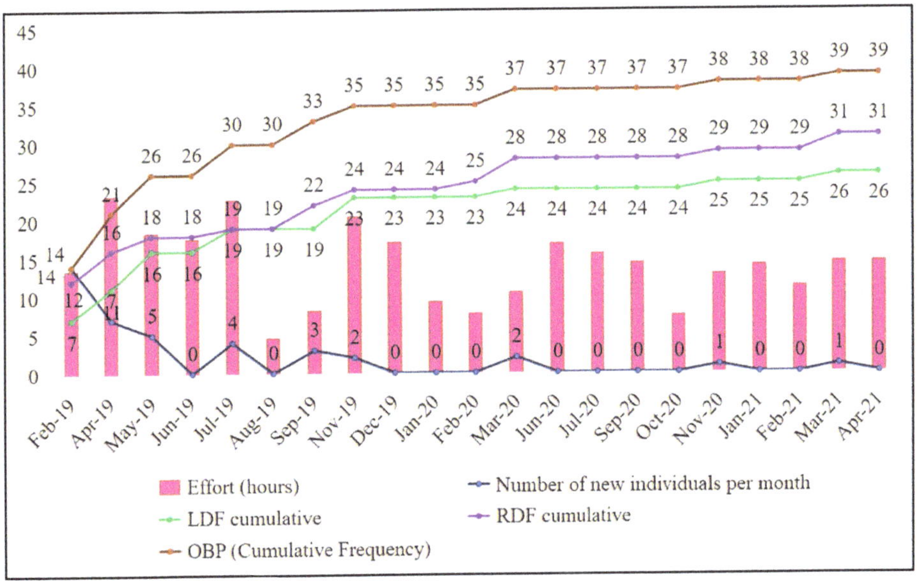

Figure 4. Discovery curve showing the accumulation of newly identified individual Irrawaddy dolphins during the period of study from February 2019 to April 2021, related to the effort in hours per month when sightings were recorded. LDF = left dorsal fin; RDF = right dorsal fin; OBP = regardless of side.

In the surveys conducted between February 2019 and April 2021, a total of 39 new individuals were identified based on either the left dorsal fin or the right dorsal fin. Upon checking the first catalog developed for Irrawaddy dolphins in West Penang for thirty individuals [11], only three individuals, namely, OBP003, OBP010, and OBP016, were rediscovered at this time. A new catalog will be developed using the current data from 2019, 2020, and 2021. From the current data, it appears that recaptures were only seen for 15 individual Irrawaddy dolphins (Table 1).

Adverse weather conditions led to the cancelation of several field days either halfway through a survey day or a whole day before the survey. Photo identification became increasingly difficult because of the decreased sighting ability of the observers during bad weather and avoidance behavior from the dolphins, which were only sighted once or twice and did not surface again.

Table 1. Individual Irrawaddy dolphins that were recaptured during the field surveys in 2019, 2020, and 2021.

No	Individual's Name	Months They Were Encountered					
		First Encounter	Second Encounter	Third Encounter	Fourth Encounter	Fifth Encounter	Sixth Encounter
1	OBP044	February 2019	April 2019				
2	OBP047	February 2019	May 2019				
3	OBP051	February 2019	March 2020				
4	OBP053	April 2019	May 2019				
5	OBP054	April 2019	November 2019	July 2020	November 2020		
6	OBP057	April 2019	March 2020				
7	OBP061	May 2019	March 2020				
8	OBP062	May 2019	November 2020				
9	OBP064	May 2019	March 2020				
10	OBP066	July 2019	February 2020				
11	OBP067	July 2019	November 2019				
12	OBP068	November 2019	March 2020	March 2021			
13	OBP070	September 2019	November 2019	November 2020	March 2021		
14	OBP003	July 2019	September 2019	November 2019	March 2020	November 2020	March 2021
15	OBP016	March 2020	March 2021				

3.2. Resighting Pattern

Data on the resighting patterns of Irrawaddy dolphins were limited because few recaptures were obtained throughout the survey period; however, the information so far indicates that the common areas where the Irrawaddy dolphin can be observed appear to be in Sungai Pinang and Sungai Burung (Figure 5).

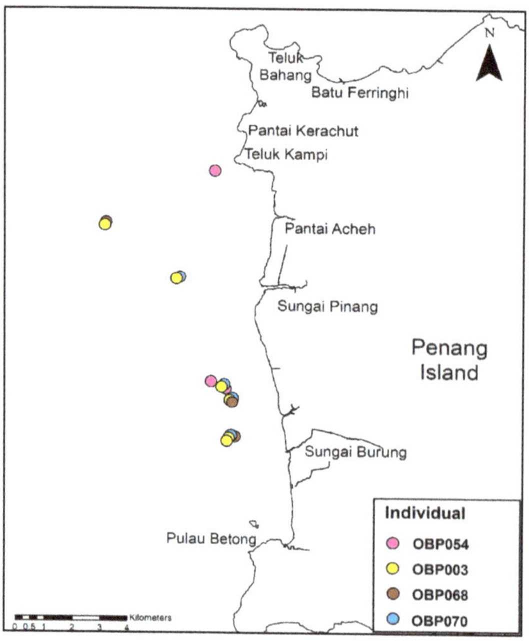

Figure 5. Resighting patterns of Irrawaddy dolphins OBP054, OBP003, OBP068, and OBP070.

Irrawaddy dolphins named OBP054, OBP068, OBP003 and OBP070 were observed more than once (see Table 1). OBP068 was seen three times, while OBP054 and OBP070 were seen four times. There were six occasions of sightings of OBP003 (Table 1), while the rest of the individuals were resighted only once during the survey (Table 1).

3.3. Minimal Distances Traveled by Individual Dolphins

The Irrawaddy dolphin individuals that were resighted more than once (OBP003, OBP054, OBP068, and OBP070) were seen between Sungai Burung, Sungai Pinang, Pantai Acheh, and Teluk Kampi (Figure 5). OBP054 traveled a greater distance, where it was observed in Teluk Kampi, 10.1 km away from the Sungai Burung region (Figure 6a, Table 2). This individual was seen first in Sungai Pinang in April 2019, then in Sungai Burung (November 2019), Teluk Kampi (July 2020), and then back in Sungai Pinang again in November 2020 (Figure 6a, Table 1).

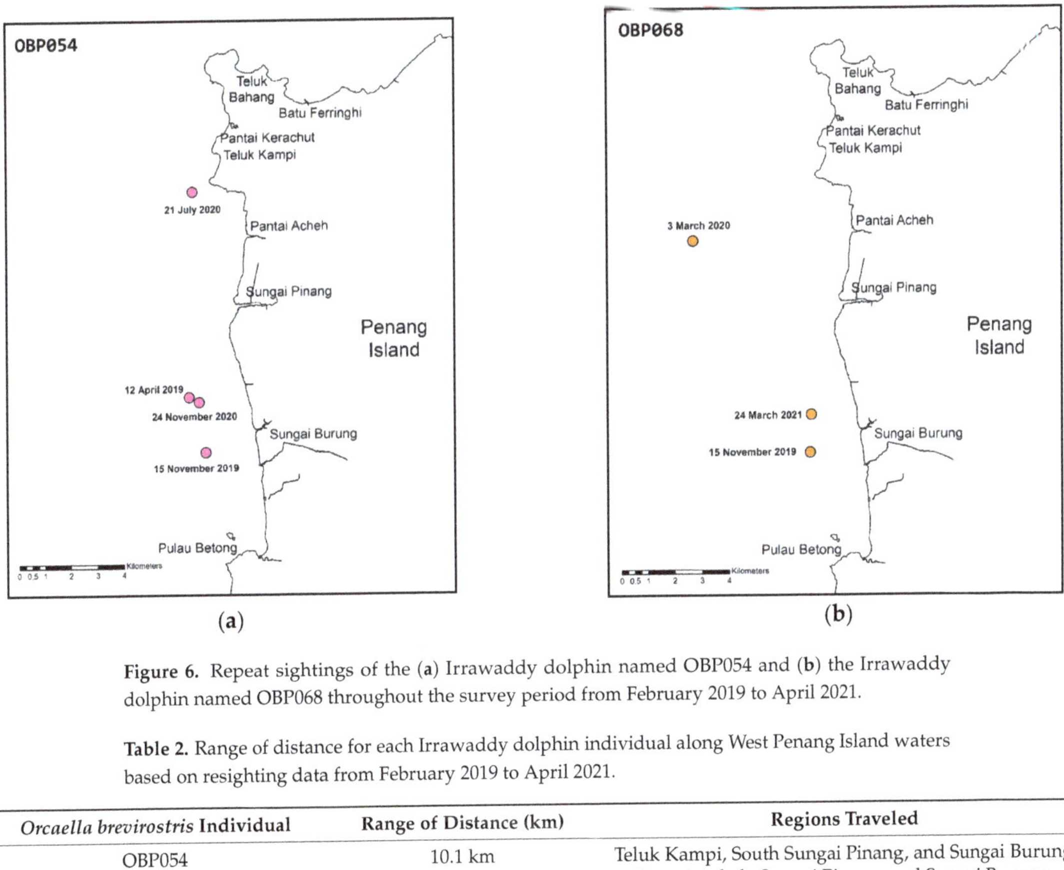

Figure 6. Repeat sightings of the (**a**) Irrawaddy dolphin named OBP054 and (**b**) the Irrawaddy dolphin named OBP068 throughout the survey period from February 2019 to April 2021.

Table 2. Range of distance for each Irrawaddy dolphin individual along West Penang Island waters based on resighting data from February 2019 to April 2021.

Orcaella brevirostris Individual	Range of Distance (km)	Regions Traveled
OBP054	10.1 km	Teluk Kampi, South Sungai Pinang, and Sungai Burung
OBP068	9.2 km	Pantai Acheh, Sungai Pinang, and Sungai Burung
OBP070	6.3 km	Sungai Pinang and Sungai Burung
OBP003	9.5 km	Pantai Acheh, Sungai Pinang, and Sungai Burung

The Irrawaddy dolphin named OBP068 had a slightly shorter moving distance of 9.2 km and traveled from Pantai Acheh to the Sungai Burung region (Figure 6b, Table 2). It was sighted in Sungai Burung in November 2019 and then in Pantai Acheh in March 2020, followed by Sungai Pinang in March 2021 (Figure 6b, Table 2).

Individual OBP070, which was seen four times, had the shortest range of movement, which totaled up to 6.3 km from Sungai Pinang to Sungai Burung (Figure 7a, Table 2). This dolphin was seen in the Sungai Pinang region in September 2019, followed by the Sungai Burung region in November 2019, November 2020, and March 2021.

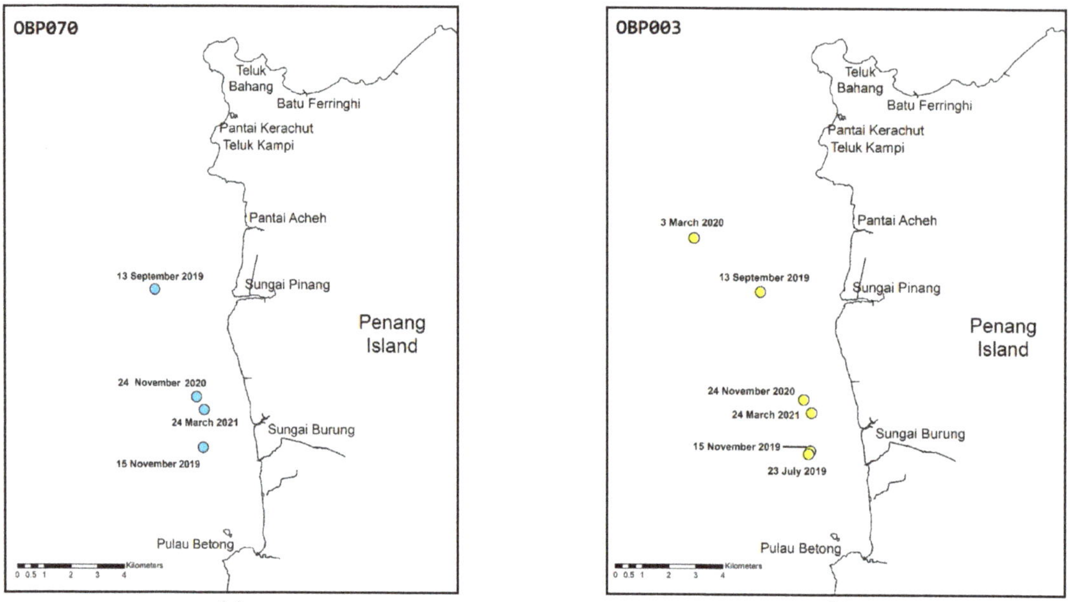

(a) (b)

Figure 7. Repeat sightings of the (**a**) Irrawaddy dolphin named OBP070 and (**b**) the Irrawaddy dolphin named OBP003 throughout the survey period from February 2019 to April 2021.

The Irrawaddy dolphin named OBP003 traveled 9.5 km (Figure 7b, Table 2). This dolphin was first spotted in July 2019 in Sungai Burung, in Sungai Pinang in September, in Sungai Burung in November, in Pantai Acheh in March, and then again in Sungai Burung in November 2020 and March 2021.

3.4. Sighting Rates and Residence Index (RI)

Among the 39 individually identified dolphins, 24 individuals had a sighting rate of 2.56%, followed by 11 individuals with a sighting rate of 5.13% (see Table 3, Table A2). One individual (OBP068) had a sighting rate of 7.69%, two (OBP054 and OBP070) had a rate of 10.26%, and one (OBP003) had a sighting rate of 15.38% (Figure 8). This result shows that only three individuals had moderate sighting rates, while the majority (36 of the individuals) had low sighting rates.

The majority of the individuals were transients (32 out of 39, 82.1%), followed by occasional individuals (4 out of 39, 10.3%), namely, OBP016, OBP062, OBP066, and OBP068. Only three individuals (OBP054, OBP070, and OBP003) accounted for all the residents (3 out of 39, or 7.7%).

The majority of the identified individuals (24 animals) had a low residence index of 0.01 (Table 4). Eleven individuals had an RI of 0.04 (individuals were found in two occasions in two months), one individual had an RI of 0.09 (discovered in three sightings in three months), two individuals had an RI of 0.16 (found in four sightings in four months), and one individual had an RI 0.36 (found in six sightings in six months).

Table 3. Number of marked Irrawaddy dolphins, percentage, sighting rates, and category of sightings of the 39 marked Irrawaddy dolphins.

Sighting Rate (%)	Category of Sighting Rate (LSR, MSR, and HSR *)	No. of Individual Marked Irrawaddy Dolphin	Percentage (%)
2.56	LSR	24	61.54
5.13	LSR	11	28.21
7.69	LSR	1	2.56
10.26	MSR	2	5.13
15.38	MSR	1	2.56
	Total	39	100.00

* LSR = low sighting rate, MSR = medium sighting rate, HSR = high sighting rate.

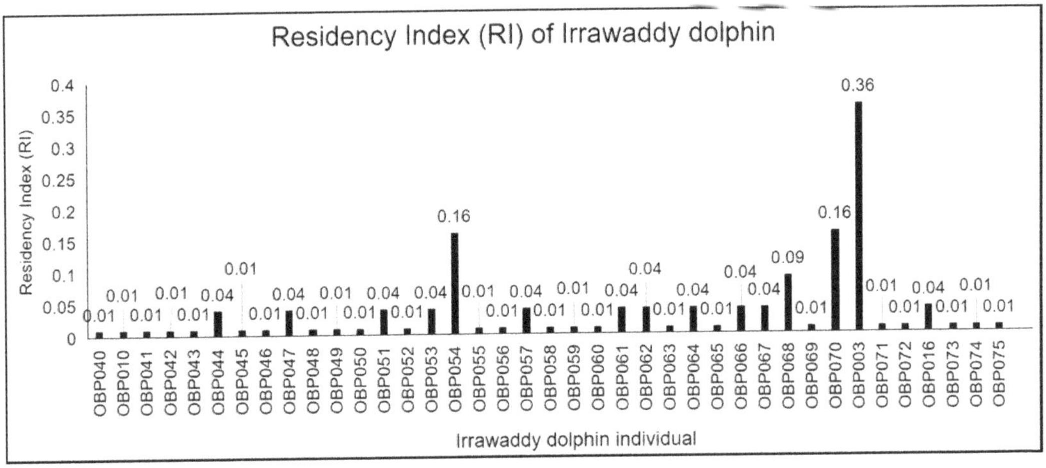

Figure 8. Residence indexes calculated for *Orcaella brevirostris* individuals discovered in 2019, 2020, and 2021 in West Penang Island.

Table 4. Number and percentage of Irrawaddy dolphins and their respective residence index (RI) values.

Residence Index (RI)	No. of Marked Individual Irrawaddy Dolphins	Percentage (%)
0.01	24	61.54
0.04	11	28.21
0.09	1	2.56
0.16	2	5.13
0.36	1	2.56

3.5. Population Size

We estimated the population size using mark–recapture analyses of photo-identified dorsal fins using the MARK software (version 10.1), which has been successfully used by researchers to estimate Irrawaddy dolphin population sizes [16,19,24].

3.6. Mark–Recapture Analysis

Based on the discovery curve (Figure 4), we discovered 39 marked Irrawaddy dolphins and 40 dolphins with unmarked fins (Grade 3). The proportion of marked individual photos was 0.706.

The MARK software was used to analyse OBP, LDF and RDF in a close population model. The selected model M(b) which describes the capture probabilities that varied with the dolphin's behaviour in response to being captured is the best fit model for OBP because it had the lowest AIC value (Akaike's Information Criterion). This model provided an estimate of 37 individuals (3.88 SE) (95% CI = 36 to 55). For the LDF site, with a population estimate of N-hat = 30 individuals (8.88 SE, 95% CI = 24 to 70), M(b) was the best fit. The RDF's population estimate, N-hat = 37 individuals (2.12 SE, 95% CI = 24 to 36), also identified M(b) as the most fit model. Meanwhile, in the open-population POPAN formulation for Jolly–Seber estimations, the phi(.), p(t), pent(N) model was the most fit model for OBP (Table 5). The population was estimated to be 45 individuals, with a standard error of 4.55 (95% CI: 45 to 190). For the LDF side, the best-fitting model for the open population estimated was phi(.), p(.), pent(N). The population estimate for LDF was 81 $\pm$ 34.94 SE (95% confidence interval = 36 to 182). In the case of RDFs, the phi(.), p(.), pent(N) model had the lowest AIC value. The model estimated 67 individuals $\pm$ 17.93 SE (95% CI: 40 to 112) (see Table 5).

The corrected population estimates ($N_{corrected}$), which included the unmarked individuals, were calculated using the OBP, LDF, and RDF N-hat results from mark–recapture analyses (Table 5). For the closed population, the corrected population estimated for OBP was 52 individuals (95% CI: 37 to 74). For the left fin side photos (LDF), adjusted for proportion, the estimated population ($N_{corrected}$) was 40 individuals (95% CI = 30 to 53). The corrected population estimate for RDF was 37 individuals, with a 95% CI of 25 to 54 (Table 5). Meanwhile, for the open population, the corrected population estimate for OBP was 64 (95% CI = 45 to 90). For LDF, the $N_{corrected}$ population estimate was 107 individuals (95% CI = 81 to 142), while for RDF, the $N_{corrected}$ was 99 individuals (95% CI = 67 to 146) (Table 5).

Table 5. Best model for closed and open populations for left dorsal fin side, right dorsal fin side, and regardless of side (OBP).

	Closed Population Model													
	Model	**AICc**	**Delta AICc**	**AICc Weight**	**Model Likelihood**	**Parameter**	**Deviance**	**−2log(L)**	**N-hat**	**$N_{corrected}$**	**SE**	**CV**	**95% CI**	**95% CI Corrected**
OBP	M(b)	−46.2429	0.0315	0.47198	0.9844	3	18.5972	−52.419	37	52	3.879	0.07	36 to 55	37 to 74
LDF	M(b)	−33.5257	0	0.64613	1	2	5.7949	−37.661	30	40	8.881	0.22	24 to 70	30 to 53
RDF	M(b)	−26.2571	0	0.87046	1	3	7.2159	−32.518	25	37	2.119	0.06	24 to 36	25 to 54
	Open population model													
	Model	**AICc**	**Delta AICc**	**AICc Weight**	**Model Likelihood**	**Parameter**	**Deviance**	**−2log(L)**	**N-hat**	**$N_{corrected}$**	**SE**	**CV**	**95% CI**	**95% CI corrected**
OBP	phi(.) p(t) pent(N)	117.8479	0	0.95366	1	8	−36.1515	99.8197	45	64	4.554	0.07	37 to 54	45 to 90
LDF	phi(.) p(.) pent(N)	62.9629	0	0.74121	1	6	−27.0058	47.3108	81	107	34.938	0.33	36 to 182	81 to 142
RDF	phi(.) p(.) pent(N)	86.2406	0	0.90352	1	5	−33.9626	74.4759	67	99	17.931	0.18	40 to 112	67 to 146

4. Discussion

The current results indicate that there is a small population of Irrawaddy dolphins that is probably transient in West Penang.

Out of a total of 39 individuals, there were resightings of only 15 Irrawaddy dolphins. Out of these fifteen sightings, only five individuals were sighted more than three times, indicating that the general population is not a residential one. Only four individuals were resighted more than once, namely, OBP068, OBP054, OBP070, and OBP003, indicating that these individuals visit West Penang more often.

Individual OBP003 was discovered in 2013 during a previous study [11] with a residency index of 0.04, as compared with now, which was 0.36. This means that previously OBP003 was not resident in West Penang but currently is. On the other hand, OBP010 had a residency index of 0.24 [11] compared with 0.01 currently. Individual OBP016 had a residency index of 0.02 [11] compared with 0.04 currently, indicating that its residency status was the same from 2013 to 2021, as far as the evidence shows. This also indicates that the animal occasionally visits West Penang throughout the year. A residence index was developed to overcome biases when the dolphins were often seen in a short period of time [20], which was not true in our case. However, we calculated the index to show a crude estimate of the degree of dolphin residency.

The fact that the resighting rate and residence index were low indicates that the population does not stay for long periods of time, with the majority of the 39 individuals being transients. They may frequently move in and out of the study area in search of prey. The encounter rate was estimated to be 0.19 sightings per hour (or 1.4 sightings per 100 km), which indicates a low encounter rate [13]. This encounter rate is slightly lower than the 0.25 sightings per hour reported in 2013 [3], which does not indicate any significant changes in relative abundance.

A study conducted simultaneously with the current study showed that milling [13] was more predominant than feeding behavior for Irrawaddy dolphins, indicating that the current environment is possibly less suitable for finding prey. Compared with recent surveys conducted from 2019 to 2021, feeding was not the predominant behavior for the Irrawaddy dolphin, as opposed to 2013, where feeding was the predominant behavior [3]. Avoidance behavior could also lead to poor resighting rates, as avoidance behavior was recorded 13 times for Irrawaddy dolphins in the period from 2019 to 2021 [13]. However, it is unclear to what extent avoidance behavior affects the resighting of animals.

From marked Irrawaddy dolphins, the estimated population size obtained from both the closed and open population models was 52 and 64 individuals respectively. Even though the population size is slightly higher compared with the year 2013 (32 to 51 individuals) [3], the result is incomparable, as the current survey area size was larger than the previous one even though both studies were in West Penang. However, this still indicates that there is a viable population of Irrawaddy dolphins, and from that, we can observe if the population will increase or decrease in the future. The population is small and comparable to other regions such as in Cowie Bay, Sabah, in 2010, with a population of 28 individuals (95% confidence limits, CL = 28–34) [4]. Large population sizes have been estimated for open estuarine waters in Bangladesh with 5383 individuals (CV = 40%) [25].

Largely, these results indicate that there exists a population of Irrawaddy dolphins that has low site fidelity, i.e., has a low sighting rate and low residence, indicating the population is open rather than closed. The current study probably only investigated a small portion of a larger metapopulation (that is, more or less permanently resident), in which case, it may extend to unknown boundaries and be underestimated in open population estimate modeling. Another factor is that some individuals may show avoidance behavior to boats, thus decreasing the chance of resightings and obtaining Grade 3 photos.

There is no kind of modeling (including one based on behavior) that can assess or explain their absence during a survey. Therefore, the study could have left a portion of the local population unstudied. More survey effort is needed to ascertain population size changes in the years to come. For instance, if the population is indeed open, it may be necessary to know if these individuals are seen in other estuarine systems close to West

Penang, such as Kuala Muda (north of Penang) and Pulau Aman (south of Penang). Further photo-identification studies can be conducted in the future in these areas. Research into the avoidance behavior of Irrawaddy dolphins should also be considered. However, at present, decisions can be made based on the current results for their conservation. Since the region of West Penang is an important area for Irrawaddy dolphins, a suitable management plan that incorporates local fisher and dolphin needs should be established.

Supplementary Materials: The following supporting information can be downloaded at: https://www.mdpi.com/article/10.3390/oceans4040029/s1, Dolphin catalogue (LDF and RDF) of each Irrawaddy dolphin individual in west Penang Island.

Author Contributions: Conceptualization, N.F.A. and L.R.; methodology, N.F.A. and L.R.; investigation, N.F.A. and L.R.; writing—original draft preparation, N.F.A.; writing—review and editing, L.R.; visualization, N.F.A.; supervision, L.R.; project administration, L.R.; funding acquisition, L.R. All authors have read and agreed to the published version of the manuscript.

Funding: We would like to acknowledge the Ocean Park Conservation Foundation, Hong Kong (OPCFHK), for graciously funding our research project in Penang, Malaysia.

Data Availability Statement: The data presented in this study are available upon request from the corresponding author.

Acknowledgments: We would like to acknowledge and express our appreciation to the staff at the Centre for Marine and Coastal Studies (CEMACS) at Universiti Sains Malaysia, as well as the dedicated staff and students who contributed to the fieldwork and data analysis. Our heartfelt thanks also go to the volunteers who provided invaluable assistance during the boat surveys.

Conflicts of Interest: The authors declare no conflict of interest. The funders had no role in the design of the study; in the collection, analyses, or interpretation of the data; in the writing of the manuscript; or in the decision to publish the results.

Appendix A

Table A1. Months, Dates, Distance, and Hours Traveled on Effort in 2019, 2020, and 2021.

Year	Month	Date of Survey	Distance Traveled on Effort (km)	Effort Hours (h)
2019	February	18, 19 and 20	96	11.38
	April	12, 15, 16, 17 and 18	230	20.64
	May	13, 14, 15 and 16	182	18.75
	June	24, 25, 26, 27 and 28	239	19.23
	July	19, 22, 23, 24 and 25	219	21.49
	August	20, 26 and 27	119	6.17
	September	12, 13 and 18	189	10.56
	November	13, 15, 18 and 19	251	12.84
	December	4, 12, 13 and 15	242	10.88
2020	January	29, 30 and 31	113	9.33
	February	1, 24 and 25	116	7.73
	March	16 and 17	143	10.50
	June	22, 23, 24, 25 and 26	192	16.88
	July	16, 17, 20, 21 and 22	184	15.48
	September	23, 25, 28, 29 and 30	147	14.33
	October	19, 20, 21 and 22	135	7.4
	November	21, 22, 24, 25 and 26	183	12.86
2021	January	25, 26, 27 and 28	167	13.96
	February	22, 23, 24, 25, 26 and 27	255	11.27
	March	24, 25, 26, 27 and 28	198	14.4
	April	22, 26, 28 and 29	145	14.35
		Total	3745	280.43

Appendix B

Table A2. Sighting Rate, Residency Index Value, Category of Sighting Rate, and Category of Residency for Each Irrawaddy Dolphin Individual Discovered.

Irrawaddy Dolphin Individual	Sighting Rate (%)	Residency Index (RI)	Category of Sighting Rate	Category of Residency
OBP040	2.56	0.01	LSR	Transient
OBP010	2.56	0.01	LSR	Transient
OBP041	2.56	0.01	LSR	Transient
OBP042	2.56	0.01	LSR	Transient
OBP043	2.56	0.01	LSR	Transient
OBP044	5.13	0.04	LSR	Transient
OBP045	2.56	0.01	LSR	Transient
OBP046	2.56	0.01	LSR	Transient
OBP047	5.13	0.04	LSR	Transient
OBP048	2.56	0.01	LSR	Transient
OBP049	2.56	0.01	LSR	Transient
OBP050	2.56	0.01	LSR	Transient
OBP051	5.13	0.04	LSR	Transient
OBP052	2.56	0.01	LSR	Transient
OBP053	5.13	0.04	LSR	Transient
OBP054	10.26	0.16	MSR	Seasonal Resident
OBP055	2.56	0.01	LSR	Transient
OBP056	2.56	0.01	LSR	Transient
OBP057	5.13	0.04	LSR	Transient
OBP058	2.56	0.01	LSR	Transient
OBP059	2.56	0.01	LSR	Transient
OBP060	2.56	0.01	LSR	Transient
OBP061	5.13	0.04	LSR	Transient
OBP062	5.13	0.04	LSR	Occasional
OBP063	2.56	0.01	LSR	Transient
OBP064	5.13	0.04	LSR	Transient
OBP065	2.56	0.01	LSR	Transient
OBP066	5.13	0.04	LSR	Occasional
OBP067	5.13	0.04	LSR	Transient
OBP068	7.69	0.09	LSR	Occasional
OBP069	2.56	0.01	LSR	Transient
OBP070	10.26	0.16	MSR	Seasonal Resident
OBP003	15.38	0.36	MSR	Seasonal Resident
OBP071	2.56	0.01	LSR	Transient
OBP072	2.56	0.01	LSR	Transient
OBP016	5.13	0.04	LSR	Occasional

Notes: Some Irrawaddy dolphins are transient, occurring only in one season, while others are occasional, seen in all seasons. Both transient and occasional dolphins have low sighting rates.

References

1. Minton, G.; Smith, B.D.; Braulik, G.T.; Kreb, D.; Sutaria, D.; Reeves, R. *Orcaella brevirostris* (errata version published in 2018). *IUCN Red List Threat. Species* **2017**, *2017*, e.T15419A123790805.
2. Rajamani, L.; Porter, L.; Dolar, L.; Rodriguez-Vargas, L.H.; Yobe, M. Marine Mammals of Coastal Penang Island, Malaysia. *Aquat. Mamm.* **2018**, *44*, 319–327. [CrossRef]
3. Rodríguez-Vargas, L.H.; Rajamani, L.; Dolar, L.; Porter, L. Population size, distribution and daylight behaviour of Irrawaddy dolphins (*Orcaella brevirostris*) in Penang Island, Malaysia. *Raffles Bull. Zool.* **2019**, *67*, 671–683.
4. Teoh, S.W.; Jaaman, S.; Palaniappan, P.M. A preliminary study of population size of Irrawaddy Dolphins (*Orcaella brevirostris*) in Cowie Bay, Sabah, Malaysia. *J. Trop. Biol. Conserv.* **2013**, *10*, 23–26.
5. Van Bressem, M.F.; Minton, G.; Sutaria, D.; Kelkar, N.; Peter, C.; Zulkarnaen, M.; Mansur, R.M.; Porter, L.; Vargas, L.H.R.; Rajamani, L. Cutaneous nodules in Irrawaddy dolphins: An emerging disease in vulnerable populations. *Diseases Aquat. Org.* **2014**, *107*, 181–189. [CrossRef] [PubMed]
6. Minton, G.; Peter, C.; Poh, A.N.Z.; Ngeian, J.; Braulik, G.; Hammond, P.S.; Tuen, A.A. Population estimates and distribution patterns of Irrawaddy dolphins (*Orcaella brevirostris*) and Indo-Pacific Finless porpoises (*Neophocaena phocaenoides*) in the Kuching Bay, Sarawak. *Raffles Bull. Zool.* **2013**, *61*, 877–888.
7. Hoffman, J.M.; Ponnampalam, L.S.; Araújo, C.C.; Wang, J.Y.; Kuit, S.H.; Hung, S.K. Comparison of Indo-Pacific humpback dolphin (*Sousa chinensis*) whistles from two areas of western Peninsular Malaysia. *J. Acoust. Soc. Am.* **2015**, *138*, 2829–2835. [CrossRef] [PubMed]
8. Smith, B.D.; Beasley, I. Orcaella brevirostris (Malampaya Sound subpopulation). Available online: https://dx.doi.org/10.2305/IUCN.UK.2004.RLTS.T44187A10858619.en (accessed on 6 September 2018).
9. Kreb, D.; Budiono. Cetacean diversity and habitat preferences in tropical waters of East Kalimantan, Indonesia. *Raffles Bull. Zool.* **2005**, *53*, 149–155.
10. Smith, B.D. Irrawaddy Dolphin: *Orcaella brevirostris*. In *Encyclopedia of Marine Mammal*, 2nd ed.; Perrin, W.F., Würsig, B., Thewissen, J.G.M., Eds.; Academic Press: San Diego, CA, USA, 2009; pp. 638–641.
11. Rodriguez-Vargas, L.H. The Distribution of Aquatic Mammals in the Water of Penang Island, Malaysia, with a Focus on the Ecology and Conservation of the Irrawaddy Dolphin *Orcaella brevirostris* (Owen in Grey, 1866). Master's Thesis, University Sains Malaysia, Penang, Malaysia, 2015.
12. Rosli, N.R.M.; Yahya, K. Trends of sediment loading in catchment areas of Pinang river in Malaysia. *APCBEE Procedia* **2013**, *5*, 128–133. [CrossRef]
13. Ali, N.F.; Rajamani, L.; Abd Rahman, A.; Porter, L.; Sim, Y.K.; Fadzly, N. Habitat use and behaviour of the Irrawaddy dolphin, *Orcaella brevirostris* and the Indo-Pacific finless porpoise, *Neophocaena phocaenoides* off the west coast of Penang Island, Malaysia. *Raffles Bull. Zool.* **2023**, *71*, 169–194.
14. Urian, K.; Gorgone, A.; Read, A.; Balmer, B.; Wells, R.S.; Berggren, P.; Durban, J.; Eguchi, T.; Rayment, W.; Hammond, P.S. Recommendation for photo-identification methods used in capture-recapture models with cetaceans. *Mar. Mammal Sci.* **2015**, *31*, 298–321. [CrossRef]
15. Hammond, P.S. Mark-recapture. In *Encyclopedia of Marine Mammal*, 2nd ed.; Perrin, W.F., Würsig, B., Thewissen, J.G.M., Eds.; Academic Press: San Diego, CA, USA, 2009; pp. 705–709.
16. Sutaria, D.; Marsh, H. Abundance estimates of Irrawaddy dolphins in Chilika Lagoon, India, using photo-identification based mark- recapture methods. *Mar. Mammal Sci.* **2011**, *27*, 338–348. [CrossRef]
17. Otis, D.L.; Burnham, K.P.; White, G.C.; Anderson, D.R. Statistical inference from capture data on closed animal populations. *Wildl. Monogr.* **1978**, *62*, 3–135.
18. Cooch, E.; White, G.C. Program MARK: A Gentle Introduction. 2018 Eighteenth Edition, p. 1151. Available online: http://www.phidot.org/software/mark/docs/book/ (accessed on 6 September 2018).
19. Kreb, D.; Lhota, S.; Porter, L.; Redman, A.; Susanti, I.; Lazecky, M. Long-term population and distribution dynamics of an endangered Irrawaddy dolphin population in Balikpapan Bay, Indonesia in response to coastal development. *Front. Mar. Sci.* **2020**, *7*, 746. [CrossRef]
20. Karczmarski, L. Group dynamics of humpback dolphins (*Sousa chinensis*) in the Algoa Bay region, South Africa. *J. Zool.* **1999**, *249*, 283–293. [CrossRef]
21. Parra, G.J.; Corkeron, P.J.; Marsh, H. Population sizes, site fidelity and residence patterns of Australian snubfin and Indo-Pacific humpback dolphins: Implications for conservation. *Biol. Conserv.* **2006**, *129*, 167–180. [CrossRef]
22. Fury, C.A.; Harrison, P.L. Abundance, site fidelity and range patterns of Indo-Pacific bottlenose dolphins (*Tursiops aduncus*) in two Australian subtropical estuaries. *Mar. Freshw. Res.* **2008**, *59*, 1015–1027. [CrossRef]
23. Zolman, E.S. Residence patterns of bottlenose dolphins (*Tursiops truncatus*) in the Stono River estuary, Charleston County, South Carolina, USA. *Mar. Mammal Sci.* **2002**, *18*, 879–892. [CrossRef]

24. Beasley, I.; Pollock, K.; Jefferson, T.A.; Arnold, P.; Morse, L.; Yim, S.; Lor Kim, S.; Marsh, H. Likely future extirpation of another Asian river dolphin: The critically endangered population of the Irrawaddy dolphin in the Mekong River is small and declining. *Mar. Mammal Sci.* **2013**, *29*, E226–E252. [CrossRef]
25. Smith, B.D.; Ahmed, B.; Mowgli, R.M.; Strindberg, S. Species occurrence and distributional ecology of nearshore cetaceans in the Bay of Bengal, Bangladesh, with abundance estimates for Irrawaddy dolphins *Orcaella brevirostris* and Finless porpoises *Neophocaena phocaenoides*. *J. Cetacean Res. Manag.* **2008**, *10*, 45–58. [CrossRef]

oceans

Article

Plastic, It's What's for Dinner: A Preliminary Comparison of Ingested Particles in Bottlenose Dolphins and Their Prey

Leslie B. Hart [1,2,*], Miranda Dziobak [1,3], Randall S. Wells [4], Elizabeth Berens McCabe [4], Eric Conger [5], Tita Curtin [1], Maggie Knight [6] and John Weinstein [7]

1 Department of Health and Human Performance, School of Health Sciences, College of Charleston, Charleston, SC 29424, USA; dziobakmk@cofc.edu (M.D.); curtinte@g.cofc.edu (T.C.)
2 Center for Coastal Environmental and Human Health, College of Charleston, Charleston, SC 29424, USA
3 Department of Environmental Health Sciences, Arnold School of Public Health, University of South Carolina, Columbia, SC 29208, USA
4 Chicago Zoological Society's Sarasota Dolphin Research Program, c/o Mote Marine Laboratory, Sarasota, FL 34236, USA; rwells@mote.org (R.S.W.); emccabe@mote.org (E.B.M.)
5 Department of Biology, School of Sciences, Mathematics, and Engineering, College of Charleston, Charleston, SC 29424, USA; congerej@g.cofc.edu
6 Graduate Program in Marine Biology, Grice Marine Laboratory, College of Charleston, Charleston, SC 29424, USA; knightma@g.cofc.edu
7 Department of Biology, The Citadel, Charleston, SC 29409, USA; weinsteinj@citadel.edu
* Correspondence: hartlb@cofc.edu

Abstract: Microplastic ingestion was reported for common bottlenose dolphins (*Tursiops truncatus*) inhabiting Sarasota Bay, FL, USA, a community that also has prevalent exposure to plasticizers (i.e., phthalates) at concentrations higher than human reference populations. Exposure sources are currently unknown, but plastic-contaminated prey could be a vector. To explore the potential for trophic exposure, prey fish muscle and gastrointestinal tract (GIT) tissues and contents were screened for suspected microplastics, and particle properties (e.g., color, shape, surface texture) were compared with those observed in gastric samples from free-ranging dolphins. Twenty-nine fish across four species (hardhead catfish, *Ariopsis felis*; pigfish, *Orthopristis chrysoptera*; pinfish, *Lagodon rhomboides*; and Gulf toadfish, *Opsanus beta*) were collected from Sarasota Bay during September 2022. Overall, 97% of fish (n = 28) had suspected microplastics, and GIT abundance was higher than muscle. Fish and dolphin samples contained fibers and films; however, foams were common in dolphin samples and not observed in fish. Suspected tire wear particles (TWPs) were not in dolphin samples, but 23.1% and 32.0% of fish muscle and GIT samples, respectively, contained at least one suspected TWP. While some similarities in particles were shared between dolphins and fish, small sample sizes and incongruent findings for foams and TWPs suggest further investigation is warranted to understand trophic transfer potential.

Keywords: plastic pollution; OneHealth; contaminant; plasticizers; seafood safety; marine mammal; cetacean; fish

1. Introduction

Marine plastic pollution is a massive environmental concern. According to a recent study, the world's oceans contain approximately 171 trillion plastic particles, and considering global dependence on plastic items, inadequate waste management practices, and policy inaction, the amount of marine plastic pollution could nearly triple by 2040 [1,2]. Additionally, the properties that garnered popular interest in plastic (e.g., durability, affordability) contribute to the pervasiveness and ubiquity of this pollutant. Plastic pollution has been identified in every ocean on the planet [3] and across multiple marine taxa [4,5].

Marine plastic debris is often categorized by size [6]. Macro- and mesoplastics (≥5 mm diameter), which may enter the marine environment directly as waste, can lethally and sublethally impact the health of large marine vertebrates, such as sea turtles [7] and marine mammals [8–10], due to entanglement or ingestion. Microplastics (<5 mm in diameter) can enter the environment indirectly via inadequate wastewater filtration or fragmentation of larger plastic items [6,11], and the small particle size makes marine fauna particularly vulnerable to microplastic exposure and toxicity. Plastic particle uptake by fish can occur via direct/active consumption, unintentional ingestion, or branchial intrusion [12,13], and particles have been detected in the gills, gastrointestinal tract (contents and digested tissues), liver, and muscle of exposed fish [5,12,14,15]. Microplastic exposure in fish has been widely reported both geographically and taxonomically [1,5,16], but individual, population, and species-level differences in exposure can be influenced by food availability, feeding behavior, degree and nature of plastic pollution, and foraging sense (i.e., visual vs. chemosensory cues; [13]). Given the widespread prevalence of plastic particle detection in fish and other lower trophic organisms, trophic transfer of microplastics to apex predators seems likely. For example, Romeo et al. (2015) observed microplastics in 30–75% of stomachs from several apex pelagic fish species considered to be specialist feeders, suggesting exposure from active consumption of contaminated prey [17].

Recently, microplastic ingestion was documented in free-ranging common bottlenose dolphins (*Tursiops truncatus*) inhabiting Sarasota Bay, FL, USA [18]. This dolphin community also has prevalent exposure to plasticizers (i.e., phthalates; [19,20]) at levels higher than human reference populations [21]. The source of Sarasota Bay bottlenose dolphin plastic and plasticizer exposure remains unknown, but we hypothesize that plastic-contaminated prey fish could be a possible vector. The objective of this study was to screen for microplastics in muscle and gastrointestinal tracts from bottlenose dolphin prey fish collected in Sarasota Bay and compare particle properties (e.g., color, shape, surface texture) to those observed in gastric samples from bottlenose dolphins. In addition to providing insight on a source of xenobiotic exposure in dolphins, this study may also reveal risks to seafood safety due to trophic concurrence with coastal human communities, given that dolphins and humans are at similar trophic levels [22].

2. Materials and Methods

2.1. Fish Collection and Sampling

Common prey fish for Sarasota Bay bottlenose dolphins have been identified through examination of the stomach contents of stranded dolphins and field observation [23–25]. Resident dolphins are considered selective feeders, choosing sound-producing fish disproportionately relative to their availability, as well as non-soniferous species [24,25]. For this study, we sought prey fish of 11 different species: Gulf toadfish (*Opsanus beta*), spot (*Leiostomus xanthurus*), spotted seatrout (*Cynoscion nebulosus*), pinfish (*Lagodon rhomboides*), sheepshead (*Archosargus probatocephalus*), striped mullet (*Mugil cephalus*), ladyfish (*Elops saurus*), pigfish (*Orthopristis chrysoptera*), Atlantic thread herring (*Opisthonema oglinum*), scaled sardine (*Harengula jaguana*), and hardhead catfish (*Ariopsis felis*). Additionally, the collected fish were between 12 cm and 64 cm in total length to ensure sufficient tissue volume for analyses. Fish samples for this project came from seasonal fish surveys in Sarasota Bay during September 2022, as part of long-term abundance monitoring [26]. Sampling occurred via state-approved licensing (Florida Fish and Wildlife Conservation Commission Special Activity License nos. 19-0809A-SR) and according to protocols approved by Mote Marine Laboratory's Institutional Animal Care and Use Committee (IACUC). Whole fish were wrapped in solvent-rinsed foil and stored frozen at −20 °C until dissection. Prior to dissection, all fish were thawed and removed from the aluminum foil. Total length (cm), total mass (g), gastrointestinal tract mass (g), and muscle mass (g) were collected. For each fish, the entire gastrointestinal tract (GIT) and muscle tissue from one lateral side were removed using stainless steel instruments and stored separately in glass jars at −20 °C. Dissection trays, petri dishes, and instruments were rinsed three times with purified water (Milli-Q®) (Millipore, Molsheim, France) prior to dissection and tissue collection.

2.2. Dolphin Sampling

Bottlenose dolphin gastric sampling and analysis were previously described in [18]. Briefly, samples were collected during catch-and-release health assessments, in which dolphins were encircled using a net in shallow water, temporarily restrained, and brought on board a specialized, padded, and shaded sampling vessel for morphological and physical examination by veterinarians. During the examination, gastric samples were collected by passing a small veterinary feeding tube through the esophagus into the stomach [27–29]. Samples were stored in glass jars and frozen at −20 °C until particle analysis. Sarasota Bay bottlenose dolphin health assessments were conducted under permit from the National Oceanic and Atmospheric Administration's (NOAA) National Marine Fisheries Service (NMFS) and IACUC-approved protocols.

2.3. Microplastic Screening

Microplastic screening of fish tissues followed methods used in [18,30]. Briefly, muscle and gastrointestinal (i.e., stomach and intestine) tissue from each fish were placed into a glass beaker, and organic (non-plastic) material in the samples was digested by adding a 10% potassium hydroxide (KOH) solution and incubated at 60 °C for 24–72 h [31]. Following digestion, samples were vacuum filtered onto GF/A 1.6 µm glass fiber filters in a fume hood and left to dry in covered glass petri dishes.

Particles of at least 35 µm were characterized visually using a dissection microscope (Leica EZ4, magnification 8–35×) according to physical attributes including shape (e.g., fiber, film, fragment, foam), surface texture (e.g., smooth, rough, rubber), and color (e.g., transparent, blue, black; [32]). Various parameters were used to identify potential plastic materials. For example, suspected plastic fibers were indicated by a smooth, uniform surface with a length that exceeded the width [33]. Suspected plastic fragments were characterized by smooth or angular edges that appeared to be broken from a larger piece of debris [33]. Fragments were further characterized as suspected tire wear particles (TWPs) if they were black, cylindrical, had a rubbery surface texture, and maintained their shape when manipulated with forceps [34]. Suspected foam particles were characterized by a round shape and honeycomb-like porosity [35]. Suspected fiber bundles were characterized by 20 or more fibers tangled together in a way that prevented them from being separated [36,37]. All particles at least 100 µm in size and with characteristics previously described were tested with a hot needle (250 °C) and suspected to be of plastic origin if the needle left a mark on or melted the particle surface [34,38,39]. Fourier transform infrared (FTIR) spectroscopy was available for polymer identification of particles ranging from 500 µm to 5 mm.

2.4. Quality Assurance/Quality Control

Rigorous precautions were taken while handling and processing samples. A 100% cotton lab coat and nitrile gloves were worn during fish dissections and laboratory analyses. All tools and glassware were carefully rinsed with deionized water prior to use. For QA/QC purposes, analysis of one "laboratory blank" (i.e., non-tissue sample) was performed simultaneously with each set of sample digestions to correct for potential procedural contamination, and three positive controls with commercially purchased polyethylene, polystyrene, and polyester microplastic particles were used to determine recovery efficiency. Mean recovery percentages for fish were 60% for film, 83% for foam, and 85% for fibers, while mean recovery percentages for dolphins were 90% for film, 87% for foam, and 90% for fibers [18]. Additionally, during fish dissections, an open petri dish was kept on the top of the bench to collect potential environmental contaminants and processed using the same methods as the fish tissues ("dissection blank").

2.5. Statistical Methods

Descriptive statistics were used to calculate the percentage of fish with suspected plastics detected in either muscle or GIT samples, overall, by species, and by tissue type (i.e., muscle vs. GIT). The particle load for fish was calculated as the number of suspected plastic particles per gram of tissue, and these concentrations were compared between tissue types and species. Similarly, suspected plastic attributes (i.e., color and shape) were compared between tissue types and species. Finally, particle attributes were qualitatively compared with ingested microplastics in bottlenose dolphins reported in [18].

3. Results

In total, 29 fish across four species (hardhead catfish (n = 2); pigfish (n = 12); pinfish (n = 10); and Gulf toadfish (n = 5)) were collected from two sampling stations in Sarasota Bay, FL, during September 2022 (Figure 1). All fish were screened for microplastics in either muscle (n = 4), GIT tissue (n = 3), or both (n = 22). Ninety-seven percent of the fish screened (n = 28) had evidence of at least one suspected plastic particle in either muscle or GI tissue. Species counts and mean tissue mass are reported in Table 1.

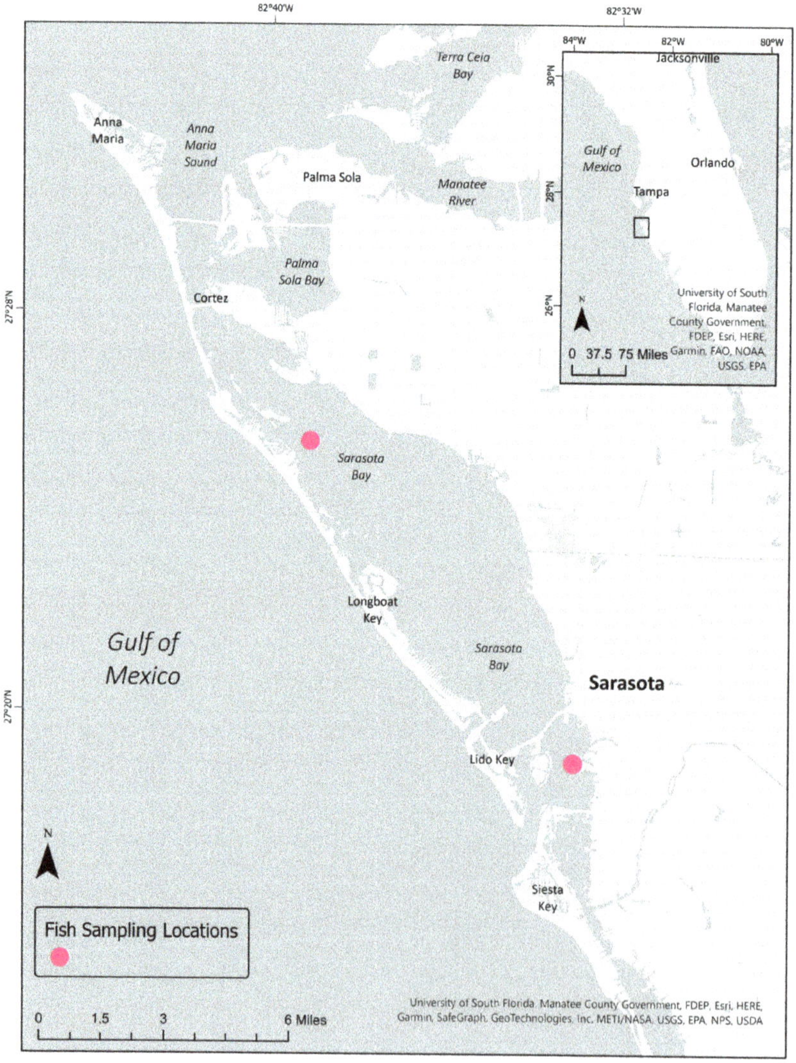

Figure 1. Locations of fish sampling in Sarasota Bay, FL, USA.

Table 1. Mean muscle and GIT mass (g) by fish species.

Species	Sample Size	Muscle Mass (g) Mean (s.d.)	GIT Mass (g) Mean (s.d.)
Hardhead catfish (*Ariopsis felis*)	2	16.3–23.2 [1]	12.8 [1]
Pigfish (*Orthopristis chrysoptera*)	12	2.8 (0.8)	6.4 (2.1)
Pinfish (*Lagodon rhomboides*)	10	3.8 (1.3)	12.0 (7.0)
Gulf toadfish (*Opsanus beta*)	5	4.1 (3.0)	4.8 (1.9)

[1] Actual values reported due to small sample size.

3.1. QA/QC Results

Suspected plastic particles were identified in both laboratory and dissection blanks (Table 2). Single fibers dominated the particles observed in the dissection blanks (*n* = 115), and major colors included transparent (*n* = 45), blue (*n* = 28), yellowed (*n* = 13), and pink/purple (*n* = 9) or orange (n = 9). "Yellowed" refers to fibers with a darker yellow appearance, likely discolored during the digestion process or particle degradation [40]. Transparent (*n* = 1) and yellowed (*n* = 1) films were also observed in dissection blanks, as well as yellowed (*n* = 1) non-TWP fragments. Single fibers (*n* = 60), one yellow film, and one red, non-TWP fragment were observed in laboratory blanks, with most fibers transparent (*n* = 19), orange (*n* = 12), or blue in color. Suspected plastic particles identified in muscle and GIT tissue samples were corrected for dissection and laboratory blank contamination by removing particles of the same shape and color from total particle counts.

Table 2. Suspected plastic particles observed in laboratory and dissection blanks.

Blank Type	n	Single Fibers	Fiber Bundles	Films	Fragments (TWP)	Fragments (Non-TWP)	Foams
Dissection	6	115	0	2	0	1	0
Laboratory	11	60	0	1	0	1	0

3.2. Suspected Plastics in Prey Fish Muscle Tissue

Among blank-corrected muscle samples (*n* = 26), suspected plastic particles were observed in 77% of the fish evaluated (*n* = 20). Overall, the number of observed particles per fish ranged from 0 to 40; the highest count was observed among pigfish. The average number of particles per species was: hardhead catfish (3.0), pigfish (8.0), pinfish (3.8), and Gulf toadfish (5.6). Pigfish had the highest average suspected plastic particle load (2.9 particles/g tissue), while catfish had the lowest (0.2 particle/g tissue; Table 3). Of the 172 total particles observed in muscle tissue across all species, 90.1% were single fibers (Table 3). Other suspected particle types observed in muscle tissue included films (*n* = 1), non-TWP fragments (*n* = 1), and TWP fragments (*n* = 15; Table 3). Foam and fiber bundles were not observed in the muscle tissue of any fish examined. Of the 155 fibers observed across all species, most were either bright yellow in color (34.2%) or transparent that yellowed with time or processing (27.7%). The majority of total tire wear particles (*n* = 15) were observed in catfish (26.7%) or Gulf toadfish (60.0%; Table 3). All observed suspected particles in muscle tissue were less than 500 μm in diameter; thus, FTIR was not conducted to determine polymer composition.

Table 3. Suspected plastic particle counts by shape in blank-corrected fish muscle samples (n = 26). Particle load reported as number of particles (#) per gram of tissue.

Species	Sample Size	Total Particles	Particle Load (#/g Tissue)	Single Fibers	Fiber Bundles	Films	Non-TWP Fragments	TWP Fragments
Hardhead catfish (*Ariopsis felis*)	2	6	0.2	2	0	0	0	4
Pigfish (*Orthopristis chrysoptera*)	11	88	2.9	86	0	0	1	1
Pinfish (*Lagodon rhomboides*)	8	50	1.6	49	0	0	0	1
Gulf toadfish (*Opsanus beta*)	5	28	1.4	18	0	1	0	9
Total	26	172	-	155	0	1	1	15

3.3. *Suspected Plastics in Prey Fish Gastrointestinal Tracts*

Among blank-corrected GIT samples (n = 25), 88.0% of fish examined had at least one suspected plastic particle. Among GIT samples, suspected plastic particle counts ranged from 1 to 67, with pigfish having the highest count. By species, the average number of particles was hardhead catfish (9.0, n = 1), pigfish (16.1), pinfish (10.8), and Gulf toadfish (15.2). Gulf toadfish had the highest average suspected plastic particle load (3.2 particles/g tissue), while hardhead catfish and pinfish had the lowest (0.7 and 0.9 particle/g tissue, respectively; Table 4). Across all fish, 343 suspected particles were observed, and most were films (36.4%), followed by fiber bundles (29.2%), single fibers (28/3%), TWP fragments (2.9%), and non-TWP fragments (1.7%; Table 4). Similarly to muscle samples, foams were not observed in any fish GI tissue. Most films were transparent (82.4%), and other colors observed were blue (n = 1), red (n = 8), and yellowed (n = 13). TWP fragments were most common among pigfish (50% of observed) and not observed in the GI tract of the single catfish or most pinfish samples (Table 4). Most observed suspected particles in muscle tissue were less than 500 µm in diameter; thus, FTIR was not able to be conducted to determine polymer composition.

Table 4. Suspected plastic particle counts by shape in fish gastrointestinal samples (n = 25). Particle load reported as number of particles (#) per gram of tissue.

Species	Sample Size	Total Particles	Particle Load (#/g Tissue)	Single Fibers	Fiber Bundles	Films	Non-TWP Fragments	TWP Fragments
Hardhead catfish (*Ariopsis felis*)	1	9	0.7	1	0	3	0	0
Pigfish (*Orthopristis chrysoptera*)	10	161	2.5	32	44	77	3	5
Pinfish (*Lagodon rhomboides*)	9	97	0.9	36	34	23	2	2
Gulf toadfish (*Opsanus beta*)	5	76	3.2	22	28	22	1	3
Total	25	343		97	100	125	6	10

3.4. *Comparison of Suspected Plastics in Fish and Bottlenose Dolphins*

Suspected plastic particles in muscle and GIT samples from fish collected from Sarasota Bay, FL, USA, were compared with ingested particles from bottlenose dolphins inhabiting the same area, as reported in [18]. In general, all sampled dolphins (n = 7) had at least one suspected plastic particle detected in gastric fluid, and particle counts exceeding 50 for an individual dolphin were common [18]. Dolphin samples contained primarily fibers, films, and foams (Figure 2a); however, foams were not observed in muscle or GIT tissues from fish (Figure 2b,c). Fibers and fragments were the dominant particle shapes observed in fish muscle (Figure 2b), while fish GIT samples contained a mixture of fibers, films, and fragments (non-TWP and TWP; Figure 2c). TWP fragments were not observed in bottlenose dolphin gastric samples, but 23.1% and 32.0% of muscle and GIT samples, respectively, contained at least one TWP fragment.

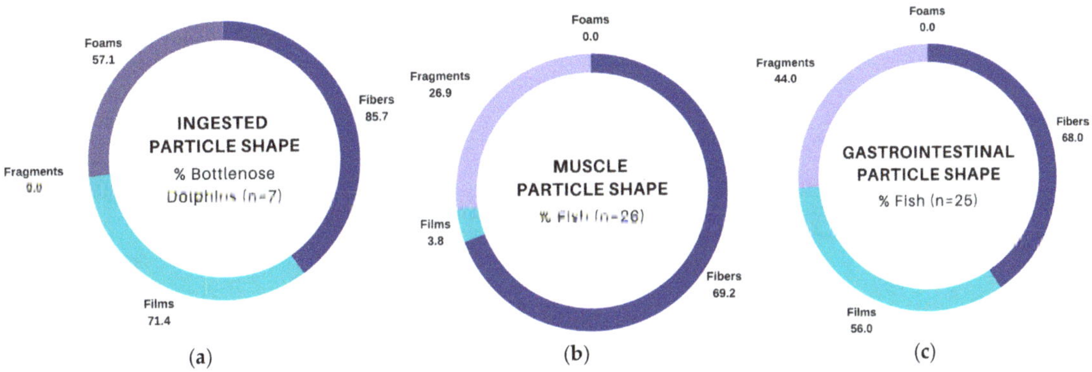

Figure 2. Comparison of suspected plastic particles (blank-corrected) in fish and bottlenose dolphins (*Tursiops truncatus*) sampled in Sarasota Bay, FL, USA: (**a**) percentage of dolphin gastric samples with at least one of the listed particle shapes, data from [18]; (**b**) percentage of fish muscle samples with at least one of the listed particle shapes; (**c**) percentage of fish gastrointestinal tract samples with at least one of the listed particle shapes.

Fibers were common in bottlenose dolphin gastric samples and both types of fish tissues (Figure 2a–c); however, there were differences in the distribution of colors. Transparent fibers were most common in dolphin samples (42.9% of samples; Figure 3a), while brighter colors such as yellow/yellowed, red, and orange were more abundant among fish samples (Figure 3b,c). Single fiber counts were more numerous in fish (Tables 3 and 4) compared to dolphins [18], and fiber bundles that were present in fish GIT samples (Table 4) were not reported in bottlenose dolphins [18].

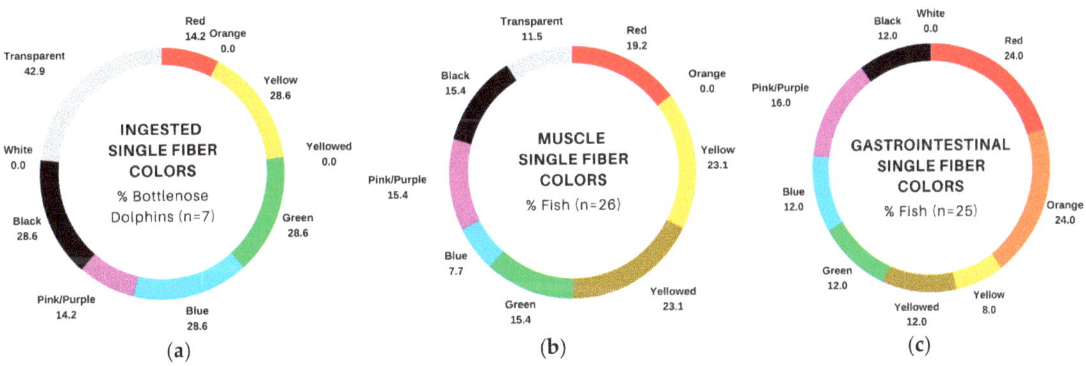

Figure 3. Comparison of single fiber colors (blank-corrected) in fish and bottlenose dolphins (*Tursiops truncatus*) sampled in Sarasota Bay, FL, USA: (**a**) percentage of dolphin gastric samples with at least one of the listed fiber colors, data from [18]; (**b**) percentage of fish muscle samples with at least one of the listed fiber colors; (**c**) percentage of fish gastrointestinal tract samples with at least one of the listed fiber colors.

Films were also observed in dolphin gastric samples (71.4%; Figure 2a) and both fish tissue types, although they were much less abundant in muscle (3.8%; Figure 2b) than in GIT (56.0%; Figure 2c). Most films in bottlenose dolphin gastric samples and fish samples were transparent (Figure 4a–c), although 20.0% of fish GIT samples also contained yellowed films, which were not observed in bottlenose dolphins (Figure 4a,c).

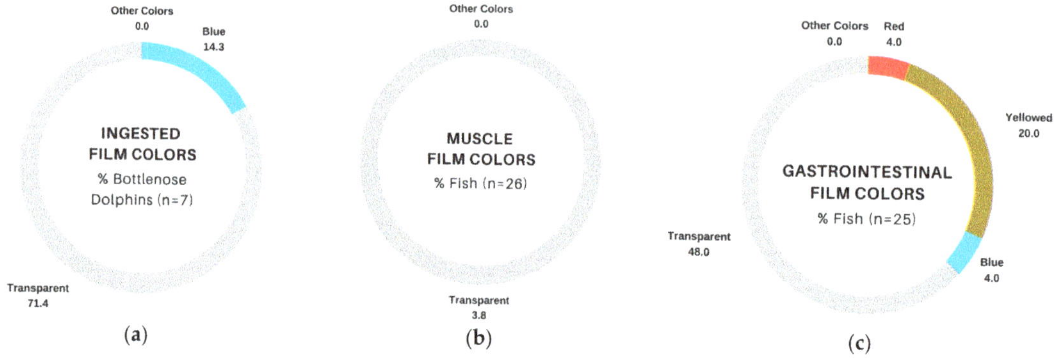

Figure 4. Comparison of film colors (blank-corrected) in fish and bottlenose dolphins (*Tursiops truncatus*) sampled in Sarasota Bay, FL, USA: (**a**) percentage of dolphin gastric samples with at least one of the listed film colors, data from [18]; (**b**) percentage of fish muscle samples with at least one of the listed film colors; (**c**) percentage of fish gastrointestinal tract samples with at least one of the listed film colors.

4. Discussion

This study provides additional evidence of exposure to suspected plastic particles among marine fauna inhabiting an urban estuary in Sarasota Bay, FL, USA. Sarasota Bay is located on the central, west coast of Florida and is a semi-closed lagoon system with minimal tidal exchange [41]. As an urban watershed, this region houses multiple residential, industrial, and commercial centers [42]. Findings from this study are not surprising as plastic ingestion in marine fauna has been well documented across multiple taxa, including bivalves [43–45], fish [46–49], and marine mammals (e.g., pinnipeds, [50–52]; mysticetes, [8,53,54]; odontocetes, [55–57]). Additionally, waters surrounding urban centers are generally more polluted than rural areas [58–60], thereby increasing exposure to plastic debris for aquatic animals that live in close proximity to urbanized areas [48,61–63].

Among the four fish species investigated in this study, suspected plastic particles were observed in both muscle and GIT tissue/contents, with particles more abundant in GIT. We expected to find suspected plastic particles in the gastrointestinal tract based on previous work [62,64–67], similar to researchers who have studied intentional/selective [68,69] and unintentional [70] ingestion. Observations of particles in muscle tissue were also not surprising based on previous studies in wild-caught fish from the Persian Gulf [71,72], India [73], and the Mediterranean Sea [74], as well as experimental evidence by [75], who demonstrated in juvenile seabass (*Dicentrarchus labrax*) that particles can be translocated to muscle tissue. These authors suggest that translocation to muscle tissue likely occurs via lymphatic or vascular systems, is restricted to small particle sizes, and likely explains the differences in total particle counts observed between muscle and GIT tissues [75].

Our comparison of ingested particles in dolphins with suspected particles observed in four prey species (hardhead catfish, pinfish, pigfish, and Gulf toadfish) revealed both similarities and differences in particle attributes. Single fibers were common in bottlenose dolphin samples and fish tissues, but the distribution of colors differed, with transparent fibers dominating dolphin samples and bright colors dominating single fibers in fish samples. Among marine fauna, ingestion of microplastics can be both indirect (i.e., ingesting microplastic-contaminated prey; [50,76,77]) and direct (i.e., ingesting plastics from the water column; [13,46,68]). Planktivorous fish species that hunt via sight can mistake microplastics for prey that are similar in size and color [13]. For example, in [46], red particles were frequently observed in omnivorous fish that commonly consume red algae, and [68] demonstrated blue particle selectivity among amberstripe scads (*Decapterus muroadsi*), which commonly eat blue copepods. It also seems possible that the color of ingested fibers could change or fade over time due to chemical degradation [40] occurring in the digestive tract. In other words, yellowed or transparent fibers could have originated as a different color. With respect to the potential for trophic transfer of fibers between fish and dolphins, slow egestion rates of fibers in fish [78] could render them available for incidental consumption by foraging dolphins.

Transparent films were also commonly observed in dolphin and fish samples. This particle shape has been widely reported in previous studies of fish exposed to microplastics [79–83], and given evidence of trophic transfer of other microplastic shapes [50], it seems plausible that contaminated fish could be the source of transparent films for bottlenose dolphins. TWP fragments were not observed in any dolphin sample, but they were observed in approximately 23.1% and 32.0% of fish muscle and GIT tissues, respectively. TWP fragments have been observed in multiple matrices (e.g., air, water, sediment, and biota (reviewed by [84])) and geographic locations (e.g., Charleston Harbor (USA) [34], San Francisco Bay (USA) [85], and the Seine (France), Chesapeake (USA), and Yoda (Japan) watersheds [86]). So, given the preponderance of these fragments in fish tissues, it is somewhat surprising that they were not observed in the gastric fluids of bottlenose dolphins from Sarasota Bay [18]. One possible explanation is that TWP fragments have a higher propensity to become entrapped in the mucus lining the gastrointestinal tract than other microplastic particles. Supporting this notion are observations regarding the fate of nanoparticles in isolated intestines, which found the bulk of particles became trapped in the mucus lining, and no particles were found in the gastric fluids of the lumen or adhered to the epithelium [87]. Certainly, further research is warranted concerning the fate of microplastic particles, including TWPs, in the gastrointestinal tracts of mammals. White foams were the most abundant particle observed in bottlenose dolphin gastric samples, sometimes in quantities exceeding 50 per sample [18]. White foams were not present in either fish tissue. Possible explanations for this difference could be the relatively small number of fish evaluated for each species or the small number of species screened; additional sampling of fish within and across more species could help to discern whether the ingested foams in dolphins could be related to contaminated prey. Alternatively, it could be possible that dolphins accidentally ingest foams if they engage in object play behavior [88,89] or physically manipulate larger, floating foam objects (e.g., food and drink containers, buoys).

One of the limitations in microplastic research is the recurring issue of sample contamination by ambient microplastics during sample processing and analysis [90]. Rigorous precautions were taken while handling and processing samples. Additionally, conservative approaches to blank-correct particle counts in fish tissues were performed such that particle shapes and colors observed in blanks were removed from the tissue counts of associated fish. Despite these stringent measures to minimize contamination, we were unable to confirm the polymer composition of particles from muscles and GITs. Fourier transform infrared spectroscopy (FTIR) is an instrumental technique used to identify functional groups using light refraction [91], but FTIR's utility is limited to particles ranging from 500 μm to 5 mm. Most microscopy-identified particles in this study were <500 μm; thus, most of the reported particle counts are based on hot needle testing [39]. Sample size is another limitation impacting our ability to trace ingested microplastics in bottlenose dolphins to particles observed in prey fish. Hart et al. (2022) evaluated gastric samples for seven individuals and observed abundant white foams, which were not present in any fish sample screened for this study. While these findings might suggest that dolphins are not exposed to white foam particles via trophic transfer, it is also possible that we did not sample enough fish, overall or within a species, to detect white foams, as they have been observed in other estuarine fish [67]. Small sample sizes also limited our ability to quantitatively compare particle counts and characteristics between species and tissue types. Additional bottlenose dolphin and fish samples will be collected to further explore evidence of trophic transfer and the species-level risk of plastic exposure for bottlenose dolphins. Recent efforts have been made to enhance our understanding of marine mammal diets using quantitative fatty acid signature analysis (QFASA; [92]). These techniques compare fatty acid profiles in tissue from a predator with a database of fatty acid profiles in common prey species so that researchers can identify the primary components of an individual's diet [92]. Blubber samples for fatty acid analyses were also collected from Sarasota Bay bottlenose dolphins, so future studies could use these techniques and species-level microplastic counts (for fish) to understand individualized diets and conduct individualized risk assessments.

Marine plastic debris is now recognized as a pollutant of international concern due to its impacts on wildlife and seafood safety. Eriksen et al. (2023) estimated that the oceans contain more than 171 trillion plastic particles, and most (92.4%; [3]) are microplastics. Given that an estimated 41% of the world's population lives within 100 km of the coast [93], this widespread marine pollutant may have substantial public health consequences. Observations of microplastics in the tissues of wild-caught fish and gastric samples of estuarine bottlenose dolphins may warn of local environmental pollution and seafood safety risks for coastal communities.

Author Contributions: Conceptualization, L.B.H.; methodology, M.D. and L.B.H.; validation, M.D. and L.B.H.; formal analysis, L.B.H. and M.D.; investigation, M.D., E.C., T.C., M.K., R.S.W. and E.B.M.; resources, L.B.H., R.S.W., E.B.M. and J.W.; data curation, L.B.H., T.C. and R.S.W.; writing—original draft preparation, L.B.H. and E.C.; writing—review and editing, M.D., M.K., T.C., R.S.W., E.B.M. and J.W.; visualization, L.B.H.; supervision, L.B.H., M.D., R.S.W. and J.W.; project administration, L.B.H. and R.S.W.; funding acquisition, L.B.H. and R.S.W. All authors have read and agreed to the published version of the manuscript.

Funding: The research reported in this publication was supported by the National Institute of Environmental Health Sciences of the National Institutes of Health under Award Number R15ES034169. The content is solely the responsibility of the authors and does not necessarily represent the official views of the National Institutes of Health. Bottlenose dolphin health assessments in Sarasota Bay were supported by Dolphin Quest, Inc., and prey fish collection was supported by the Charles and Margery Barancik Foundation.

Institutional Review Board Statement: Fish sampling for this study occurred via Florida state-approved licensing (Florida Fish and Wildlife Conservation Commission Special Activity License nos. 19-0809A-SR) and according to protocols approved by Mote Marine Laboratory's Institutional Animal Care and Use Committee (IACUC).

Informed Consent Statement: Not applicable.

Data Availability Statement: The fish data (used for this manuscript) can be accessed using this DOI link: https://doi.org/10.5061/dryad.fn2z34v1d. Bottlenose dolphin data can be accessed using the repository cited in the original paper (Hart et al., 2022), doi: 10.3389/fmars.2022.947124.

Acknowledgments: The authors sincerely appreciate the training received from Bonnie Ertel and Dana Norton for microplastic isolation and quantification. Thanks also to Christina Toms, Jonathan Crossman, and Kylee DiMaggio for their assistance with sample collection. We are also grateful for the SDRP interns and volunteers who provided crew support for fishing operations. Finally, many thanks to The Citadel, the College of Charleston's Department of Health and Human Performance, and the Center for Coastal Environmental and Human Health for laboratory and logistical support.

Conflicts of Interest: The authors declare no conflict of interest. The funding sponsors had no role in the design of this study; in the collection, analyses, or interpretation of data; in the writing of the manuscript, and in the decision to publish the results.

References

1. Alberghini, L.; Truant, A.; Santonicola, S.; Colavita, G.; Giaccone, V. Microplastics in Fish and Fishery Products and Risks for Human Health: A Review. *Int. J. Environ. Res. Public Health* **2023**, *20*, 789. [CrossRef] [PubMed]
2. Eriksen, M.; Cowger, W.; Erdle, L.M.; Coffin, S.; Villarrubia-Gómez, P.; Moore, C.J.; Carpenter, E.J.; Day, R.H.; Thiel, M.; Wilcox, C. A Growing Plastic Smog, Now Estimated to Be over 170 Trillion Plastic Particles Afloat in the World's Oceans—Urgent Solutions Required. *PLoS ONE* **2023**, *18*, e0281596. [CrossRef] [PubMed]
3. Eriksen, M.; Lebreton, L.C.M.; Carson, H.S.; Thiel, M.; Moore, C.J.; Borerro, J.C.; Galgani, F.; Ryan, P.G.; Reisser, J. Plastic Pollution in the World's Oceans: More than 5 Trillion Plastic Pieces Weighing over 250,000 Tons Afloat at Sea. *PLoS ONE* **2014**, *9*, e111913. [CrossRef] [PubMed]
4. Kühn, S.; van Franeker, J.A. Quantitative Overview of Marine Debris Ingested by Marine Megafauna. *Mar. Pollut. Bull.* **2020**, *151*, 110858. [CrossRef] [PubMed]
5. Ugwu, K.; Herrera, A.; Gómez, M. Microplastics in Marine Biota: A Review. *Mar. Pollut. Bull.* **2021**, *169*, 112540. [CrossRef] [PubMed]
6. Andrady, A.L. Microplastics in the Marine Environment. *Mar. Pollut. Bull.* **2011**, *62*, 1596–1605. [CrossRef] [PubMed]
7. Campani, T.; Baini, M.; Giannetti, M.; Cancelli, F.; Mancusi, C.; Serena, F.; Marsili, L.; Casini, S.; Fossi, M.C. Presence of plastic debris in loggerhead turtle stranded along the Tuscany coasts of the Pelagos Sanctuary for Mediterranean Marine Mammals (Italy). *Mar. Pollut. Bull.* **2013**, *74*, 225–230. [CrossRef] [PubMed]
8. Besseling, E.; Foekema, E.M.; Van Franeker, J.A.; Leopold, M.F.; Kühn, S.; Bravo Rebolledo, E.L.; Heße, E.; Mielke, L.; IJzer, J.; Kamminga, P.; et al. Microplastic in a Macro Filter Feeder: Humpback Whale *Megaptera novaeangliae*. *Mar. Pollut. Bull.* **2015**, *95*, 248–252. [CrossRef]
9. Lusher, A.L.; Hernandez-Milian, G.; O'Brien, J.; Berrow, S.; O'Connor, I.; Officer, R. Microplastic and Macroplastic Ingestion by a Deep Diving, Oceanic Cetacean: The True's Beaked Whale *Mesoplodon Mirus*. *Environ. Pollut.* **2015**, *199*, 185–191. [CrossRef]
10. Denuncio, P.; Bastida, R.; Dassis, M.; Giardino, G.; Gerpe, M.; Rodríguez, D. Plastic Ingestion in Franciscana Dolphins, *Pontoporia blainvillei* (Gervais and d'Orbigny, 1844), from Argentina. *Mar. Pollut. Bull.* **2011**, *62*, 1836–1841. [CrossRef]
11. Cole, M.; Lindeque, P.; Halsband, C.; Galloway, T.S. Microplastics as Contaminants in the Marine Environment: A Review. *Mar. Pollut. Bull.* **2011**, *62*, 2588–2597. [CrossRef]
12. Su, L.; Deng, H.; Li, B.; Chen, Q.; Pettigrove, V.; Wu, C.; Shi, H. The Occurrence of Microplastic in Specific Organs in Commercially Caught Fishes from Coast and Estuary Area of East China. *J. Hazard. Mater.* **2019**, *365*, 716–724. [CrossRef]
13. Roch, S.; Friedrich, C.; Brinker, A. Uptake Routes of Microplastics in Fishes: Practical and Theoretical Approaches to Test Existing Theories. *Sci. Rep.* **2020**, *10*, 3896. [CrossRef] [PubMed]
14. Barboza, L.G.A.; Lopes, C.; Oliveira, P.; Bessa, F.; Otero, V.; Henriques, B.; Raimundo, J.; Caetano, M.; Vale, C.; Guilhermino, L. Microplastics in Wild Fish from North East Atlantic Ocean and Its Potential for Causing Neurotoxic Effects, Lipid Oxidative Damage, and Human Health Risks Associated with Ingestion Exposure. *Sci. Total Environ.* **2020**, *717*, 134625. [CrossRef] [PubMed]
15. Zitouni, N.; Bousserrhine, N.; Missawi, O.; Boughattas, I.; Chèvre, N.; Santos, R.; Belbekhouche, S.; Alphonse, V.; Tisserand, F.; Balmassiere, L.; et al. Uptake, Tissue Distribution and Toxicological Effects of Environmental Microplastics in Early Juvenile Fish *Dicentrarchus labrax*. *J. Hazard. Mater.* **2021**, *403*, 124055. [CrossRef]
16. Chen, K.-J.; Chen, M.-C.; Chen, T.-H. Plastic Ingestion by Fish in the Coastal Waters of the Hengchun Peninsula, Taiwan: Associated with Human Activity but No Evidence of Biomagnification. *Ecotoxicol. Environ. Saf.* **2021**, *213*, 112056. [CrossRef] [PubMed]
17. Romeo, T.; Pietro, B.; Pedà, C.; Consoli, P.; Andaloro, F.; Fossi, M.C. First Evidence of Presence of Plastic Debris in Stomach of Large Pelagic Fish in the Mediterranean Sea. *Mar. Pollut. Bull.* **2015**, *95*, 358–361. [CrossRef]
18. Hart, L.B.; Dziobak, M.; Wells, R.S.; Ertel, B.; Weinstein, J. Microplastics in Gastric Samples from Common Bottlenose Dolphins *(Tursiops truncatus)* Residing in Sarasota Bay FL (USA). *Front. Mar. Sci.* **2022**, *9*, 947124. [CrossRef]
19. Hart, L.B.; Beckingham, B.; Wells, R.S.; Alten Flagg, M.; Wischusen, K.; Moors, A.; Kucklick, J.; Pisarski, E.; Wirth, E. Urinary Phthalate Metabolites in Common Bottlenose Dolphins (*Tursiops truncatus*) from Sarasota Bay, FL, USA. *GeoHealth* **2018**, *2*, 313–326. [CrossRef]
20. Dziobak, M.K.; Wells, R.S.; Pisarski, E.C.; Wirth, E.F.; Hart, L.B. Demographic Assessment of Mono(2-Ethylhexyl) Phthalate (MEHP) and Monoethyl Phthalate (MEP) Concentrations in Common Bottlenose Dolphins (*Tursiops truncatus*) from Sarasota Bay, FL, USA. *GeoHealth* **2021**, *5*, e2020GH000348. [CrossRef]
21. Hart, L.B.; Dziobak, M.K.; Pisarski, E.C.; Wirth, E.F.; Wells, R.S. Sentinels of Synthetics—A Comparison of Phthalate Exposure between Common Bottlenose Dolphins (*Tursiops truncatus*) and Human Reference Populations. *PLoS ONE* **2020**, *15*, e0240506. [CrossRef] [PubMed]
22. Backer, L.C.; Bolton, B.; Litz, J.A.; Trevillian, J.; Kieszak, S.; Kucklick, J. Environmental Contaminants in Coastal Populations: Comparisons with the National Health and Nutrition Examination Survey (NHANES) and Resident Dolphins. *Sci. Total Environ.* **2019**, *696*, 134041. [CrossRef] [PubMed]
23. Barros, N.B.; Wells, R.S. Prey and Feeding Patterns of Resident Bottlenose Dolphins (*Tursiops truncatus*) in Sarasota Bay, Florida. *J. Mammal.* **1998**, *79*, 1045–1059. [CrossRef]
24. Berens McCabe, E.J.; Gannon, D.P.; Barros, N.B.; Wells, R.S. Prey Selection by Resident Common Bottlenose Dolphins (*Tursiops truncatus*) in Sarasota Bay, Florida. *Mar. Biol.* **2010**, *157*, 931–942. [CrossRef]

25. Wells, R.; McHugh, K.; Douglas, D.; Shippee, S.; Berens McCabe, E.; Barros, N.; Phillips, G. Evaluation of Potential Protective Factors Against Metabolic Syndrome in Bottlenose Dolphins: Feeding and Activity Patterns of Dolphins in Sarasota Bay, Florida. *Front. Endocrinol.* **2013**, *4*, 139. [CrossRef]
26. Gannon, D.P.; McCabe, E.J.B.; Camilleri, S.A.; Gannon, J.G.; Brueggen, M.K.; Barleycorn, A.A.; Palubok, V.I.; Kirkpatrick, G.J.; Wells, R.S. Effects of *Karenia brevis* harmful algal blooms on nearshore fish communities in southwest Florida. *Mar. Ecol. Prog. Ser.* **2009**, *378*, 171–186. [CrossRef]
27. Wells, R.S.; Rhinehart, H.L.; Hansen, L.J.; Sweeney, J.C.; Townsend, F.I.; Stone, R.; Casper, D.R.; Scott, M.D.; Hohn, A.A.; Rowles, T.K. Bottlenose Dolphins as Marine Ecosystem Sentinels: Developing a Health Monitoring System. *EcoHealth* **2004**, *1*, 246–254. [CrossRef]
28. Wells, R.S. Learning from Nature: Bottlenose Dolphin Care and Husbandry. *Zoo Biol.* **2009**, *28*, 635–651. [CrossRef] [PubMed]
29. Barratclough, A.; Wells, R.S.; Schwacke, L.H.; Rowles, T.K.; Gomez, F.M.; Fauquier, D.A.; Sweeney, J.C.; Townsend, F.I.; Hansen, L.J.; Zolman, E.S.; et al. Health Assessments of Common Bottlenose Dolphins (*Tursiops truncatus*): Past, Present, and Potential Conservation Applications. *Front. Vet. Sci.* **2019**, *6*, 444. [CrossRef]
30. Lusher, A.L.; Welden, N.A.; Sobral, P.; Cole, M. Sampling, Isolating and Identifying Microplastics Ingested by Fish and Invertebrates. *Anal. Methods* **2017**, *9*, 1346–1360. [CrossRef]
31. Karami, A.; Golieskardi, A.; Choo, C.K.; Romano, N.; Ho, Y.B.; Salamatinia, B. A high-performance protocol for extraction of microplastics in fish. *Sci. Total Environ.* **2017**, *578*, 485–494. [CrossRef] [PubMed]
32. Shim, W.J.; Hong, S.H.; Eo, S. Identification methods in microplastic analysis: A review. *Anal. Methods* **2017**, *9*, 1384–1391. [CrossRef]
33. Lusher, A.L.; Bråte, I.L.N.; Munno, K.; Hurley, R.R.; Welden, N.A. Is It or Isn't It: The Importance of Visual Classification in Microplastic Characterization. *Appl. Spectrosc.* **2020**, *74*, 1139–1153. [CrossRef] [PubMed]
34. Leads, R.R.; Weinstein, J.E. Occurrence of Tire Wear Particles and Other Microplastics within the Tributaries of the Charleston Harbor Estuary, South Carolina, USA. *Mar. Pollut. Bull.* **2019**, *145*, 569–582. [CrossRef] [PubMed]
35. Manikanda Bharath, K.; Muthulakshmi, A.L.; Natesan, U. Microplastic Contamination around the Landfills: Distribution, Characterization and Threats: A Review. *Curr. Opin. Environ. Sci. Health* **2023**, *31*, 100422. [CrossRef]
36. Rochman, C.M.; Brookson, C.; Bikker, J.; Djuric, N.; Earn, A.; Bucci, K.; Athey, S.; Huntington, A.; McIlwraith, H.; Munno, K.; et al. Rethinking microplastics as a diverse contaminant suite. *Environ. Toxicol. Chem.* **2019**, *38*, 703–711. [CrossRef] [PubMed]
37. Unsworth, K.F.R.; Higgs, A.; Walter, B.; Cullen-Unsworth, L.C.; Inman, I.; Jones, B.L. Canopy accumulation: Are seagrass meadows a sink of microplastics? *Oceans* **2021**, *2*, 162–178. [CrossRef]
38. De Witte, B.; Devriese, L.; Bekaert, K.; Hoffman, S.; Vandermeersch, G.; Cooreman, K.; Robbens, J. Quality Assessment of the Blue Mussel (Mytilus Edulis): Comparison between Commercial and Wild Types. *Mar. Pollut. Bull.* **2014**, *85*, 146–155. [CrossRef]
39. Devriese, L.I.; van der Meulen, M.D.; Maes, T.; Bekaert, K.; Paul-Pont, I.; Frère, L.; Robbens, J.; Vethaak, A.D. Microplastic Contamination in Brown Shrimp (*Crangon crangon*, Linnaeus 1758) from Coastal Waters of the Southern North Sea and Channel Area. *Mar. Pollut. Bull.* **2015**, *98*, 179–187. [CrossRef]
40. Corcoran, P.L. Degradation of microplastics in the environment. In *Handbook of Microplastics in the Environment*; Springer International Publishing: Cham, Switzerland, 2022; pp. 531–542.
41. Sarasota Bay Comprehensive Conservation and Management Plan Update & State of the Bay Report 2014. Available online: https://www.epa.gov/sites/production/files/2015-09/documents/ccmp.stateofthebay-for-website-august2014.pdf (accessed on 14 August 2023).
42. USF 2020. Sarasota Bay Watershed: Geography and Land Use—Sarasota.WaterAtlas.Org. Available online: https://www.sarasota.wateratlas.usf.edu/watershed/geography.asp?wshedid=5&wbodyatlas=watershed (accessed on 14 August 2023).
43. Aswathy, J.; Sharma, S.R.K.; Mini, K.G. Microplastic contamination in commercially important bivalves from the southwest coast of India. *Environ. Pollut.* **2022**, *305*, 119250.
44. Rojas-Jimenez, K.; Villalobos-Rojas, F.; Gatgens-García, J.; Rodríguez-Arias, M.; Hernández-Montero, N.; Wehrtmann, I.S. Presence of Microplastics in Six Bivalve Species (*Mollusca, Bivalvia*) Commercially Exploited at the Pacific Coast of Costa Rica, Central America. *Mar. Pollut. Bull.* **2022**, *183*, 114040. [CrossRef]
45. Rios-Fuster, B.; Alomar, C.; González, G.P.; Martínez, R.M.G.; Rojas, D.L.S.; Hernando, P.F.; Deudero, S. Assessing microplastic ingestion and occurrence of bisphenols and phthalates in bivalves, fish and holothurians from a Mediterranean marine protected area. *Environ. Res.* **2022**, *214*, 114034. [CrossRef] [PubMed]
46. Mizraji, R.; Ahrendt, C.; Perez-Venegas, D.; Vargas, J.; Pulgar, J.; Aldana, M.; Patricio Ojeda, F.; Duarte, C.; Galbán-Malagón, C. Is the Feeding Type Related with the Content of Microplastics in Intertidal Fish Gut? *Mar. Pollut. Bull.* **2017**, *116*, 498–500. [CrossRef] [PubMed]
47. Zhang, S.; Wang, N.; Gong, S.; Gao, S. The Patterns of Trophic Transfer of Microplastic Ingestion by Fish in the Artificial Reef Area and Adjacent Waters of Haizhou Bay. *Mar. Pollut. Bull.* **2022**, *177*, 113565. [CrossRef] [PubMed]
48. Costa, L.L.; da Costa, I.D.; da Silva Oliveira, A.; Zalmon, I.R. Microplastic Ecology: Testing the Influence of Ecological Traits and Urbanization in Microplastic Ingestion by Sandy Beach Fauna. *Estuar. Coast. Shelf Sci.* **2023**, *290*, 108406. [CrossRef]
49. Huang, L.; Li, Q.P.; Li, H.; Lin, L.; Xu, X.; Yuan, X.; Koongolla, J.B.; Li, H. Microplastic contamination in coral reef fishes and its potential risks in the remote Xisha areas of the South China Sea. *Mar. Pollut. Bull.* **2023**, *186*, 114399. [CrossRef] [PubMed]

50. Nelms, S.E.; Galloway, T.S.; Godley, B.J.; Jarvis, D.S.; Lindeque, P.K. Investigating Microplastic Trophic Transfer in Marine Top Predators. *Environ. Pollut.* **2018**, *238*, 999–1007. [CrossRef]
51. Donohue, M.J.; Masura, J.; Gelatt, T.; Ream, R.; Baker, J.D.; Faulhaber, K.; Lerner, D.T. Evaluating Exposure of Northern Fur Seals, Callorhinus Ursinus, to Microplastic Pollution through Fecal Analysis. *Mar. Pollut. Bull.* **2019**, *138*, 213–221. [CrossRef]
52. McIvor, A.J.; Pires, R.; Lopes, C.; Raimundo, J.; Campos, P.F.; Pais, M.P.; Canning-Clode, J.; Dinis, A. Assessing Microplastic Exposure of the Critically Endangered Mediterranean Monk Seal (*Monachus monachus*) on a Remote Oceanic Island. *Sci. Total Environ.* **2023**, *856*, 159077. [CrossRef]
53. Fossi, M.C.; Panti, C.; Guerranti, C.; Coppola, D.; Giannetti, M.; Marsili, L.; Minutoli, R. Are Baleen Whales Exposed to the Threat of Microplastics? A Case Study of the Mediterranean Fin Whale (*Balaenoptera physalus*). *Mar. Pollut. Bull.* **2012**, *64*, 2374–2379. [CrossRef]
54. Zantis, L.J.; Bosker, T.; Lawler, F.; Nelms, S.E.; O'Rorke, R.; Constantine, R.; Sewell, M.; Carroll, E.L. Assessing Microplastic Exposure of Large Marine Filter-Feeders. *Sci. Total Environ.* **2022**, *818*, 151815. [CrossRef] [PubMed]
55. Hernandez-Gonzalez, A.; Saavedra, C.; Gago, J.; Covelo, P.; Santos, M.B.; Pierce, G.J. Microplastics in the Stomach Contents of Common Dolphin (*Delphinus delphis*) Stranded on the Galician Coasts (NW Spain, 2005–2010). *Mar. Pollut. Bull.* **2018**, *137*, 526–532. [CrossRef]
56. Moore, R.C.; Loseto, L.; Noel, M.; Etemadifar, A.; Brewster, J.D., MacPhee, S.; Bendell, L.; Ross, P.S. Microplastics in Beluga Whales (*Delphinapterus leucas*) from the Eastern Beaufort Sea. *Mar. Pollut. Bull.* **2020**, *150*, 110723. [CrossRef]
57. Zhang, X.; Luo, D.; Yu, R.-Q.; Xie, Z.; He, L.; Wu, Y. Microplastics in the Endangered Indo-Pacific Humpback Dolphins (*Sousa chinensis*) from the Pearl River Estuary, China. *Environ. Pollut.* **2021**, *270*, 116057. [CrossRef] [PubMed]
58. Merga, L.B.; Redondo-Hasselerharm, P.E.; Van den Brink, P.J.; Koelmans, A.A. Distribution of Microplastic and Small Macroplastic Particles across Four Fish Species and Sediment in an African Lake. *Sci. Total Environ.* **2020**, *741*, 140527. [CrossRef] [PubMed]
59. Wardlaw, C.M.; Corcoran, P.L.; Neff, B.D. Factors Influencing the Variation of Microplastic Uptake in Demersal Fishes from the Upper Thames River Ontario. *Environ. Pollut.* **2022**, *313*, 120095. [CrossRef] [PubMed]
60. Kwon, O.Y.; Kang, J.-H.; Hong, S.H.; Shim, W.J. Spatial Distribution of Microplastic in the Surface Waters along the Coast of Korea. *Mar. Pollut. Bull.* **2020**, *155*, 110729. [CrossRef]
61. Peters, C.A.; Bratton, S.P. Urbanization Is a Major Influence on Microplastic Ingestion by Sunfish in the Brazos River Basin, Central Texas, USA. *Environ. Pollut.* **2016**, *210*, 380–387. [CrossRef]
62. Garcia, T.D.; Cardozo, A.L.P.; Quirino, B.A.; Yofukuji, K.Y.; Ganassin, M.J.M.; dos Santos, N.C.L.; Fugi, R. Ingestion of Microplastic by Fish of Different Feeding Habits in Urbanized and Non-Urbanized Streams in Southern Brazil. *Water Air Soil Pollut.* **2020**, *231*, 434. [CrossRef]
63. Lato, K.A.; Thorne, L.H.; Fuirst, M.; Brownawell, B.J. Microplastic Abundance in Gull Nests in Relation to Urbanization. *Mar. Pollut. Bull.* **2021**, *164*, 112058. [CrossRef]
64. Peters, C.; Thomas, P.; Rieper, K.; Bratton, S. Foraging Preferences Influence Microplastic Ingestion by Six Marine Fish Species from the Texas Gulf Coast. *Mar. Pollut. Bull.* **2017**, *124*, 82–88. [CrossRef] [PubMed]
65. Hastuti, A.R.; Lumbanbatu, D.T.; Wardiatno, Y. The Presence of Microplastics in the Digestive Tract of Commercial Fishes off Pantai Indah Kapuk Coast, Jakarta, Indonesia. *Biodiversitas J. Biol. Divers.* **2019**, *20*, 1233–1242. [CrossRef]
66. de Vries, A.N.; Govoni, D.; Árnason, S.H.; Carlsson, P. Microplastic Ingestion by Fish: Body Size, Condition Factor and Gut Fullness Are Not Related to the Amount of Plastics Consumed. *Mar. Pollut. Bull.* **2020**, *151*, 110827. [CrossRef] [PubMed]
67. Parker, B.W.; Beckingham, B.A.; Ingram, B.C.; Ballenger, J.C.; Weinstein, J.E.; Sancho, G. Microplastic and Tire Wear Particle Occurrence in Fishes from an Urban Estuary: Influence of Feeding Characteristics on Exposure Risk. *Mar. Pollut. Bull.* **2020**, *160*, 111539. [CrossRef] [PubMed]
68. Ory, N.C.; Sobral, P.; Ferreira, J.L.; Thiel, M. Amberstripe Scad Decapterus Muroadsi (*Carangidae*) Fish Ingest Blue Microplastics Resembling Their Copepod Prey along the Coast of Rapa Nui (Easter Island) in the South Pacific Subtropical Gyre. *Sci. Total Environ.* **2017**, *586*, 430–437. [CrossRef] [PubMed]
69. Yin, X.; Wu, J.; Liu, Y.; Chen, X.; Xie, C.; Liang, Y.; Li, J.; Jiang, Z. Accumulation of Microplastics in Fish Guts and Gills from a Large Natural Lake: Selective or Non-Selective? *Environ. Pollut.* **2022**, *309*, 119785. [CrossRef] [PubMed]
70. Bowen, L.W.; Liu, Q.-X.; Fu, S.; Ma, C.; Chen, Q.; Su, L.; Craig, N.J.; Shi, H. Fish ingest microplastics unintentionally. *Environ. Sci. Technol.* **2021**, *55*, 10471–10479.
71. Abbasi, S.; Soltani, N.; Keshavarzi, B.; Moore, F.; Turner, A.; Hassanaghaei, M. Microplastics in Different Tissues of Fish and Prawn from the Musa Estuary, Persian Gulf. *Chemosphere* **2018**, *205*, 80–87. [CrossRef]
72. Akhbarizadeh, R.; Moore, F.; Keshavarzi, B. Investigating Microplastics Bioaccumulation and Biomagnification in Seafood from the Persian Gulf: A Threat to Human Health? *Food Addit. Contam. Part A* **2019**, *36*, 1696–1708. [CrossRef]
73. Daniel, D.B.; Ashraf, P.M.; Thomas, S.N. Microplastics in the Edible and Inedible Tissues of Pelagic Fishes Sold for Human Consumption in Kerala, India. *Environ. Pollut.* **2020**, *266*, 115365. [CrossRef]
74. Di Giacinto, F.; Di Renzo, L.; Mascilongo, G.; Notarstefano, V.; Gioacchini, G.; Giorgini, E.; Bogdanović, T.; Petričević, S.; Listeš, E.; Brkljača, M.; et al. Detection of Microplastics, Polymers and Additives in Edible Muscle of Swordfish (*Xiphias Gladius*) and Bluefin Tuna (*Thunnus thynnus*) Caught in the Mediterranean Sea. *J. Sea Res.* **2023**, *192*, 102359. [CrossRef]
75. Zeytin, S.; Wagner, G.; Mackay-Roberts, N.; Gerdts, G.; Schuirmann, E.; Klockmann, S.; Slater, M. Quantifying Microplastic Translocation from Feed to the Fillet in European Sea Bass *Dicentrarchus labrax*. *Mar. Pollut. Bull.* **2020**, *156*, 111210. [CrossRef]

76. Hasegawa, T.; Nakaoka, M. Trophic transfer of microplastics from mysids to fish greatly exceeds direct ingestion from the water column. *Environ. Pollut.* **2021**, *273*, 116468. [CrossRef] [PubMed]
77. Sucharitakul, P.; Pitt, K.A.; Welsh, D.T. Trophic transfer of microbeads to jellyfish and the importance of aging microbeads for microplastic experiments. *Mar. Pollut. Bull.* **2021**, *172*, 112867. [CrossRef]
78. Xiong, X.; Tu, Y.; Chen, X.; Jiang, X.; Shi, H.; Wu, C.; Elser, J.J. Ingestion and Egestion of Polyethylene Microplastics by Goldfish (*Carassius auratus*): Influence of Color and Morphological Features. *Heliyon* **2019**, *5*, e03063. [CrossRef]
79. Phillips, M.B.; Bonner, T.H. Occurrence and Amount of Microplastic Ingested by Fishes in Watersheds of the Gulf of Mexico. *Mar. Pollut. Bull.* **2015**, *100*, 264–269. [CrossRef] [PubMed]
80. Baalkhuyur, F.M.; Bin Dohaish, E.-J.A.; Elhalwagy, M.E.A.; Alikunhi, N.M.; AlSuwailem, A.M.; Røstad, A.; Coker, D.J.; Berumen, M.L.; Duarte, C.M. Microplastic in the Gastrointestinal Tract of Fishes along the Saudi Arabian Red Sea Coast. *Mar. Pollut. Bull.* **2018**, *131*, 407–415. [CrossRef]
81. Adika, S.A.; Mahu, E.; Crane, R.; Marchant, R.; Montford, J.; Folorunsho, R.; Gordon, C. Microplastic Ingestion by Pelagic and Demersal Fish Species from the Eastern Central Atlantic Ocean, off the Coast of Ghana. *Mar. Pollut. Bull.* **2020**, *153*, 110998. [CrossRef]
82. Garcés-Ordóñez, O.; Mejía-Esquivia, K.A.; Sierra-Labastidas, T.; Patiño, A.; Blandón, L.M.; Espinosa Díaz, L.F. Prevalence of Microplastic Contamination in the Digestive Tract of Fishes from Mangrove Ecosystem in Cispata, Colombian Caribbean. *Mar. Pollut. Bull.* **2020**, *154*, 111085. [CrossRef]
83. Wootton, N.; Ferreira, M.; Reis-Santos, P.; Gillanders, B.M. A Comparison of Microplastic in Fish from Australia and Fiji. *Front. Mar. Sci.* **2021**, *8*, 690991. [CrossRef]
84. Wang, H.; Luo, Z.; Yu, R.; Yan, C.; Zhou, S.; Xing, B. Tire Wear Particles: Trends from Bibliometric Analysis, Environmental Distribution with Meta-Analysis, and Implications. *Environ. Pollut.* **2023**, *322*, 121150. [CrossRef]
85. Werbowski, L.M.; Gilbreath, A.N.; Munno, K.; Zhu, X.; Grbic, J.; Wu, T.; Sutton, R.; Sedlak, M.D.; Deshpande, A.D.; Rochman, C.M. Urban Stormwater Runoff: A Major Pathway for Anthropogenic Particles, Black Rubbery Fragments, and Other Types of Microplastics to Urban Receiving Waters. *ACS EST Water* **2021**, *1*, 1420–1428. [CrossRef]
86. Unice, K.M.; Kreider, M.L.; Panko, J.M. Comparison of Tire and Road Wear Particle Concentrations in Sediment for Watersheds in France, Japan, and the United States by Quantitative Pyrolysis GC/MS Analysis. *Environ. Sci. Technol.* **2013**, *47*, 8138–8147. [CrossRef] [PubMed]
87. Sinnecker, H.; Krause, T.; Koelling, S.; Lautenschläger, I.; Frey, A. The Gut Wall Provides an Effective Barrier against Nanoparticle Uptake. *Beilstein J. Nanotechnol.* **2014**, *5*, 2092–2101. [CrossRef] [PubMed]
88. Kuczaj, S.A.; Yeater, D.B. Observations of rough-toothed dolphins (*Steno bredanensis*) off the coast of Utila, Honduras. *J. Mar. Biol. Assoc. UK* **2007**, *87*, 141–148. [CrossRef]
89. Martin, A.; da Silva, V.; Rothery, P. Object Carrying as Socio-Sexual Display in an Aquatic Mammal. *Biol. Lett.* **2008**, *4*, 243–245. [CrossRef] [PubMed]
90. Foekema, E.M.; De Gruijter, C.; Mergia, M.T.; van Franeker, J.A.; Murk, A.J.; Koelmans, A.A. Plastic in north sea fish. *Environ. Sci. Technol.* **2013**, *47*, 8818–8824. [CrossRef] [PubMed]
91. Berna, F. Fourier transform infrared spectroscopy (FTIR). In *Encyclopedia of Earth Sciences Series*; Gilbert, A.S., Ed.; Springer: Dordrecht, The Netherlands, 2017; pp. 285–286.
92. Xie, Q.; Ning, X.; He, X.; Deng, L.; Wu, Z.; Huang, B.; Gui, D.; Wu, Y. First Evaluation of Quantitative Fatty Acid Signature Analysis (QFASA) in Dolphins. *Reg. Stud. Mar. Sci.* **2022**, *50*, 102141. [CrossRef]
93. Martínez, L.M.; Intralawan, A.; Vázquez, G.; Pérez-Maqueo, O.; Sutton, P.; Landgrave, R. The coasts of our world: Ecological, economic and social importance. *Ecol. Econ.* **2007**, *63*, 254–272. [CrossRef]

oceans

MDPI

Article

Harbor Porpoise Aggregations in the Salish Sea

Dave Anderson [1,*], Laurie Shuster [1,2], Cindy R. Elliser [3], Katrina MacIver [3], Erin Johns Gless [4], Johannes Krieger [4] and Anna Hall [5]

1 Cascadia Research Collective, 218 1/2 4th Ave W, Olympia, WA 98501, USA
2 Pierce College, 9401 Farwest Dr SW, Lakewood, WA 98498, USA
3 Pacific Mammal Research, 1513 A Ave, Anacortes, WA 98221, USA; cindy.elliser@pacmam.org (C.R.E.); katrina.maciver@pacmam.org (K.M.)
4 Pacific Whale Watch Association, 355 Harris Ave #104, Bellingham, WA 98225, USA; erin@pacificwhalewatchassociation.com (E.J.G.); johannes@crystalseas.com (J.K.)
5 Sea View Marine Sciences, 4415 Spellman Place, Victoria, BC V9C 4C5, Canada; annahall@shaw.ca
* Correspondence: danderson@cascadiaresearch.org

Abstract: Harbor porpoises are typically seen in small groups of 1–3 individuals, with aggregations of 20+ individuals treated as rare events. Since the 1990s, the harbor porpoise population in the Salish Sea has seen a significant recovery, and an increased number of observed aggregations that exceed the more usual small group sizes has been observed in recent years. By combining the observational data of United States and Canadian research organizations, community scientists, and whale watch captains or naturalists, we demonstrate that harbor porpoise aggregations appear to be more common than previously known, with 160 aggregations documented in 2022 alone. Behavioral data also indicate that foraging behaviors are common and social behaviors, like mating, are seen more often during these encounters compared to small groups. Other behaviors that are considered to be rare or unknown were also observed during these encounters, including cooperative foraging and vessel approach. These aggregations are likely important foraging and social gatherings for harbor porpoises. This holistic approach integrating data from two countries and multiple sources provides a population level assessment that more effectively reflects the behavior of harbor porpoises in this region, which do not recognize the socio-political boundaries imposed upon the natural world.

Keywords: harbor porpoise; aggregation; social behavior; *Phocoena phocoena*; foraging behavior; large group; transboundary; community science; Salish Sea

1. Introduction

Harbor porpoises (*Phocoena phocoena vomerina*) were a commonly observed cetacean in the waters of Puget Sound and throughout the greater Salish Sea region during the 1940s [1], but by the 1970s, their numbers were greatly reduced throughout the Salish Sea within Washington State, USA (hereafter, Washington or WA), and were completely absent from Puget Sound [2,3]. Information on harbor porpoise abundance in the Salish Sea within British Columbia, Canada, (hereafter British Columbia, or BC) prior to the mid-1990s is sparse [4,5] (Hall unpublished data). Several systematic studies spanned the late 1990's and early 2000's that included the inland waters of southern British Columbia [6,7]. Within the last 30 years, there has been a marked increase in harbor porpoise presence in both WA and southern BC waters of the Salish Sea. A recent study integrating passive acoustic monitoring and community observation logs (historic and contemporary) has documented the frequent presence of harbor porpoises near the Port of Prince Rupert, BC, including large aggregations during winter months [8]. Aerial surveys documented harbor porpoise numbers increasing in Washington waters through the 1990s and reentering Puget Sound beginning in 2000 [9,10]. The first sighting of a small group in South Puget Sound, the southernmost area within the Salish Sea, was in September 2005 by two of the authors

(Shuster and Anderson), with regular sightings beginning in 2008 (Anderson, unpublished data). Harbor porpoises are once again the most common cetacean found throughout the Salish Sea; however, knowledge is discontinuous about the species in these waters [11].

Harbor porpoises are found in coastal waters throughout the Northern Hemisphere. Population health and conservation status of harbor porpoises vary by subspecies and region. The global population of *Phocoena phocoena* is listed as a species of Least Concern on the International Union for Conservation of Nature (IUCN) Red List [12]. The Committee on the Status of Endangered Wildlife in Canada (COSEWIC) list the Pacific harbor porpoise [13] population as Special Concern. They are not listed as a threatened or endangered species in the United States, nor classified as strategic populations in the Washington State Inland Waters [14]. Entanglement in fishing gear is considered to be the greatest threat (though this is not as prevalent in US/BC waters as other populations in their global range), with pollution, anthropogenic noise and prey depletion being additional concerns [12]. While they are not endangered or threatened, as top predators that remain in these waters year round, they are important indicator species for the health of the Salish Sea. Population recovery provides an opportunity to identify and close knowledge gaps, collect baseline data, understand habitat usage and behavior, and inform conservation recommendations for this unique region [11].

Salish Sea harbor porpoises are most often seen singly or in small groups, averaging less than three animals [10,11,15], which is typical throughout their global range [16]. However, harbor porpoises occasionally come together in larger aggregations (Figure 1 and Figure S1), where many smaller groups are in close proximity to each other. These aggregations can be with animals densely packed in a small area, or spread over several kilometers, possibly consisting of distinct subgroups that are close enough to interact with each other [17–19]. In other regions, some aggregations are thought to be related to seasonal migrations, for example, when ice forms in the Bay of Fundy [20], or herring migrate in the Baltic Sea [21]. In areas where harbor porpoise are not known to migrate, these occurrences are considered spurious and rare and it is suspected they are feeding aggregations located in an area with a high concentration of prey [17].

Figure 1. Harbor porpoise group within an aggregation of an estimated 300+ individuals on 24 February 2021. Photo credit Trevor Derie, Pacific Mammal Research.

The prevalence and importance of aggregations are often dismissed or treated as rare events in the Salish Sea of Washington and British Columbia. In their seminal work on marine mammals in Washington State, Scheffer and Slipp make no mention of larger aggregations, with the observations, "usually in groups of 2 to 5, occasionally 10 to 12", though they do note that, "rarely are more than 3 of a group in sight at one time, although several groups may gather in favored waters" [1]. A literature review of published reference works that include this region indicates that authors predominantly mentioned smaller group sizes, or did not mention group size at all [22,23], while others mention larger aggregations, usually referring to them as "rare" or "occasional" [24–27], and some works suggested seasonality as a factor influencing larger aggregations with them occurring in summer and early fall [28,29]. The recent study documenting large aggregations during winter months in the waters near the Port of Prince Rupert is an exception [8].

Several surveys for harbor porpoise have been conducted in the northeast Pacific Ocean, including both the Salish Sea and the outer coast, between 1984 and 2015. Only a few note occasional sightings of groups or aggregations of over 20 animals, with only two publications (and three total sightings of 50, 100 and 195 individuals) noted for Salish Sea waters (labeled Inland WA, Table 1). Full survey data were not available for retrospective analysis for either publication. However, the more recent data from aerial surveys by Evenson [9] were available for review and had only one off-effort sighting of an aggregation of over 20 individuals (Evenson, unpublished data). An exception to these data is on the northern BC coast, where harbor porpoise aggregations have been recorded around Prince Rupert, a busy commercial port, with the largest aggregations between December and April, including one of 200–1000 individuals [8]. It is important to note that during line or strip transect surveys, investigators might only record individual groups, not taking note of how multiple groups within a small geographic area create an aggregation.

Table 1. Summary of survey reports for northeast Pacific Ocean harbor porpoise, and whether large groups or aggregations are mentioned. "W. coast" is "West coast", "WA" is Washington, "BC" is British Columbia, "SJDF" is Strait of Juan de Fuca, "SJI" is San Juan Islands, and "SG" is Strait of Georgia.

Region	Dates	Type	Sightings	Citation
W. coast, USA	1984–1986	Ship	852 groups, 2 groups over 20 (0.23%)	[30]
W. coast, USA	1984–1985	Aerial	366 groups, group size range not mentioned	[31]
Inland WA, USA			Two aggregations (50 and 100)	[32]
San Juan Islands, WA, USA	1991–1992	Boat and shore	Four survey types: 301 groups (random boat) with 1 group of "at least 18 harbor porpoise" 125 (fixed boat), 634 (shore abundance), unknown (shore location) includes sighting of 195 individuals	[33]
OR and WA	1991	Aerial	579 groups of 1–7 individuals	[34]
Inland WA	1994–2014	Aerial	1270 groups, 1 aggregation of 28 individuals (0.08%)	[9], Evenson unpub. data
Coastal OR, WA, BC, SJDF, SJI and SG	2002	Aerial	606 groups of 1–7 individuals	[35]
Coastal OR, WA, BC, and SJDF, SJI and SG	2003	Aerial	499 groups of 1–12 individuals	[36]
Inland WA	2013–2015	Aerial	338 groups, mean group size 1.7 No max group size reported	[10]
Northwest BC, Chatham Sound, Prince Rupert	1993–2022	Shore, including community science	626 groups, including 12 aggregations over 31 individuals Largest group was 200–1000 individuals	[8]

Recent observations, however, suggest that these large aggregations may be much more common in the Salish Sea than previously documented. Harbor porpoise aggregations in these waters are not related to migration or icing up, as harbor porpoises are known to remain year-round [6,19], with long-term photo identification (photo ID) [11], genetic data [37], and tag data [38] suggesting the possibility of high site fidelity among this population. On-going photo ID studies in British Columbia are also noting positive identifications of individuals on an inter-annual basis (Porpoise Conservation Society, unpublished data). Long-term sighting data analyses (1991–2008) from British Columbia determined harbor porpoise high density aggregation data are associated with foraging and reproductive behaviors, specific habitats, and oceanographic variables related to tidal phase and mixing [19].

Harbor porpoises are opportunistic feeders, with the majority of their diet composed of small forage fish, along with some cephalopods, crustaceans and arthropods [39,40] and occasionally larger fish are also consumed [41]. A wide variety of forage fish found in the Salish Sea are known food sources for harbor porpoises including Pacific herring (*Chupea pallasi*), Pacific sand lance (*Ammodytes hexapterus*), and surf smelt (*Hypomesus pretiosus*) [42,43]. Salmon and steelhead runs are common in the many rivers entering the Salish Sea [44]. Though salmonids are not considered to be a significant portion of harbor porpoises' diet, as opportunistic feeders, porpoises may eat salmonoid smolts when available [45], and in some locations, porpoises have been observed taking adult salmon [41]. Eulachon (*Thaleichthys pacificus*), a high-fat-content fish, is found in the northern Salish Sea and could serve as an ideal high-calorie food [46]. Traditionally, Northern anchovy (*Engraulis mordax*) was not abundant in these waters until the 2014–2016 Blob event that increased offshore ocean temperatures [47,48] and led to greater abundance of anchovies, especially in South Puget Sound [46,49]. Market squid (*Doryteuthis opalescens*), another staple of harbor porpoises' diet, are also found in South Puget Sound from December through February [46,50].

Large aggregations of harbor porpoises were independently observed by the co-authors and our combined data serve as a method to investigate the occurrence of aggregations throughout the Salish Sea. We compared data from several sources, including small boat surveys, whale watch vessels, land-based marine mammal monitoring field efforts, and community/citizen scientist observers. We quantified the occurrence of these large aggregations, their relation to seasonal patterns, and the prevalence of social behaviors (including mating, fission/fusion of subgroups, coordinated feeding behavior, and willingness to approach vessels) during these groupings. We document that these aggregations occur more commonly than previously thought and suggest that they provide important feeding and socializing opportunities for Salish Sea harbor porpoises.

2. Materials and Methods

2.1. Study Location

The Salish Sea is an inland fjord-like body of water composed of many inlets, passages and bays in Washington and British Columbia (Figure 2). The major basins include the Strait of Juan de Fuca (Juan de Fuca Strait in Canada), connecting to the Pacific Ocean; the San Juan Islands, northeast of the Strait of Juan de Fuca in Washington; the Gulf Islands, in Canada north of the San Juan Islands; the Strait of Georgia, between mainland BC and Vancouver Island; and Puget Sound, south of the east end of the Strait of Juan de Fuca.

2.2. Data Collection

Four organizations contributed harbor porpoise sighting data of aggregation sizes ≥ 20 animals for this analysis between February 2017 and March 2023. Data were collected through boat and land-based surveys by local researchers, public reports from community scientists (including a public sighting app), and sighting reports from whale-watch captains and naturalists via a private sighting app.

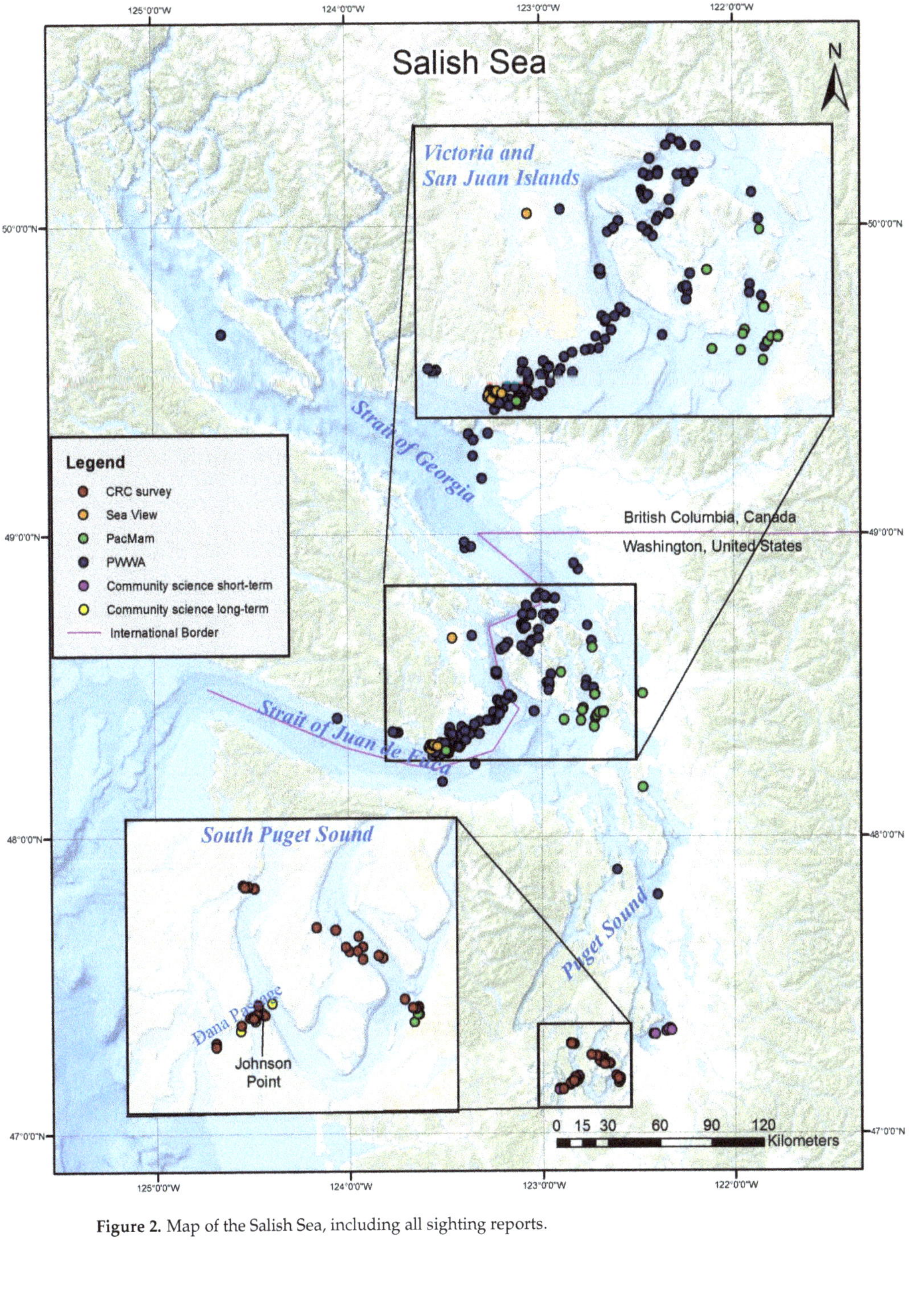

Figure 2. Map of the Salish Sea, including all sighting reports.

2.2.1. Cascadia Research Collective

Cascadia Research Collective (CRC) has conducted year-round, regular, small-boat-based (4.2 m Zodiac) surveys in South Puget Sound since summer 2016. Sightings of all marine mammals are recorded in Google Sheets. Porpoise counts are estimates of the number of animals within good sighting distance from the boat, usually around 300 m, with a Beaufort Sea State of under three. For larger aggregations, several sightings are recorded while passing through the area. Additionally, reports are collected from fishers and community scientist residents living on banks overlooking various locations of Puget Sound. Only reports from experienced observers, or those that supplied photographs or video, were included. Sightings of 31 groups ≥ 20 included in this analysis were reported between 9 January 2019 and 11 May 2023.

2.2.2. Pacific Mammal Research

Pacific Mammal Research (PacMam), based in Anacortes, Washington, is a research organization studying harbor porpoises and harbor seals (*Phoca vitulina*) through land-based, long-term photo ID and behavioral surveys. In March of 2021, a custom opportunistic sighting form (PacMam harbor porpoise project) was created using the Epicollect5 app platform through a collaboration with Kwiáht (Center for the Historical Ecology of the Salish Sea). This app allows the public to easily document opportunistic harbor porpoise sightings throughout the Salish Sea. Community science participants were recruited through public presentations and through social media. The majority of sightings are from the general public, though there are some from local researchers. Information on total group size, number of calves, Global Positioning System (GPS) location, weather, tidal phase, boat presence, gull presence, behavior, length of time watching the porpoises, expertise of the observer, and any extra notes can be documented. Data entry is not required for every field and observer expertise varies; therefore, some sighting records do not contain information about each of these factors. To date, users of this app have documented over 300 harbor porpoise sightings throughout the Salish Sea, from South Puget Sound, north to the San Juan Islands, and out the Strait of Juan de Fuca. These sightings are not restricted to large aggregations and range from 1–100+ harbor porpoises. Thus, for this study a subset of the data was used (group sizes ≥ 20, and observer expertise level of experienced or expert). Twenty-nine sightings occurring between January and November 2022 were included in this analysis.

2.2.3. Pacific Whale Watch Association

The Pacific Whale Watch Association (PWWA) is a professional association of ecotourism operators in Washington and British Columbia. Whale watching season is primarily March through November, with only a few companies operating year-round. Tours focus on finding large whales, including killer whales (*Orcinus orca*), humpback whales (*Megaptera novaeangliae*), and gray whales (*Eschrichtius robustus*), limiting their operations to areas where the larger whales are likely to be located. As of 2023, the PWWA comprises 30 member companies departing from 23 ports ranging as far south as Seattle, WA, as far north as Telegraph Cove, BC, and as far west as Port Renfrew, BC. PWWA members utilize the private PWWA App, developed by Johannes Krieger in 2018, to record wildlife sightings throughout the Salish Sea. Sightings of harbor porpoises in the region are fairly common and not routinely reported by whale watchers, but for this study, PWWA captains and naturalists were asked to document "large aggregations" of harbor porpoises, groups of 10 or more individuals, beginning in April 2021. Included in this analysis were 146 sightings of groups ≥ 20 occurring between February 2021 and December 2022. Sightings records in the PWWA App include species, group size, travel direction (if known), time, date, and GPS location of the sighting.

2.2.4. Sea View

Sea View Marine Sciences (Sea View) specializes in marine mammal research, monitoring, and mitigation. Sea View is on Vancouver Island near Victoria, British Columbia and has worked extensively in southern BC waters conducting numerous field assessments and research projects with professional biologists and observers. From 2017 to 2023, harbor porpoise group size and behavioral data were collected by Sea View as part of a larger Marine Mammal Monitoring Program of the Canadian Department of National Defence (DND) training operations in the Salish Sea. Field efforts and data collection were conducted entirely in Canadian waters and were conducted according to the DND schedule. Sightings of 14 groups $\geq$ 20 between August 2018 and September 2022 were included in this analysis.

2.2.5. All Data

Duplication of sightings across platforms was possible; therefore, care was taken to remove these from the data. Sighting reports from PWWA vessels, which were duplicated in the PacMam data, using criteria of same day, location and reporting party, were deleted from the PacMam data. Reports to CRC that matched PacMam data were deleted from CRC data. Potential duplicates in PacMam and PWWA data were removed by checking for reports on the same day with locations within 2 km (20–49 individuals), 4 km (50–99 individuals), or 8 km (>100 individuals), without islands between reported locations, where only one of the two or more sightings were kept for analysis. The encounter with the highest estimate of individuals was kept, as the aggregations could grow over time. No duplications were found in the PacMam data, and 20 sightings were removed from the PWWA data.

Large aggregations can cover several square kilometers, so even accurate GPS locations taken within the aggregation do not represent the extent of the entire aggregation. Map of sightings was generated using ArcMAP 10.8.2. (Figure 2).

The timing of these aggregations can vary, and we differentiate between long- and short-term events. Long-term aggregations are defined as harbor porpoises remaining in the same area, in large numbers (20+), lasting at least one week, as documented by observers recording the presence of the aggregation on multiple days with good sighting conditions, with few gaps of more than a few days. Short-term aggregations are defined as large numbers of harbor porpoises (20+), usually lasting for a few hours, or up to a few days at most.

To determine the temporal occurrence of these events, results from each dataset were analyzed to determine if aggregations occurred throughout the year. Additional analysis of possible variation in occurrence relating to seasonality was not conducted, as much variability is due to bias in observer effort. (For example, opportunistic data are highly biased toward summer and early fall, when weather is best and more boats are on the water).

All contributing groups recorded porpoise behavior, paying particular attention to those rarely seen outside of these aggregations, especially social and unique foraging behaviors not possible in smaller groups. CRC and PacMam data were analyzed for the frequency of occurrence of behaviors in large versus small groups. CRC data were from dedicated boat surveys and have more detailed descriptions of behavior. Frequencies of probable foraging, traveling, probable mating, vessel approach, vessel avoidance, fission–fusion, logging, side surface feeding, and general splashing (non-mating breaching, tail slaps, chases, porpoising) were compared between large and small groups. For PacMam app data, the options for behavior are more general, as the app is for public use. Frequencies of "traveling in one direction" (travel), "play, socializing, leaping" (social), and dives/surface chases, probable feeding (foraging) were compared between large and small groups. In both datasets, some sightings had more than one behavior documented. In these cases the sighting was scored for each of those behaviors, so a sighting with both foraging and travel behaviors would be scored as a foraging sighting and a traveling sighting to accurately analyze the frequency of behaviors. Because of this, the total number of sightings used in the frequency analysis is more than the actual total number of sightings. Analysis of this type could not be conducted for the PWWA App data because those users were specifically asked to document larger groups (10 or more individuals); thus, the data are biased to larger groups and does not include enough smaller groups for comparison. Sea View's research occurred during DND training operations. Data presented in this publication were used with DND authorization and approval for analysis of harbor porpoise high-density aggregations observed during the Marine Mammal Monitoring Program. DND approval was granted for analysis for large aggregations only, so comparison with smaller groups was not possible with these data.

3. Results

3.1. CRC Data

CRC has conducted 97 surveys since 2016, covering 8790 km in the South and Central basins of the Puget Sound. During this period, harbor porpoises were encountered 450 times. Aggregations of 20 or more individuals were encountered 31 times (6.9% of encounters) from January 2017 through March 2023, all within South Puget Sound (Table S1). Six of the included sightings were in long-term aggregations, to verify reports by community scientists. Due to the visibility limitations experienced by observers in a small boat, obtaining a count that fully represents the extent or number of animals present in the larger aggregations is not possible because of the difficulty in viewing the entire extent of the aggregation.

There have been 16 community science reports of short-term aggregations of 20 or more animals, including two reports of aggregations of 100 or more, submitted to CRC and included in this analysis (Table S3). The first report of a large aggregation in South Puget Sound was submitted in 2012 by a marina employee. Fishers in Case Inlet encountered an estimated 200 individuals and observed breaching, chasing, and wake riding.

Of particular note are large, long-term aggregations around Johnson Point in 2019, 2020, and 2021. Community scientists used high-power binoculars from their hilltop home overlooking Dana Passage and observed porpoises when winds were light and the Beaufort Sea State was less than 3. Aggregations formed in early winter and persisted through March or April, with an additional aggregation in October and November 2021. Animals were sighted daily when winds were light and the aggregation frequently shifted position within the observation area of approximately 12 square km. A wide variety of behaviors were reported, including many foraging dives, cooperative feeding in bait balls, following slow-moving vessels, and breaches (Table S2).

3.2. PacMam Data

Reports to PacMam included 22 unique reports from January through November 2022 of short-term aggregations of groups of 20 or more individuals, including three groups of 100 or more in 2022 (Table S3). These data were extracted from a dataset of 286 reports, representing 7.7% of sightings. Travel, foraging, and social behavior states were observed. Specific behaviors observed included directional surfacing (travel), surface chases (foraging), and mating attempts (social).

3.3. PWWA Data

After accounting for potential duplicates, PWWA naturalists and captains logged 33 unique, short-term aggregations of 20 or more individuals, including five groups of 100 or more individuals in 2021. There were short term aggregations of 93 groups of 20 or more individuals, including 15 aggregations of 100 or more individuals in 2022 (Table S4). Only aggregations 10+ animals were recorded; therefore, it is unknown what percentage these aggregations represent of PWWA operations.

3.4. Sea View Data

Sea View documented 14 porpoise aggregation sightings occurring between August 2018 and September 2022 representing 13.3% of the observational dataset for this time period. These data were extracted from a larger dataset of 105 sightings of individuals and groups of harbor porpoise with a minimum total of 1542 porpoise observed across all sightings and years. Thirteen were short-term aggregations with 12 in Strait of Juan de Fuca, and one in Saanich Inlet, BC. Two events (10 August 2018 and 6 June 2019) were observed with the number of animals in the aggregation increasing throughout the day, to 20+ and 100+ animals, respectively. Foraging behaviors were observed during both events.

Site fidelity for aggregations was noted for a nearshore habitat in Strait of Juan de Fuca on 28 April 2021, 19 May 2021 and 8 June 2021 with a consistent group size of ~15–20 animals. Behaviors noted during these three events included foraging and socializing.

A three-day aggregation occurred in June 2022, during which foraging and reproductive behaviors were documented by Sea View. During this event, less commonly observed harbor porpoise behaviors occurred, including wake-riding and multiple aerial behaviors. These observations in Strait of Juan de Fuca are spatially consistent with the high-density aggregations reported by Hall [19], suggesting long-term habitat use that spans decades in this part of British Columbia.

3.5. All Data

Harbor porpoise aggregations were encountered in every month of the year (Table 2). Higher sightings during March-October are likely attributed to greater observation efforts for all datasets due to increased day length, better weather, and good sighting conditions during these months. This is especially true for the PWWA data as these are the primary months whale-watching vessels are on the water (only a few are active November–February).

Table 2. Distribution of sightings by month. Sightings of large aggregations occurred every month of the year. The seasonal variation documented is likely due to increased observer efforts during warmer months, that have longer days, and more days with good sighting conditions. Community scientist reports of long-term aggregations were recorded one time per aggregation per month.

	Jan	Feb	Mar	Apr	May	Jun	Jul	Aug	Sep	Oct	Nov	Dec
CRC survey	3	4	4	1	6	0	4	5	2	0	2	0
Community science long-term (CRC)	2	3	3	1	1	0	0	0	0	1	2	1
Community science short-term (CRC)	1	2	1	4	4	4	0	0	0	0	0	0
PacMam app	0	3	17	1	2	12	5	6	2	1	1	0
PWWA	0	4	16	5	8	27	20	13	14	16	2	1
Sea View	0	0	0	1	3	6	1	2	1	0	0	0

Across all these data collection platforms, behaviors recorded during these encounters could include synchronized group foraging, traveling, mating (which often occurs with males breaching as they attempt to mate—this is a consistent behavior seen in harbor porpoise populations world-wide) (Figure 3) [51], surface chases (Figure 4), spyhops, non-mating breaches, porpoising, fusion–fission of groups, and logging. Due to inconsistent recording between platforms, non-mating social behaviors were grouped into "Social behavior/splashing" (Table 3) (Figure 5). Sub-groups often experience fusion/fission during foraging, coming together for a series of dives, before splitting up again, not always in the same groups. Wake riding was also observed in several cases, with one short-duration bow ride recorded in Dana Passage. Porpoises will often approach slow-moving or stopped vessels during these aggregations. In the short- and long-term aggregations of 100+ individuals, mating attempts can be common, sometimes numbering in the dozens over a relatively short direct observation period (1–2 h). One aggregation had over 40 mating attempts, recorded by an experienced observer with PacMam.

Figure 3. Behavior typical of a mating practice or attempt. Photographed during an aggregation of approximately 20 individuals on 31 August 2018. Photo credit: Laurie Shuster and David Anderson, Cascadia Research Collective, taken under National Marine Fisheries Service (NMFS) permit number 20605.

Figure 4. Activity typical of a chase. Photographed during aggregation of over 100 individuals on 21 January 2019. Photo credit: Laurie Shuster and David Anderson, Cascadia Research Collective, taken under National Marine Fisheries Service (NMFS) permit number 20605.

Table 3. Behaviors recorded during encounters. Not all encounters had recorded behavior, as encounters can be short in duration and/or at a distance where it is difficult to determine. Multiple behaviors can occur during a single encounter, especially with larger groups. * One aggregation had over 40 mating attempts, recorded by an experienced observer.

	CRC		PacMam	
	Aggregations (Total 31)	**Small Group (Total 398)**	**Aggregation (Total 56)**	**Small Group (Total 274)**
Foraging	29 (93.5%)	202 (50.8%)	22 (39.3%)	91 (33.2%)
Traveling	2 (6.5%)	80 (20.1%)	6 (10.7%)	107 (39.1%)
Mating	5 (16.1%)	11 (2.8%)	2 (3.6%) *	3 (1.2%)
Social behavior/splashing	13 (41.9%)	23 (5.8%)	4 (7.1%)	10 (3.6%)
Vessel approach	5 (16.1%)	25 (6.3%)	--	--
Fission–fusion	4 (12.9%)	16 (4.0%)	--	--

Figure 5. Splash during aggregation with four porpoises in foreground. Photographed during aggregation of approximately 200 individuals on 2 June 2022. Photo credit: Anna Hall, Sea View Marine Sciences.

Based on the prevalence of foraging activities during most, if not all, of these aggregations, it is likely that they are primarily foraging opportunities. However, social activities are also quite common, and it is equally likely that these aggregations provide opportunities for increased social interactions.

4. Discussion

Reports of harbor porpoise aggregations in the Salish Sea are considered rare events; however, these transboundary data demonstrate that aggregations are more common than previously documented and occur year-round (Table 2) in the Salish Sea. This conflicts with previous studies that suggest the aggregations were seasonal and limited to summer and early fall [28,29]. In other locations, harbor porpoise populations are known to migrate seasonally [20,21] and it is possible that previous Salish Sea researchers assumed a similar seasonal pattern. However, historical knowledge of Salish Sea harbor porpoises is sparse [52], due to the limited amount and breadth of research studies conducted. Therefore, it is unlikely those limited studies had the ability to accurately document these grouping patterns. In addition, when the harbor porpoise numbers were depleted, there may not have been enough animals or aggregations to be able to accurately determine seasonality.

The recovery of the Salish Sea harbor porpoise population [9] has likely been a driver behind increased occurrence of aggregations. At the same time, Dall's porpoise (*Phocoenoides dalli*) populations have substantially decreased and are seen infrequently in the Salish Sea [9]. Dall's may avoid areas with harbor porpoise presence [4] and their departure could also provide increased opportunities for resources such as prey and preferred habitat.

The availability of food is likely a dominant influence in the formation of these aggregations. Foraging behaviors were frequently documented, with most individuals in the aggregations exhibiting regular foraging dives or surface-level chases. The lower percentages documented in the PacMam data are likely due to observer bias. These data were generally collected by the public, and for behavior, "unknown" was an option for entry. For example, about 21% for both large and small groups included unknown behavior. Thus, the lower documented foraging behavior in the PacMam data is likely an underrepresentation. Fusion/fission behavior was observed, with subgroups of 10 or more animals synchronizing their dives then splitting up again after a series of dives. Synchronized feeding on surface bait balls, including where individual porpoises were seen on their sides at the surface, swimming in a circle were observed. This could be a way to condense a bait ball before making a feeding pass, as has been documented in other cetacean species [53–55]. This foraging behavior is rarely seen outside large groups, [19] (Anderson and Shuster, unpublished data Cooperative feeding during high-density aggregations was noted by Hall [19] in southern British Columbia.

The large number of harbor porpoise aggregations documented in this study indicate prey abundance sufficient to support both short- and long-term events. The high metabolic rate of harbor porpoises [56–58] would require substantial food resources to support aggregations of over 100 porpoises that inhabit an area for several months. Porpoises have been documented foraging in the presence of other marine mammals and birds in many locations. During winter months in 2019 and 2020, sufficient prey were present in nearby Case Inlet in South Puget Sound to feed several hundred California sea lions (*Zalophus californianus*) (Jefferies, pers. com.) along with a large aggregation of harbor porpoises and other marine species such as harbor seals, three long-beaked common dolphins (*Delphinus capensis*), and large flocks of marine birds that were frequently seen in the area. Regional oceanographic elements, such as shoals, fronts, and high tidal flows can concentrate plankton and forage fish. Numerous spawning events in these waters create a potential plentiful food source for harbor porpoises. Five species of fish that are key components of harbor porpoise diet are present in the Salish Sea: Pacific herring, Pacific sand lance, surf smelt, eulachon, northern anchovy [6,39]. Thus, there are a variety of prey species that may concentrate in large enough numbers to support short- and longer-term harbor porpoise aggregations, along with other species. More research is needed to understand which forage species are more important for driving these events.

Social behaviors and interactions appear to be important components of harbor porpoise aggregations (Table 3). While social interactions are occasionally observed in smaller groups, they were over twice as likely to occur in larger groups. Mating attempts are seen year-round in small groups [59] (Elliser unpublished data, Anderson unpublished data, Hall unpublished data), but typically only one or two attempts are observed, whereas dozens of attempts have been observed in some aggregations.

During aggregation events, porpoises are also more likely to interact with slow-moving or stopped vessels, are less likely to substantially relocate to avoid fast moving vessels and have been observed wake riding or following the prop wash of boats. Porpoises' proximity and interaction with boats was documented in 16.1% of CRC survey observations and noted in some community science reports. In several South Puget Sound aggregation encounters, harbor porpoises have approached the research vessel in a small sub-group, diving under the boat, and then reemerging on the other side while porpoising away at high speed (Shuster and Anderson, unpublished data.). Whale watch personnel have documented porpoises approaching their vessels during encounters with larger aggregations and noted an increased likelihood of porpoises wake riding or following the prop wash of slow-moving vessels. However, vessel-approach behaviors are not limited to larger aggregations. Land-based observers at Burrows Pass in northern Washington have also documented animals in smaller groups approaching vessels (Table S4), wake riding, and interacting with prop wash (Elliser, unpublished data), as have boat-based observers in South Puget Sound (Shuster and Anderson, unpublished data).

While harbor porpoises are generally considered relatively solitary, formation into larger groups provides numerous opportunities to interact and may promote a diverse repertoire of social interactions. Abundance of food is a potential impetus for aggregation formation; therefore, porpoises may spend less time and energy foraging, enabling a shift of resources toward social behaviors.

The social structure of harbor porpoises has not been well investigated and is unknown at this time. Due to their vocalization patterns (e.g., lack of whistles normally attributed to communicative calls) and tendency towards very small groups (1–3 individuals), it has been thought that they do not have very strong social ties. However, indications of sociality are present in the literature. Breaching and splashing in wild harbor porpoises have been attributed as social behaviors. A previous review of harbor porpoise social behaviors, from wild and captive settings, noted these to be well developed and set within a context of individual and group relationships [19].

Harbor porpoises have been observed using complex, cooperative foraging behaviors with role specialization that is rarely seen in animals [60,61]. In the Firth of Clyde (Scotland), a common dolphin was found to change its vocalizations to match local harbor porpoises [62], and harbor porpoise clicks have been shown to be used in communicative contexts, not just foraging [63,64]. Although little is known about their associations, there is early evidence through photo ID that shows at least some individuals are often sighted together over weeks to months at a time (Elliser, unpublished data). It is likely that social interactions are more important to this species than what is observed in the limited social encounters observed at the surface [63]. These large aggregations may be important aspects of their social structure. The importance of larger groups is seen in other species, such as the Southern Resident killer whales (SRKW). In the Salish Sea, SRKW are usually found in tight, matrilineal pods but periodically join to form superpods where the individuals mix and socialize with members of other pods [65]. Large aggregations may provide similar opportunities for individual harbor porpoises to socialize with others in their community or population, and facilitate genetic diversity. Further research is needed to determine the role of these aggregations in harbor porpoise society.

Documentation of the occurrence and behavior patterns of large harbor porpoise aggregations was achievable via the coordinated efforts of researchers, community scientists, and whale-watching crews. Diverse participants of this transboundary project provided valuable scientific data, irrespective of the socio-political boundaries, and created an observation network that surpassed the scope of resources available to one small research organization.

Community engagement conversations and the use of mobile apps expanded data collection considerably in 2021 and 2022 (67% of aggregation reports). The creation and promotion of a quick and easy sighting app expanded opportunities for the public to submit sightings and encouraged people to observe animal behavior in an intentional and systematic way while completing the entry form. The PWWA request for whale-watching crews to record sightings of large harbor porpoise aggregations yielded 146 entries submitted by 21 boats. Sightings of harbor porpoises are not usually entered in the PWWA private app, yet crews regularly contributed data once they were aware of the research request. During 2021–2022, most of the sighting entries (56%) were submitted via PWWA. These examples of community interest, engagement, and promotion illustrate a potential road map for future collaborative projects that expand the reach of small research organizations.

5. Conclusions

It is clear from the results presented here that large harbor porpoise aggregations are more common in the Salish Sea than previously realized. It is likely that the aggregations documented here are a small portion of the ones actually occurring throughout these waters. Due to the behaviors observed, these aggregations are likely both important foraging and socialization opportunities for harbor porpoises. Aggregation events may also play a vital role in the reproduction of the species, as noted by the long-term habitat selection and occurrence of mating behavior commonly observed in southern BC [19]. Understanding when and why these aggregations are occurring can help us better understand the foraging ecology, behavioral ecology, and social structure of this enigmatic species.

Moreover, this may also assist in the identification of important habitats that are vital for the long-term survival of Salish Sea harbor porpoise population(s), and potential conflict with net fisheries and associated bycatch risks. However, there is a distinct knowledge gap about harbor porpoise habitat use and behavior in this region [52] that could hamper conservation efforts. The repatriation of a species inspires a wealth of research questions that could illuminate ecosystem factors relevant to numerous species; allocating additional resources to this research could have widespread impacts. Monitoring their populations can provide critical data on ecosystem health. The decline and recovery of harbor porpoises in the Salish Sea is not isolated, and is mirrored in the population in San Francisco Bay [66] around the same time. This highlights the importance of harbor porpoises as a sentinel species for the health of local ecosystems and the importance of understanding their behavioral and foraging ecology for their conservation and that of their ecosystem.

Community collaboration has become an essential component of environmental research like this and fostering additional community partnerships will be important to expanding these aggregation data. Transboundary collaborations of researchers, industry, and community are essential to examining, understanding, and protecting harbor porpoises and the Salish Sea as a habitat region. Environmental and human-influenced factors such as prey availability, water temperature, and pollution, defy international boundaries in our shared waters and collaboration is critical for the conservation of harbor porpoises and the numerous species of the Salish Sea.

To understand the function and frequency of these large harbor porpoise aggregations, the authors recommend further transboundary studies focused on identifying specific food sources that may be involved in attracting these aggregations, detailed behavioral analyses, bathymetric and environmental factors that may be conducive to aggregations, and determining any trends or patterns that may affect the location and timing of repeated aggregations. This information is critical to fully understanding and conserving this important indicator species in the Salish Sea.

Supplementary Materials: The following supporting information can be downloaded at: https://www.mdpi.com/article/10.3390/oceans4030019/s1, Table S1: Cascadia Research Collective South Puget Sound survey results; Table S2: Cascadia Research Collective long-term aggregation reports; Table S3: Cascadia Research Collective short-term aggregation reports; Table S4: Pacific Mammal Research (PacMam) aggregation reports; Table S5: Pacific Whale Watch Association (PWWA) aggregation reports; Table S6: Sea View survey results; Figure S1: Harbor porpoise in foreground, with multiple other groups in the background within large aggregation of 200+ individuals; Figure S2: Additional image of a mating attempt during an aggregation of 100+ individuals; Video S1: Compilation of videos from Cascadia Research Collective encounters.

Author Contributions: Conceptualization, D.A. and C.R.E.; software, J.K.; investigation, D.A., L.S. and A.H.; resources, D.A. and L.S.; data curation, D.A., L.S., C.R.E., K.M., E.J.G., J.K. and A.H.; writing—original draft preparation, D.A.; writing—review and editing, L.S., C.R.E., K.M., E.J.G. and A.H. All authors have read and agreed to the published version of the manuscript.

Funding: CRC research was funded by the authors. PacMam research was funded by Marathon Petroleum Foundation, Inc. and public donations. PWWA research was funded by PWWA member companies. Sea View monitoring data was funded by the Canadian Department of National Defence during training operations that required the Marine Mammal Monitoring Program to be implemented.

Institutional Review Board Statement: Not applicable.

Informed Consent Statement: Not applicable.

Data Availability Statement: The substantial aggregation data used in the preparation of this paper are available in the Supplemental Materials. Data on small groups are being retained by their respective owners, who should be contacted to negotiate availability of that data.

Acknowledgments: CRC vessel surveys were conducted under NOAA-NMFS permit #20605. We would like to thank all the owners of the whale watch companies for supporting this effort, and all the captains and naturalists for submitting harbor porpoise reports. Our thanks also go out to all the community scientists who have submitted reports, with special thanks to Pam and Charlie Rhubart, Michelle Roy and Kim Merriman, who have monitored marine mammal activity from their homes overlooking South Puget Sound waters for several years. Gratitude is also extended to the Canadian Department of National Defence for supporting data sharing for this study.

Conflicts of Interest: The authors declare no conflict of interest.

References

1. Scheffer, V.B.; Slipp, J.W. The Whales and Dolphins of Washington State with the Key to the Cetaceans of the West Coast of North America. *Am. Midl. Nat.* **1948**, *39*, 257–337. [CrossRef]
2. Everitt, R.D.; Fiscus, C.H.; DeLong, R.L. *Northern Puget Sound Marine Mammals*; Department of Commerce, National Oceanic and Atmospheric Administration: Seattle, WA, USA, 1980.
3. Osmek, S.; Calambokidis, J.; Laake, J.; Gearin, P.; Delong, R. *Assessment of the Status of Harbor Porpoise (Phocoena phocoena) in Oregon and Washington*; Department of Commerce, NOAA, National Marine Fisheries Service, Alaska Fisheries Center: Seattle, WA, USA, 1996.
4. Cowan, I.M. The Dall Porpoise, *Phocoenoides dalli* (True), of the Northern Pacific Ocean. *J. Mammal.* **1944**, *25*, 295–306. [CrossRef]
5. Pike, G.C.; MacAskie, I.B. *Marine Mammals of British Columbia*; Royal BC Museum: Victoria, British Columbia, 1969; Volume 171.
6. Hall, A.M. Seasonal Abundance and Prey Distribution of Harbour Porpoise (*Phocoena phocoena*) in Southern Vancouver Island Waters. Ph.D. Thesis, University of British Columbia, Victoria, British Columbia, 2004; p. 100.

7. Laake, J.; DeLong, R.L.; Calambokidis, J.; Osmek, S. Abundance and Distribution of Marine Mammals in Washington and British Columbia inside Waters, 1996. In *Marine Mammal Protection Act and Endangered Species Act Implementation Program 1996*; AFSC Processed Report 97-10; National Marine Fisheries Service: Seattle, WA, USA, 1997; pp. 67–73.
8. Dracott, K.; Robinson, C.V.; Brown-Dussault, A.; Birdsall, C.; Barrett-Lennard, L. Porpoises after Dark: Seasonal and Diel Patterns in Pacific Harbour Porpoise (*Phocoena phocoena vomerina*) Aggregations at One of North America's Fastest Growing Ports. *Front. Mar. Sci.* **2022**, *9*, 1955. [CrossRef]
9. Evenson, J.R.; Anderson, D.; Murphie, B.L.; Cyra, T.A.; Calambokidis, J. *Disappearance and Return of Harbor Porpoise to Puget Sound: 20 Year Pattern Revealed from Winter Aerial Surveys*; Technical Report; Washington Department of Fish and Wildlife, Wildlife Program and Cascadia Research Collective: Olympia, WA, USA, 2016.
10. Jefferson, T.A.; Smultea, M.A.; Courbis, S.S.; Campbell, G.S. Harbor Porpoise (*Phocoena phocoena*) Recovery in the Inland Waters of Washington: Estimates of Density and Abundance from Aerial Surveys, 2013–2015. *Can. J. Zool.* **2016**, *94*, 505–515. [CrossRef]
11. Elliser, C.R.; MacIver, K.H.; Green, M. Group Characteristics, Site Fidelity, and Photo-Identification of Harbor Porpoises, *Phocoena phocoena*, in Burrows Pass, Fidalgo Island, Washington. *Mar. Mammal Sci.* **2018**, *34*, 365–384. [CrossRef]
12. Braulik, G.; Minton, G.; Amano, M.; Bjørge, A. *Phocoena phocoena*. The IUCN Red List of Threatened Species. 2020, e.T17027A50369903. Available online: https://dx.doi.org/10.2305/IUCN.UK.2020-2.RLTS.T17027A50369903.en (accessed on 11 June 2023).
13. Committee on the Status of Endangered Wildlife in Canada (COSEWIC). Assessment and Status Report Harbour Porpoise (Pacific Ocean Population) *Phocoena phocoena vomerina* in Canada. 2016. Available online: https://wildlife-species.canada.ca/species-risk-registry/virtual_sara/files/cosewic/sr_HarbourPorpoise_2016_e.pdf (accessed on 11 June 2023).
14. Carretta, J.V.; Forney, K.A.; Oleson, E.M.; Weller, D.W.; Lang, A.R.; Baker, J.; Muto, M.M.; Hanson, B.; Orr, A.J.; Huber, H.; et al. *Pacific Marine Mammal Stock Assessments: 2016*; NOAA Southwest Fisheries Science Center: San Diego, CA, USA, 2017. [CrossRef]
15. Raum-Suryan, K.L.; Harvey, J.T. Distribution and Abundance of and Habitat Use by Harbor Porpoise, *Phocoena phocoena*, off the Northern San Juan Islands, Washington. *Fish. Bull.* **1998**, *96*, 808–822.
16. Leatherwood, S.; Reeves, R.R. Bottlenose Dolphin Tursiops Truncatus and Other Toothed Cetaceans. In *Wild Mammals of North America, Biology, Management, Economics*; Chapman, J.A., Feldhamer, G.A., Eds.; John Hopkins University Press: Baltimore, MD, USA, 1982; pp. 368–414.
17. Hoek, W. An Unusual Aggregation of Harbor Porpoises (*Phocoena phocoena*). *Mar. Mammal Sci.* **1992**, *8*, 152–155. [CrossRef]
18. Butler, R.W.; MacVicar, R.S.; Hollick-Kenyon, S. Observation of A Super Pod of Pacific Harbor Porpoises (*Phocoena phocoena vomerina*) in the Salish Sea. *Northwest. Nat.* **2017**, *98*, 137–138. [CrossRef]
19. Hall, A. Foraging Behaviour and Reproductive Season Habitat Selection of Northeast Pacific Porpoises. Ph.D. Thesis, The University of British Columbia (Vancouver), Kelowna, British Columbia, 2011.
20. Neave, D.J.; Wright, B.S. Seasonal Migrations of the Harbor Porpoise (*Phocoena phocoena*) and Other Cetacea in the Bay of Fundy. *J. Mammal.* **1968**, *49*, 259–264. [CrossRef]
21. Koschinski, S. Current Knowledge on Harbour Porpoises (*Phocoena phocoena*) in the Baltic Sea. *Ophelia* **2001**, *55*, 167–197. [CrossRef]
22. Cresswell, G.; Walker, D.; Pusser, T. Harbor Porpoise *Phocoena phocoena*. In *Whales & Dolphins of the North American Pacific: Including Seals and other Marine Mammals*; Harbour Publishing: Madeira Park, BC, USA, 2007; pp. 158–159.
23. Angell, T.; Balcomb, K.C. Porpoises. In *Marine Birds and Mammals of Puget Sound*; University of Washington Press: Seattle, WA, USA, 1982; pp. 122–123.
24. Bjorge, A.; Tolley, K.A. Harbor Porpoise: *Phocoena phocoena*. In *Encyclopedia of Marine Mammals*; Perrin, W.F., Würsig, B., Thewissen, J.G.M., Eds.; Academic Press: Burlington, MA, USA, 2009; pp. 530–533.
25. Jefferson, T.A.; Webber, M.A.; Pitman, R.L. Harbor Porpoise—*Phocoena phocoena*. In *Marine Mammals of the World: A Comprehensive Guide to Their Identification*; Elsevier: Amsterdam, The Netherlands, 2015; pp. 326–329.
26. Allen, S.G.; Mortenson, J.; Webb, S. Harbor Porpoise *Phocoena phocoena*. In *Field Guide to Marine Mammals of the Pacific Coast*; University of California Press: Berkeley, CA, USA, 2011; pp. 377–385.
27. Ford, J.K.B. Harbour Porpoise *Phocoena phocoena*. In *Marine Mammals of British Columbia*; Royal BC Museum: Victoria, BC, USA, 2014; pp. 329–337.
28. Osborne, R.; Calambokidis, J.; Dorsey, E.M. Harbor Porpoise *Phocoena phocoena*. In *A Guide to Maine Mammals of Greater Puget Sound*; Island Publishers: Anacortes, WA, USA, 1988; pp. 25–28.
29. Reeves, R.R.; Stewart, B.S.; Clapham, P.J.; Powell, J.A. Harbor Porpoise. In *Guide to Marine Mammals of the World*; Alfred A. Knopf: New York, NY, USA, 2002; pp. 460–463.
30. Barlow, J. Harbor Porpoise, *Phocoena phocoena*, Abundance Estimation for California, Oregon, and Washington: I. Ship Surveys. *Fish. Bull.* **1988**, *86*, 417–432.
31. Barlow, J.; Oliver, C.W.; Jackson, T.D.; Taylor, B.L. Harbor Porpoise, *Phocoena phocoena*, Abundance Estimation for California, Oregon, And Washington: II. Aerial Surveys. *Fish. Bull.* **1988**, *86*, 433–444.
32. Calambokidis, J.; Speich, S.M.; Peard, J.; Steiger, G.H.; Cubbage, J.C.; Fry, D.M.; Lowenstine, L.J. *Biology of Puget Sound Marine Mammals and Marine Birds: Population Health and Evidence of Pollution Effects*; US Department of Commerce, National Oceanic and Atmospheric Administration, National Ocean Service: Rockville, MD, USA, 1985.

33. Raum-Suryan, K.L. *Distribution, Abundance, Habitat Use, and Respriation Patterns of Harbor Porpoise (Phocoena Phocoena) of the Northern SanJuan Islands*; San Jose State University: Washington, DC, USA, 1995.
34. Calambokidis, J.; Evenson, J.R.; Cubbage, J.C.; Gearin, P.J.; Osmek, S.D. *Harbor Porpoise Distribution and Abundance off Oregon and Washington from Aerial Surveys in 1991*; Cascadia Research: Seattle, WA, USA, 1992.
35. Chandler, T.; Calambokidis, J. *2002 Aerial Surveys for Harbor Porpoise and Other Marine Mammals off Oregon, Washington and British Columbia*; Contract Report to National Marine Mammal Laboratory: Seattle WA, USA, 2003.
36. Chandler, T.; Calambokidis, J. *2003 Aerial Surveys for Harbor Porpoise and Other Marine Mammals off Oregon, Washington, and British Columbia*; Contract Report to National Marine Mammal Laboratory: Seattle, WA, USA, 2003; p. 25.
37. Chivers, S.J.; Hanson, B.; Laake, J.; Gearin, P.; Muto, M.M.; Calambokidis, J.; Duffield, D.; McGuire, T.; Hodder, J.; Greig, D.; et al. *Additional Genetic Evidence for Population Structure of Phocoena phocoena off the Coasts of California, Oregon, and Washington*; Southwest Fisheries Science Center: La Jolla, CA, USA, 2007.
38. Hanson, M.B. *Seasonal Movements and Habitat Use of Dall's and Harbor Porpoises in the Inland and Coastal Waters of Washington State as Determined by Radiotelemetry*; NMFS-F/SPO-82; US Department of Commerce, National Oceanic and Atmospheric Administration Technical Memorandum: Seattle, WA, USA, 2007; pp. 53–54.
39. Nichol, L.M.; Hall, A.M.; Ellis, G.M.; Stredulinsky, E.; Boogaards, M.; Ford, J.K.B. Dietary Overlap and Niche Partitioning of Sympatric Harbour Porpoises and Dall's Porpoises in the Salish Sea. *Prog. Oceanogr.* **2013**, *115*, 202–210. [CrossRef]
40. Smith, G.J.D.; Gaskin, D.E. The Diet of Harbor Porpoises (*Phocoena phocoena* (L.)) in Coastal Waters of Eastern Canada, with Special Reference to the Bay of Fundy. *Can. J. Zool.* **1974**, *52*, 777–782. [CrossRef]
41. Elliser, C.R.; Calambokidis, J.; D'Alessandro, D.N.; Duffield, D.A.; Huggins, J.L.; Rice, J.; Szczepaniak, I.; Webber, M. Prey-Related Asphyxiation in Harbor Porpoises (*Phocoena phocoena*) along the U.S. West Coast: Importance of American Shad (*Alosa sapidissima*) on Adult Female Harbor Porpoise Mortality. *Oceans* **2020**, *1*, 94–108. [CrossRef]
42. Penttila, D. *Marine Forage Fishes in Puget Sound*; Seattle District, US Army Corps of Engineers: Seattle, WA, USA, 2007; 23p.
43. Quinn, T.; Krueger, K.; Pierce, K.; Penttila, D.; Perry, K.; Hicks, T.; Lowry, D. Patterns of Surf Smelt, *Hypomesus pretiosus*, Intertidal Spawning Habitat Use in Puget Sound, Washington State. *Estuaries Coasts* **2012**, *35*, 1214–1228. [CrossRef]
44. Quinn, T.P.; Losee, J.P. Diverse and Changing Use of the Salish Sea by Pacific Salmon, Trout, and Char. *Can. J. Fish. Aquat. Sci.* **2022**, *79*, 1003–1021. [CrossRef]
45. D'alessandro, D.N.; Duffield, D.A. Salmonid Passive Integrated Transponder Tags and Coded Wire Tags Found in the Forestomach of a Harbor Porpoise (*Phocoena phocoena*) in Southwestern Washington. *Fish. Bull.* **2019**, *117*, 303–307. [CrossRef]
46. Burger, M.; Sandell, T.; Fanshier, C.; Lindquist, A.; Biondo, P.; Lowry, D. *Findings of the 2016-17 Southern Salish Sea Acoustic Mid-Water Trawl Survey*; Washington Department of Fish and Wildlife: Olympia, WA, USA, 2020.
47. Morgan, C.A.; Beckman, B.R.; Weitkamp, L.A.; Fresh, K.L. Recent Ecosystem Disturbance in the Northern California Current. *Fisheries* **2019**, *44*, 465–474. [CrossRef]
48. Holser, R.R.; Keates, T.R.; Costa, D.P.; Edwards, C.A. Extent and Magnitude of Subsurface Anomalies During the Northeast Pacific Blob as Measured by Animal-Borne Sensors. *J. Geophys. Res. Ocean.* **2022**, *127*, e2021JC018356. [CrossRef]
49. Duguid, W.D.P.; Boldt, J.L.; Chalifour, L.; Greene, C.M.; Galbraith, M.; Hay, D.; Lowry, D.; McKinnell, S.; Neville, C.M.; Qualley, J.; et al. Historical Fluctuations and Recent Observations of Northern Anchovy *Engraulis mordax* in the Salish Sea. *Deep. Res. Part II Top. Stud. Oceanogr.* **2019**, *159*, 22–41. [CrossRef]
50. Maupin, S. Market Squid, Loligo Opalescens, in the Northeastern Pacific. In *Species Synopses, Life Histories of Selected Fish and Shellfish of the Northeast Pacific and Bering Sea*; University of Washington: Seattle, WA, USA, 1988; pp. 83–93.
51. Keener, W.; Webber, M.A.; Szczepaniak, I.D.; Markowitz, T.M.; Orbach, D.N. The Sex Life of Harbor Porpoises (*Phocoena phocoena*): Lateralized and Aerial Behavior. *Aquat. Mamm.* **2018**, *44*, 620–632. [CrossRef]
52. Elliser, C.R.; Hall, A. Return of the Salish Sea Harbor Porpoise, *Phocoena phocoena*: Knowledge Gaps, Current Research, and What We Need to Do to Protect Their Future. *Front. Mar. Sci.* **2021**, *8*, 534. [CrossRef]
53. Ramos, E.A.; Santoya, L.; Verde, J.; Walker, Z.; Castelblanco-Martínez, N.; Kiszka, J.J.; Rieucau, G. Lords of the Rings: Mud Ring Feeding by Bottlenose Dolphins in a Caribbean Estuary Revealed from Sea, Air, and Space. *Mar. Mammal Sci.* **2022**, *38*, 364–373. [CrossRef]
54. Wiley, D.; Ware, C.; Bocconcelli, A.; Cholewiak, D.; Friedlaender, A.; Thompson, M.; Weinrich, M. Underwater Components of Humpback Whale Bubble-Net Feeding Behaviour. *Behaviour* **2011**, *148*, 575–602. [CrossRef]
55. Vaughn, R.; Würsig, B.; Packard, J. Dolphin Prey Herding: Prey Ball Mobility Relative to Dolphin Group and Prey Ball Sizes, Multispecies Associates, and Feeding Duration. *Mar. Mammal Sci.* **2010**, *26*, 213–225. [CrossRef]
56. Gallagher, C.A.; Stern, S.J.; Hines, E. The Metabolic Cost of Swimming and Reproduction in Harbor Porpoises (*Phocoena phocoena*) as Predicted by a Bioenergetic Model. *Mar. Mammal Sci.* **2018**, *34*, 875–900. [CrossRef]
57. Wisniewska, D.M.M.; Johnson, M.; Teilmann, J.; Rojano-Doñate, L.; Shearer, J.; Sveegaard, S.; Miller, L.A.A.; Siebert, U.; Madsen, P.T.T. Ultra-High Foraging Rates of Harbor Porpoises Make Them Vulnerable to Anthropogenic Disturbance. *Curr. Biol.* **2016**, *26*, 1441–1446. [CrossRef]
58. Rojano-Donãte, L.; McDonald, B.I.; Wisniewska, D.M.; Johnson, M.; Teilmann, J.; Wahlberg, M.; Højer-Kristensen, J.; Madsen, P.T. High Field Metabolic Rates of Wild Harbour Porpoises. *J. Exp. Biol.* **2018**, *221*, jeb185827. [CrossRef] [PubMed]

59. Webber, M.A.; Keener, W.; Wahlberg, M.; Elliser, C.R.; MacIver, K.H.; Torres Ortiz, S.; Jakobsen, F.; Hamel, H.; Rieger, A.; Siebert, U.; et al. Sexual Behavior and Anatomy in Harbor Porpoises. In *Sex in Cetaceans: Morphology, Behavior and the Evolution of Sexual Strategies*; Würsig, B., Orbach, D.N., Eds.; Springer Nature: Cham, Switzerland, 2023.
60. Fink, B.D. Observation of Porpoise Predation on a School of Pacific Sardines. *Calif. Fish Game* **1959**, *45*, 216–217.
61. Torres Ortiz, S.; Stedt, J.; Midtiby, H.S.; Egemose, H.D.; Wahlberg, M. Group Hunting in Harbour Porpoises (*Phocoena phocoena*). *Can. J. Zool.* **2021**, *99*, 511–520. [CrossRef]
62. Cosentino, M.; Nairn, D.; Coscarella, M.; Jackson, J.C.; Windmill, J.F.C. I Beg Your Pardon? Acoustic Behaviour of a Wild Solitary Common Dolphin Who Interacts with Harbour Porpoises. *Bioacoustics* **2022**, *31*, 517–534. [CrossRef]
63. Sørensen, P.M.; Wisniewska, D.M.; Jensen, F.H.; Johnson, M.; Teilmann, J.; Madsen, P.T. Click Communication in Wild Harbour Porpoises (*Phocoena phocoena*). *Sci. Rep.* **2018**, *8*, 9702. [CrossRef] [PubMed]
64. Clausen, K.T.; Wahlberg, M.; Beedholm, K.; DeRuiter, S.; Madsen, P.T. Click Communication in Harbour Porpoises *Phocoena phocoena*. *Bioacoustics* **2010**, *20*, 1–28. [CrossRef]
65. Ford, J.K.B.; Ellis, G.M.; Balcomb, K.C. *Killer Whales: The Natural History and Genealogy of Orchinus orca in British Columbia and Washington*, 2nd ed.; UBC Press: Vancouver, BC, USA, 2000.
66. Stern, S.J.; Keener, W.; Szczepaniak, I.D.; Webber, M.A. Return of Harbor Porpoises (*Phocoena phocoena*) to San Francisco Bay. *Aquat. Mamm.* **2017**, *43*, 691–702. [CrossRef]

oceans MDPI

Article

Comparative Microbial Community Analysis of Fur Seals and Aquaculture Salmon Gut Microbiomes in Tasmania

Erin D'Agnese [1,2,*,†], Ryan J. McLaughlin [3], Mary-Anne Lea [4], Esteban Soto [1], Woutrina A. Smith [1] and John P. Bowman [2]

1. School of Veterinary Medicine, University of California, 1 Shields Avenue, Davis, CA 95616, USA; sotomartinez@ucdavis.edu (E.S.)
2. Tasmania Institute of Agriculture, University of Tasmania, Sandy Bay, TAS 7001, Australia; john.bowman@utas.edu.au
3. Practical Informatics, Pittsburgh, PA 15213, USA; mclaughlin@practical-informatics.com
4. Institute for Marine and Antarctic Studies, University of Tasmania, Battery Point, TAS 7005, Australia; maryanne.lea@utas.edu.au

* Correspondence: erin.dagnese@wildecohealth.com

† Current address: Wild EcoHealth, Tacoma, Washington, DC 98465, USA.

Abstract: In Tasmania, Australian fur seals (*Arctocephalus pusillus doriferus*) regularly interact with Atlantic salmon (*Salmo salmar* L.) aquaculture lease operations and opportunistically consume fish. The microbial communities of seals and aquaculture salmon were analyzed for potential indicators of microbial sharing and to determine the potential effects of interactions on wild seal microbiome composition. The high-throughput sequencing of the V1–V3 region of the 16S rRNA genes from the gut microbial communities of 221 fur seals was performed: 41 males caught at farms, 50 adult scats from haul-outs near farms, 24 necropsied seals, and controls from Bass Strait breeding colonies, encompassing 56 adult scats and 50 pup swabs. QIIME2 and R Studio were used for analysis. Foraging at or near salmon farms significantly shifted seal microbiome biodiversity. Taxonomic analysis showed a greater divergence in *Bacteroidota* representatives in male seals captured at farms compared to all other groups. Pathogens were identified that could be monitoring targets. Potential indicator amplicon sequence variants were found across a variety of taxa and could be used as minimally invasive indicators for interactions at this interface. The diversity and taxonomic shifts in the microbial communities of seals indicate a need to further study this interface for broader ecological implications.

Keywords: Australian fur seal; microbiome; aquaculture; Tasmania; One Health; microbial source tracking; Atlantic salmon

1. Introduction

In marine and coastal systems, humans have historically been directly connected to the wild through the fisheries of wild fish stocks and direct interaction with wild animals via recreational activities, fisheries' competition for fish stocks, and aquaculture [1–3]. However, the more indirect connections are often more insidious and easily overlooked. The interconnectedness of the wild system to humans, both directly and indirectly, has become more evident as aquaculture management in nearshore systems is improved upon.

In Tasmania, an island state of Australia, the farmed Atlantic salmon (*Salmo salar* L.) industry produced 51,298 tonnes of salmon in 2016–2017, and production has continued to increase annually. The industry in Tasmania alone produces around AUD $739 million of profit and has become the single most important aquaculture commodity in Australia [4,5]. Meanwhile, in the same region, the Australian fur seal (*Arctocephalus pusillus doriferus*) populations have rebounded since the sealing era [6]. It is the most abundant breeding fur seal in the region, and males regularly attempt to forage on the farmed salmon in the estuarine net pens and also forage on wild native fish species in the region, such as jack mackerel (*Trachurus declivis*) and rebait (*Emmelichthys nitidus*) [7,8]. The population remains reduced post-sealing, and their continued recovery remains tenuous as they face modern human influences on their ecosystem.

One of these is that the interactions between male fur seals and aquaculture leads directly to increased risk for the safety of fish, human, and marine mammals in many similar systems around the world [3,9,10]. Often, the Australian fur seals in Tasmania predate on farmed salmon through the nets of the net pens holding salmon in the estuarine habitat. This can lead to a reduction in salmon stock, risk to farm workers, and occasionally, these interactions can lead to the injury, entanglement, and mortality of the seals [11,12]. Direct impacts (entanglement, seal–human interactions, etc.) are currently being mitigated through regulation and by the industry improving pen infrastructure; however the extent of indirect effects, such as bio-contamination risks, are largely unknown. At this interface, little is known about the indirect effects, such as the microbial connections between aquaculture and wildlife, as no research has been conducted on fur seals' microbial communities in the regions around farms.

Microbial source tracking (MST) systems have been employed to follow sources of pollution and pathogens into the aquatic and marine ecosystems [13,14]. Historically, MST systems have relied on fecal coliforms and other indicator bacteria, such as *E. coli*, to follow terrestrial contamination through aquatic and marine systems, as well as determine water quality/safety [13,15]. However, these techniques are labor-intensive and have many limitations in their ability to discern point sources or the interplay between multiple sources, so the use of genetic markers and genetic libraries have been adopted as being more precise in measuring point sources as molecular methodologies continue to advance [16–19]. With advances in technology, the ability to track multiple taxa in microbial communities through systems has become possible [20–23]. Metacommunity analysis utilizes the knowledge of the diversity of species and changes in community structures to elucidate the extent of interactions within and between species assemblages, including microbial communities found in nature [16,24–27]. Comparing full microbial communities using microbiome analysis between geographic regions and trophic levels as a way of analyzing bacterial movements is an up-and-coming field of research [23,28,29]. Such work at the interface between wild fur seals, aquaculture, and human development would be very useful in laying the groundwork for biocontamination monitoring and management.

Microbiota are communities of microbial organisms living in and interacting with a distinct environment, for example, the gut or the skin of another organism [30,31]. Microbes, such as bacteria and fungi, thrive in a wide variety of environments, depending on their specific growth requirements and characteristics. Many are shared through trophic systems, from prey to predator, and up the food chain [32,33]. Some microbes are geographically more or less abundant, and their presence in a host can indicate whether that animal has spent time in that region [34,35]. Studying microbiota differences is minimally invasive and has the capability of indicating host lifestyle, what environments the host encounters, and what pathogen exposure is occurring, all in one detailed analytical study when variables such as spatial and temporal variability are well characterized [34,36].

It is known that a shift in diet and the consistent use of new environments can alter the microbial community in the gut as well [37]. Differences in nutrition, age, and regionality are also known to affect microbiome structure and diversity in many species, including pinnipeds [38–40]. It is understood that as an animal matures through life stages, their microbiome shifts with increasing microbial input from the environment and their changing diet [41]. However, in adult animals, the microbiome is considered remarkably stable [42]. Therefore, differences seen in adult pinnipeds' microbiota in the same population are expected to be minimal and, if existent, to be related to novel prey or habitat use.

As there are no native salmonids in Tasmania, salmon are not a natural prey item for wildlife here, and their potentially novel microbial communities are now entering the ecosystem and native predators' gastrointestinal tracts [43,44]. The microbes in farmed salmon may be transferred to the ecosystem, wild animals, and people in the region through biocontamination, predation, or direct interaction with salmon farms and farmed salmon. The aquaculture of Atlantic salmon is known to introduce pathogens into natural ecosystems, so a monitoring scheme would be very useful for the early detection of pathogen spread [43,44]. As many male fur seals frequent the farms during the non-breeding seasons and haul out on nearby rocky islands to rest with other seals that are foraging in the region, it is likely that shifts in microbiota would be most pronounced in the males that frequent the farms and would be most like those found in other adults that forage and haul out in the region.

This study aimed to elucidate any potential microbial community differences in seal gut microbial communities associated with foraging on or near non-native farmed salmon. This study provides a baseline understanding of the microbial community of Australian fur seal guts across Tasmania, and what potential life history and external variables lead to changes in microbiomes in this recovering pinniped population. The microbiomes of Australian fur seals found in Tasmania were compared across age class, location of sampling, and proximity to salmon farms to determine diversity differences and detect the amplicon sequence variants (ASV)s, which drive the community diversity changes found. Differentially abundant ASVs in the seal microbiomes were identified using compositional analysis and then compared to a representative sampling of the farmed salmon gut microbiome ASVs to determine whether seals in regions around farms shared ASVs with salmon gut microbiomes. The identified ASVs could prove to be useful indicator species for interactions between seals and salmon farms. The data from this project will aid in future pinniped microbiome studies and metacommunity analyses.

2. Materials and Methods

2.1. Sample Collection

Male fur seals ($n = 41$) were trapped at three salmon farm leases in the southern region of the state by the farm managers (who were permitted to do so) following the protocol set by the state government animal ethics committee and were then handled between May and November in both 2017 and 2018. Only males were trapped, as they are the only demographic that forage at the farms. Seals were trapped both inside and outside of salmon pens using baited pens, which close after a seal has grabbed the bait at the back of the trap, and were held overnight or up to 36 h in pens on land until sampling occurred under permits from the Department of Primary Industries, Parks, Water and Environment (2017/18: FA 17158 and 2018/19: FA 18114), and all handling of animals occurred under University of Tasmania Animal ethics (UTAS AEC project number: A0016754). At the time of collection, seals were sedated with Hypnovel (5 mg/mL Midazolam) (Roche products Pty Limited, Sydney, NSW, Australia) at a dose rate of 0.35–0.6 mg/kg, depending on the size of animal and its state of excitement. Rectal swabs were taken from each male, two were placed in RNAlater-like preservation buffer (40 mL of 0.5 M disodium EDTA (pH 8.0), 25 mL 1 M sodium citrate, 700 g ammonium sulfate, and 935 mL of sterile distilled water filtered through a 0.2 micron filter). The swabs/feces in the preservation buffer were frozen at −20 °C for later extraction.

For control samples, we sampled feces from adult animals that were present on haul-outs both in the region around farms and at rookery islands in the Bass Strait and took rectal swabs from pups handled at those same rookery islands (Figure 1). These controls were appropriate for our study considering the limitations related to fur seal life history/behavior, permissions, and resources available. Pups were swabbed at Tenth Island in January 2018 ($n = 50$). The sample handling and storage were the same as the above for males sampled at the farms. Feces from adults on Tenth Island and Judgement Island in January 2018 were also collected ($n = 56$) in 0.9% saline and preservation buffer as per the swab methods, fecal samples were collected by swabbing the outside of fresh feces to obtain approximately 500–800 mg of feces. In September and October (2018), whole fecal samples were collected and frozen by collaborators, and then 50 random samples of fresh scats from islands in the southeast of Tasmania were subsampled in preservation buffer to be included in the dataset. Necropsies were performed on freshly dead and moderately decomposed carcasses of fur seals from around Tasmania during this study, and fecal samples were collected in preservation buffer to be included in this study ($n = 24$), these were then assigned to the appropriate seal group based on the location of stranding.

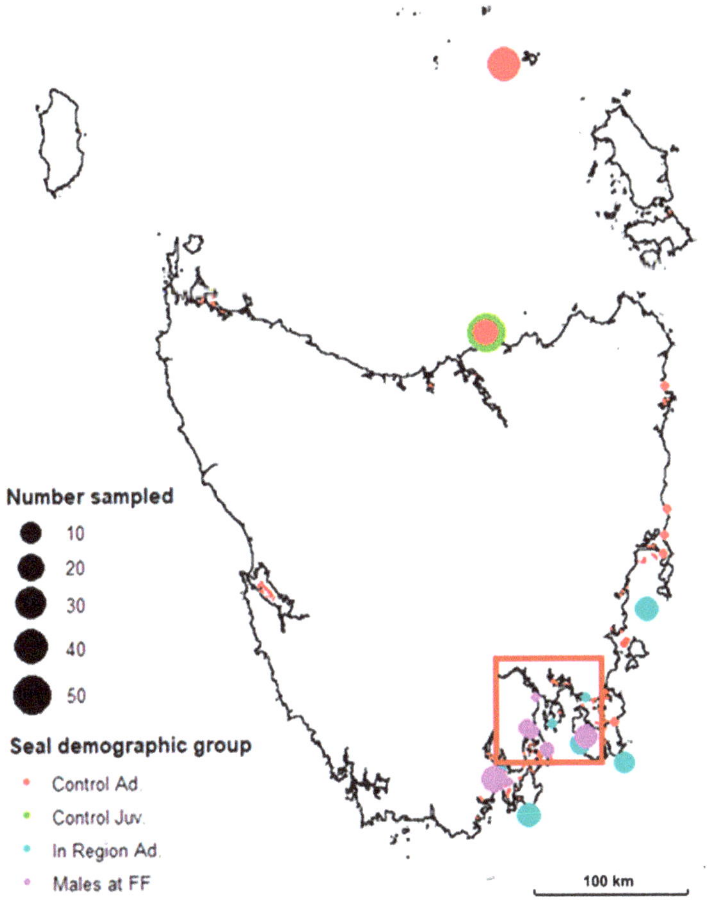

Figure 1. Map of sampling sites of fur seals around Tasmania, with colors representing seal groups sampled: control adults (Control Ad., $n = 57$), control juveniles (Control Juv., $n = 51$), adults samples from the region near farms (In Region Ad., $n = 52$), and male seals caught and samples at fish farms (Males at FF, $n = 56$). The size of the circle represents the number of animals sampled in that location. Red shapes show aquaculture leases of all types, including Atlantic salmon, other fin fish, and bivalves. The red square denotes the region studied, which includes the major urban region of Hobart, Tasmania.

2.2. DNA Extraction and Sequencing

The extraction of DNA from swab and fecal samples (180–200 mg feces/sample) was performed as per manufacturer protocol using the Qiagen DNeasy mini stool kit (Qiagen, Hilden, Germany) or the updated version of the same kit, Qiagen Fast stool mini kit (Qiagen, Hilden, Germany), using sterile techniques and was not carried out on days when salmon samples were processed in order to avoid contamination. A negative control was extracted and confirmed as negative via the PCR of the V1–V3 region of the 16S rRNA gene using 27F/519R universal primers. After extraction, DNA was stored at −20 °C until preparation for sequencing. Samples were tested for nucleic acid content and quality using the Nanodrop 8000 spectrophotometer (Thermo Scientific, Waltham, MA, USA) before sequencing. The sequencing was carried out at the Ramaciotti Centre for Genomics, Sydney, NSW, Australia, on the amplified product along with a sequencing control from implementation of the 27F/519R universal primers for the V1–V3 region of the 16S rRNA gene in order to compare the previously sequenced farmed salmon microbiomes directly, which also utilized the V1–V3 region gene target, and to increase resolution at the genus and species level in gut microbiome analysis [45,46]. The representative farmed Atlantic salmon gut microbiome sequences from the different farm leases in southern Tasmania, which are included in the analysis here, can be found in National Center for Biotechnology Information (NCBI) Sequence Read Archive under accession number SRP133157; these have been analyzed and have revealed no difference to salmon gut microbiomes from the same farms the seals were captured from and were determined to be representative of farmed salmon in Tasmania ([47] unpublished data are available through GenBank). Quality control, library preparation, and normalization was carried out before sequencing amplicons was carried out with a MiSeq v3 kit (Illumina, San Diego, CA, USA) for paired-end 300bp sequences on the Illumina MiSeq platform.

Groups for the seal and salmon combined analysis are as follows: salmon; males sampled at fish farms (FF); seals sampled in the region around salmon farms (InRegAd), which includes necropsied animals and feces collected from haul-outs in the same region of fish farms (<50 km away); control adults (feces and necropsied animals sampled from >50 km away from fish farms); and control juveniles (pups sampled at Tenth Island in the Bass Strait). The regions around FF and the sampling sites are shown on the map of Tasmania (Figure 1).

2.3. *PCR and Sanger Sequencing of Enteric Isolates for Comparison to ASV Classification Post-Processing*

The swabs and feces from the seals that were caught were lightly vortexed in 0.9% saline to suspend any bacteria on the swab in the solution, then the solution was serially diluted in sterile 0.9% saline to 10^{-1} and 10^{-2}. These dilutions were then plated on brain heart infusion (BHI) agar, marine agar, thiosulfate–citrate–bile salts–sucrose agar (TCBS), xylose lysine deoxycholate (XLD), and eosin methylene blue agar (EMB) and grown for 48 h at 37 °C. After 48 h, morphologically distinct colonies from each agar type were lifted for isolation on BHI agar or marine agar dependent on the initial growth media. Isolate DNA was extracted for PCR and identification via Sanger sequencing by aseptically transferring a pure colony into 400 µL of sterile double-distilled water and heating at 100 °C for 10 min, then centrifuged for 1 min at 12,000 rpm to pellet cell particles. The supernatant was used as the genomic DNA in PCR reactions. The bacterial 16S rRNA V1–V3 region amplicon was amplified using conventional PCR methodologies with MyTaq red mix, 27F and 519R universal primers at 10 µM, genomic DNA, and sterile double distilled water (10 µL, 1 µL, 1 µL, and 11 µL, respectively, with water added to fill reactions to 25 µL) reactions were run on a Biorad thermocycler with the following parameters: 95 °C for 1 min, then 30 cycles of 95 °C for 1 min, 55 °C for 1 min, and 72 °C for 1 min, then final extension at 72 °C for 1 min and held at 10 °C until taken off the thermocycler. Sanger sequencing was used to sequence the 16S rRNA gene PCR product at Macrogen Inc., Seoul, Republic of Korea.

2.4. *Microbiome Analysis*

All analysis of bacterial 16S rRNA gene amplicon sequence data was carried out using QIIME 2 ver. 2021.8 (https://qiime2.org/ accessed 15/06/2023) and the R packages 'tidyverse', 'qiime2R', 'vegan', 'scales', 'phyloseq', 'microbiome', 'ancom', and 'indicspecies' in RStudio (https://posit.co/products/open-source/rstudio/ accessed 15/06/2023), employing R v.3.6.1 [48,49]. Seal samples and salmon samples were processed through Qiime2 separately due to their being processed and sequenced separately. For both, the bioinformatic pipeline is as follows. Forward and reverse reads had primers and adapters removed using 'cutadapt' then were denoised, edited for quality, and joined using DADA2 (ver 1.14) before classification using 'sklearn' and the SILVA 138 16S reference database at the default confidence threshold of 70% [50]. To avoid a loss of data, no rarefaction was used—only samples with more than 1000 sequences were utilized. Alpha and beta diversity were calculated for within-sample and between-sample diversity. Seal samples were analyzed for richness and Shannon diversity metrics, and differences between factors were tested for significance using pair-wise Wilcoxon rank-sum tests. Specifically, any potential diversity difference due to sample type were tested for in order to identify any bias due to swab or feces. The beta diversity metrics calculated were Bray–Curtis distances. To visualize and cluster samples, a principal component analysis (PCoA) of Bray–Curtis distances was used and to measure between group differences, a PERMANOVA of Bray–Curtis distances was used. Due to the expected high level of diversity differences between salmon and seal microbiomes and the lack of information this would add to testing our hypothesis, we did not test the diversity differences between seals and salmon. Alpha and beta diversity metrics were calculated in R using 'vegan' and 'phyloseq' packages and visualized using 'ggplot2' and 'ggpubr' [48,51,52].

2.5. Compositional Taxonomic Diversity Analysis

The analysis of taxonomic diversity was carried out in R v.3.6.1 using the packages 'qiime2R', 'phyloseq', 'ancom', 'indicspecies', 'vegan', 'ape', 'exactRanktests', 'nlme', 'compositons', 'readr', and dependent packages. Compositional differences in microbiomes between seal groups were determined using an analysis of the composition of microbiomes (ANCOM) using ANCOM v.2 in R. This is the preferred way to handle microbiome data due to the compositional nature of the data [53]. A random intercept model adjusted for other covariates was employed for the ANCOM analysis to account for random variation within samples and for the effects of covariates in seal samples (age class, region, animal status/sample type) [54]. The ASVs identified as significantly different between seal groups were then compared to the ASVs in salmon microbiomes to find any ASVs found in the salmon microbiome that were also found as differentially abundant across seal groups.

2.6. Isolate Sequence Database Construction and Utilization

To account for the novelty witnessed through the preliminary analysis of the taxonomic classifications of the ASVs in seal gut microbiomes, a local and species specific database was constructed using the isolates cultured from a subset of the same seals used in this comparative microbiome study. The V1–V3 region sequences from the isolates cultured and sequenced via Sanger sequencing, which has low error rates and high-quality reads, were annotated via NCBI nucleotide BLAST (blastn) and the top ten best hits for the sequences were identified. These higher quality sequences and their taxonomic assignments were used to classify the ASVs that were differentially abundant between seal groups via BLAST in Python to determine whether their assignment from the global reference database was accurate for assignment compared to a reference database from cultured isolates from seal microbiomes. All further analysis was carried out in R using the following packages: 'taxonomizr', 'tibble', and 'tidyverse'. The BLAST results of the isolate sequences were reduced to one taxonomic assignment by selecting the highest percent identity, and when multiple accession numbers had the same percent identity, the least common ancestor was used. These classifications of ASVs were compared to the classification of the same ASVs via the Silva 138 reference database and sklearn classification. For each ASV, the higher percent identity taxonomic classification from the two reference databases was assigned to that ASV for the analysis of taxonomy presented for the differentially abundant ASVs.

3. Results

3.1. Seal Sample Alpha and Beta Diversity

For all seal samples (n = 216), after quality filtration, the mean of reads per sample was 11,521 reads (min: 1321; max: 65,245). The richness and the Shannon diversity indices were tested for differences between groups by testing with Wilcoxon rank-sum pairwise comparisons. There was no significant difference in richness or Shannon diversity between swabs and feces (p-values = 0.13 and 0.64, respectively), nor were there significant differences found between the scats collected at haul-outs, the live-captured seals, or the necropsy seals for either richness or Shannon diversity indices. There was a significant difference in richness between adults sampled and the juveniles sampled, as expected (p-value < 0.001), but not in the Shannon diversity index (p-value = 0.14) (Supplementary Material, Figures S1–S3). Some diversity differences were seen for richness and Shannon diversity when comparing all seal groups (control juveniles, control adults, in region adults, and males captured at farms). The Shannon diversities seen in each group are shown in Figure 2a, and the ASV richness is shown in Figure 2b. Significant differences existed between males from farms and control juveniles (p-value = 0.028) and scats collected from haul-outs in the region around farms and control adults (p-value = 0.0024) and control juveniles (p-value = 0.00027), but the in-region scats were not significantly different to swabs from males captured at the farms (p-value = 0.239) (Figure 2). As these groups were defined as distinct through the study design and the potential confounding factors showed minimal impact on the alpha diversity, we proceeded while using the groups assigned through the taxonomic diversity analysis.

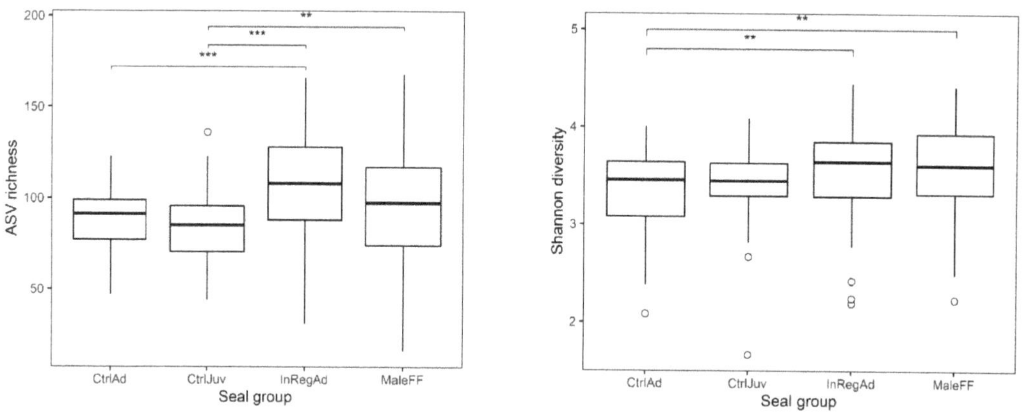

(**a**) (**b**)

Figure 2. (**a**) The Shannon diversity indices and (**b**) the ASV richness found in each group of seals sampled: males at fish farms (MaleFF, n = 56), adults/scats sampled in the region directly around salmon (InRegAd, n = 52), juveniles sampled outside of the regions around farms (CtrlJuv, n = 51), adults/scats sampled outside the regions around farms (CtrlAd, n = 57). Significant differences are represented as *** p-value < 0.001 and ** p-value < 0.05.

The PERMANOVA of Bray–Curtis distances revealed significant community differences between multiple groups of seals. Significant differences between seal groups sampled were found (pseudo-F = 8.663, p = 0.001), while seals sampled in different regions of Tasmania also had significantly different community diversity (pseudo-F = 6.391, p = 0.001). In addition, juveniles and adult seals had significantly different community diversity (pseudo-F = 8.125, p = 0.001). To account for potentially confounding effects, we tested a PERMANOVA model using backwards selection, and the final model was selected, in which the significant variables included seal group, animal status (live captured and swabbed, haul-out scat, and necropsied animal samples), month, and region; this demonstrated significant differences between groups and, to a slightly lesser degree, the other variables (seal groups: pseudo-F = 9.008, $p = 0.001$; animal status: pseudo-F = 2.243, $p = 0.001$; month: pseudo-F = 1.674, p-value = 0.001; and region: pseudo-F = 1.414, p-value = 0.028). The Bray–Curtis dissimilarity distances show community differences between groups at the ASV level rather than the taxonomic level, with many males at fish farms forming an apparent cluster on PC1, as well as overlap seen in all group of seals with juveniles specifically showing similarity to all groups sampled (Figure 3).

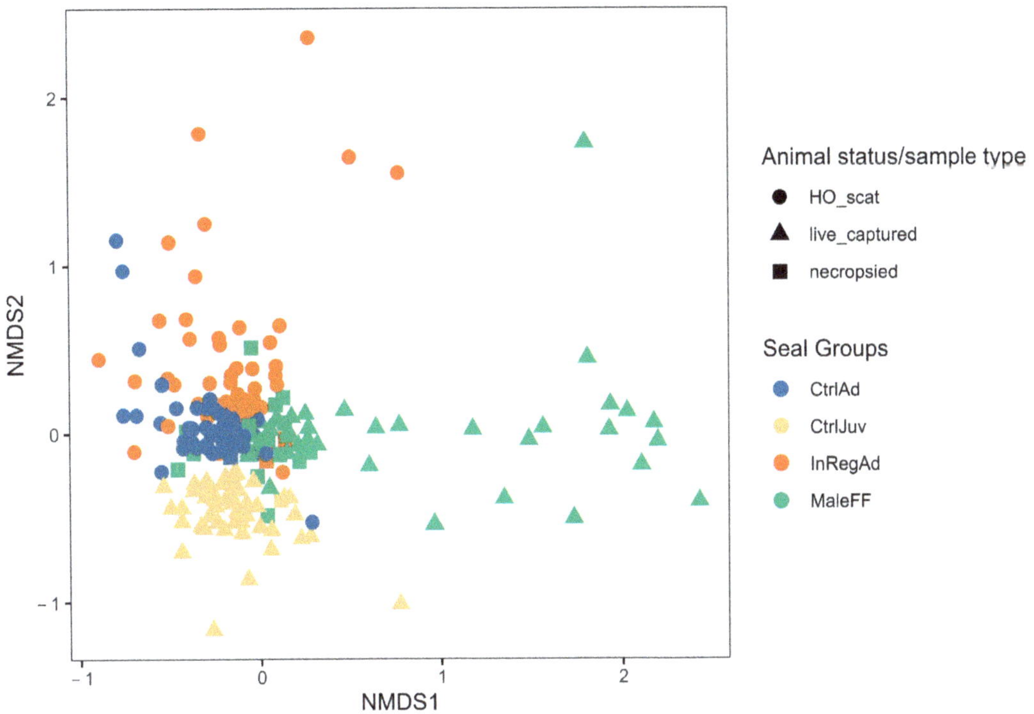

Figure 3. An NMDS plot of the Bray–Curtis distances between samples. Color groups of fur seals sampled in Tasmania: males at fish farms (MaleFF), adults/scats sampled in the region directly around salmon (InRegAd), juveniles sampled outside of the regions around farms (CtrlJuv), and adults/scats sampled outside the regions around farms (CtrlAd). Shapes represent the animal status upon sampling which accounts for different sample types.

3.2. *Taxonomic Diversity Differences within the Seals in Tasmania*

Taxonomic differences between groups of seals sampled were analyzed for common and significantly different ASVs found within and between seals sampled. As multiple subsets of adults were sampled, the relative abundance of bacterial phyla represented in the four groups of seals are shown in Figure 4. Overall, the males sampled at the farms and the animals sampled in the regions around farms had more diversity within the phylum *Bacteroidota* than the other groups of seals. There were other phyla that appeared divergent, but this phylum was the most apparently different.

The proportions of families represented in *Bacteroidota* showed that the males captured on the farms were had a larger variation in the families represented (Figure 5).

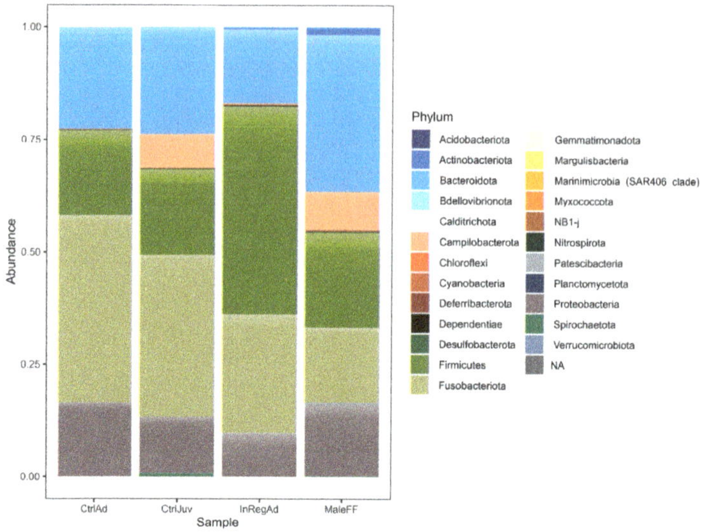

Figure 4. The relative abundance of bacterial phyla represented in the microbiomes of all four groups of fur seals sampled in Tasmania: males at fish farms (MaleFF), adults/scats sampled in the region directly around salmon (InRegAd), juveniles sampled outside of the regions around farms (CtrlJuv), adults/scats sampled outside the regions around farms (CtrlAd), NA refers to other/unidentified phylum.

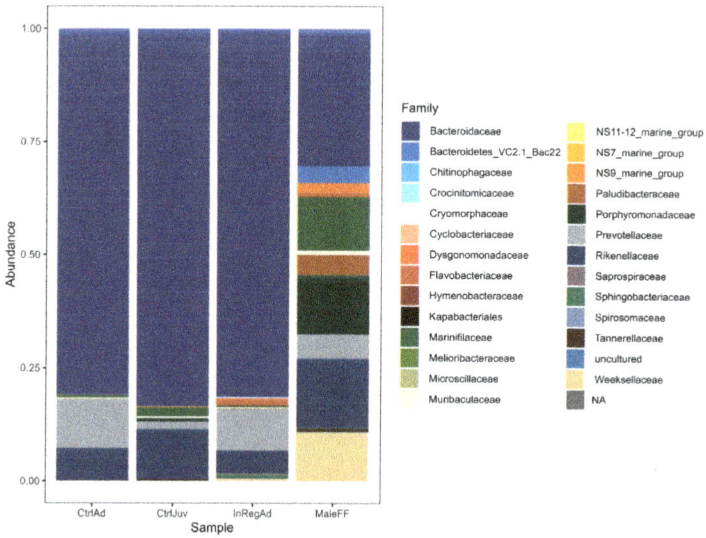

Figure 5. The families represented in the phylum in *Bacteroidota* in the microbiomes of all four groups of fur seals sampled in Tasmania: males at fish farms (MaleFF), adults/scats sampled in the region directly around salmon (InRegAd), juveniles sampled outside of the regions around farms (CtrlJuv), and adults/scats sampled outside the regions around farms (CtrlAd). NA refers to other/unidentified families in *Bacteroidota*.

The ANCOM analysis of seal microbiomes revealed 336 ASVs that were significantly different at a W-statistic threshold of 0.6 (W-statistic is a count of the sub-hypotheses passed during all of the pair-wise tests that make up the ANCOM) between seal groups when using sample-type as a covariate based on the sample-type being a confounding factor which would limit interpretation. Of the 336 ASVs, 80 ASVs were more abundant in males at fish farms than other groups of seals. Of the 202 ASVs that were differentially abundant at a stricter W-statistic threshold of 0.7, 61 were associated with males captured at the farms (Figure 6). ASVs that were found to be more significantly associated with salmon farm frequenting seals were a *Weeksellaceae* unclassified genus, *Photobacterium damselae, Neisseria* sp., *Psychrobacter* spp., *Porphyromonas* spp., *Providencia alcalifaciens,* and *Campylobacter balseri;* the rest are identified in Figure 6. The heatmap representing 79 differentially ASVs found most abundantly in adult scats sampled from the same region as the farms had several *Bacillota* (*Firmicutes*) and *Actinomycetota* (*Actinobacteria*) representatives, along with ASVs identified as *Edwardsiella tarda* (Supplementary Figure S4). From the culture results, which were used to create the local and specific reference database, a novel genus was identified. We sequenced the genome of a representative strain called Ap9-02, which is under the NCBI accession code GCA_009835005.1 (as part of bioproject PRJNA594230, the 16S rRNA sequence is on WTFG01000017.1). A full length 16S rRNA gene sequence retrieved from the draft genome confirms the closest relatives as the genera *Ornithinobacterium, Moheibacter,* and *Faecalibacter* in the family *Weeksellaceae* (phylum *Bacteroidota*) at a similarity of approximately 89–91% (using a full length 16S rRNA gene).

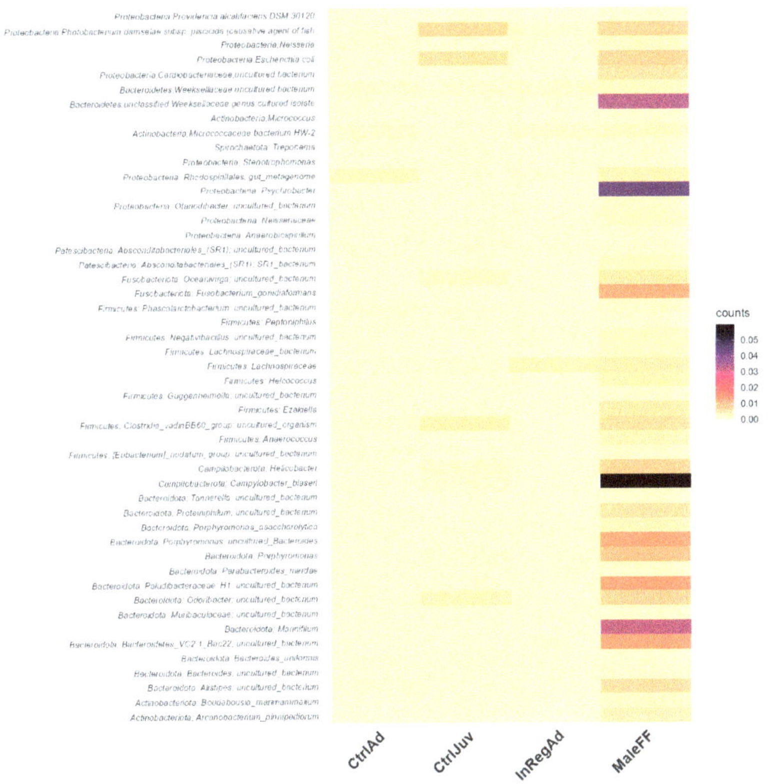

Figure 6. Heatmap of proportion of reads in each group of the 61 ASVs aggregated at the taxon level (46 taxa) found in significantly higher abundance in male seals sampled at fish farms (MaleFF) than any of the other three groups of seals tested, i.e., in region adults (InRegAd), control adults (CtrlAd), or control juveniles (CtrlJuv). These ASVs were found to be differentially abundant through the ANCOM analysis of ASVs in all seals sampled when assessing variability between groups and region as a co-variate in the model and using a W-statistic threshold of 0.7.

Six ASVs determined as differentially abundant between seal groups via ANCOM were assigned to an unclassified genus (16S rRNA gene sequence 91% similar to *Ornithobacterium rhinotracheale* (family *Weeksellaceae*, phylum *Bacteroidetes*), which had multiple isolates cultured from the seals captured at fish farms during this study; these ASVs were found significantly more frequently in the males from salmon farms (Figure 6). As the taxonomic evaluation of this unclassified *Weeksellaceae* genus (genome of representative strain Ap0902 is deposited under accession GCA_009835005.1) is ongoing due to its difficulty in being cryopreserved and revived, it will be referred to from hereon as "unclassified *Weeksellaceae* genus" in this study. All ASVs identified through ANCOM as differentially abundant that were classified as representing the unclassified *Weeksellaceae* genus were significantly related to males captured at fish farms compared to all other groups sampled, except one ASV, which was represented by 89 more reads in adult scats sampled in the same region as the farms than the males that were captured on the farms, and this ASV also had the lowest W-statistic of all six ASVs assigned to this unclassified *Weeksellaceae* genus.

3.3. *Differentially Abundant ASVs Also Present in Salmon*

Salmon microbiome sequencing data had a mean read depth of 85,953 reads/sample (range 34–239,832 reads/sample), as this mean sampling depth is much higher than the seals, as was the range, and the comparisons of diversity between seals and salmon were deemed inappropriate for extensive analysis. Briefly, after removing samples with <1000 reads/sample to compare to the seal microbiomes, there were no significant differences in richness between salmon and any group of seals; however, the Shannon diversity in salmon microbiomes were significantly lower than in all groups of seals (p-value < 0.00001) (Supplementary Figure S5). In the beta diversity measured via Bray–Curtis distances, differences were extremely apparent between salmon microbiomes and seal microbiomes, forming two distinct clusters, one of seals and one of salmon, with no overlap (Figure 7).

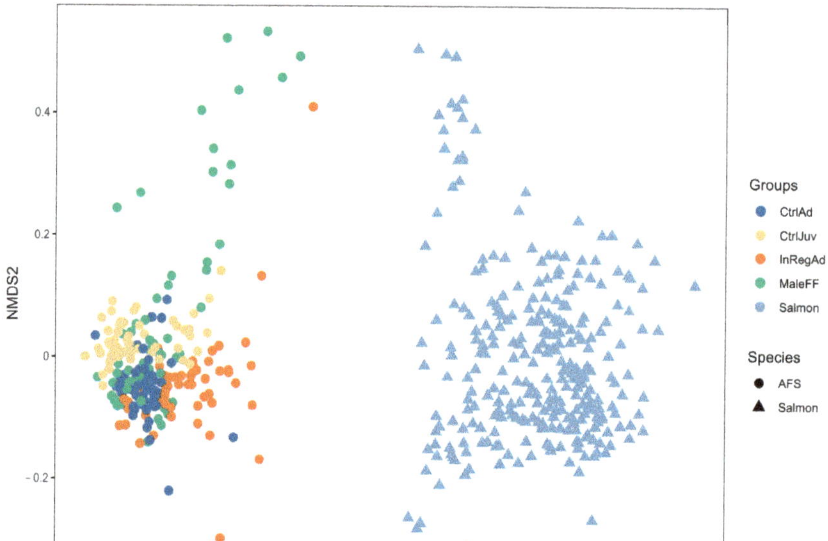

Figure 7. The nMDS plot of Bray–Curtis distances between all seal and salmon samples with colors representing seal groups and salmon: males at fish farms (MaleFF), adults/scats sampled in the region directly around salmon (InRegAd), juveniles sampled outside of the regions around farms (CtrlJuv), adults/scats sampled outside the regions around farms (CtrlAd), and salmon.

To demonstrate why this may be the case taxonomically, the relative abundance of ASVs at the phyla level in salmon and the different seal groups are shown in Figure 8. These results illustrate how different the phyla represented are in the GI tracts of these two species, and that salmon have microbiota that are dominated by *Pseudomonadota* (*Proteobacteria*) (Figure 8).

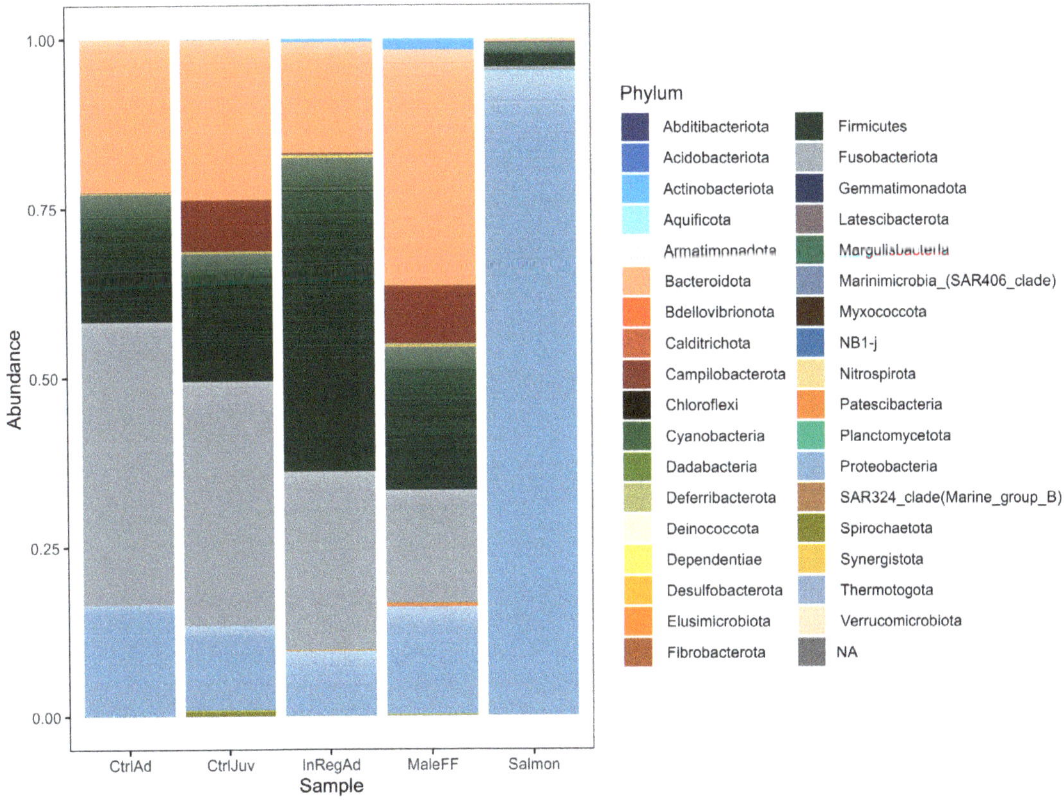

Figure 8. The relative abundance of the phyla represented in seals and farmed salmon in Tasmania for each group of seals sampled: males at fish farms (MaleFF), adults/scats sampled in the region directly around salmon (InRegAd), juveniles sampled outside of the regions around farms (CtrlJuv), adults/scats sampled outside the regions around farms (CtrlAd), and farmed salmon.

We used the ASVs, identified through the ANCOM analysis, to determine whether any were also found in the GI microbiomes of farmed salmon. There were 18 ASVs found using ANCOM that were differentially abundant in seals and were also found in salmon gut microbiomes (Table 1). Of the 18 ASVs, 11 were most associated with scats sampled from seals in the region around farms and 2 were from the males at the farms. The remaining five were split between control juveniles and control adult scats. Interestingly, there were a number of potential pathogens that were found in both species: *Providencia alcalifaciens*, *Photobacterium damselae* subsp. *damselae*, and members of *Clostridium sensu stricto 1* (Table 1). The ASVs assigned to the family *Weeksellaceae* found in salmon were not in the same genus as the novel ASVs found in seals. As the phylum *Bacteroidota* was the phylum with the greatest divergence among the groups of seals, we visualized the relative abundance of families in this phylum in the seal groups and salmon as well, which revealed more ASVs of the *Flavobacteriaceae*, *Porphyromonadaceae*, and *Dysgonomonadaceae* families in the males sampled from farms that were also found in farmed salmon (Supplementary Figure S6).

Table 1. The ASVs and their taxonomic assignment of the significantly differentially abundant ASVs between groups of seals determined through ANCOM and the number of reads assigned to them in the farmed salmon microbiome data, as well as in each group of seal, and the group which had the most reads of each ASV. W-statistic represents the number of sub-hypotheses that were significant in the ANCOM model (threshold at 0.6).

Phylum; Taxon	W *	CtrlAd	CtrlJuv	InRegAd	MaleFF	High Group	Salmon Reads
Pseudomonadota; Photobacterium damselae subsp. damselae	1073	44,141	3	5673	152	CtrlAd	664
Fusobacteriota; Cetobacterium ceti	1054	10,070	0	3029	252	CtrlAd	401
Pseudomonadota; Neisseria;uncultured bacterium	1033	244	212	3436	926	InRegAd	1751
Pseudomonadota; Escherichia-Shigella	1013	0	2776	580	18	CtrlJuv	161
Bacillota; Bacillus	1008	1381	719	1679	203	InRegAd	206
Bacillota; Clostridium sensu stricto 1	1007	0	16	1285	282	InRegAd	477
Bacillota; Bacilli	991	375	0	2757	28	InRegAd	41
Pseudomonadota; Plesiomonas	956	0	0	1568	72	InRegAd	179
Bacillota; Bacillus	932	296	365	4247	337	InRegAd	183
Pseudomonadota; Providencia alcalifaciens DSM 30120	926	32	26	1460	64	InRegAd	38
Bacillota; Peptostreptococcaceae	889	0	0	2723	0	InRegAd	707
Pseudomonadota; Providencia alcalifaciens DSM 30120	889	0	0	0	900	MaleFF	32
Bacillota; Clostridium moniliforme	819	0	0	884	207	InRegAd	63
Pseudomonadota; Escherichia coli UMEA 4075-1	767	0	118	0	0	CtrlJuv	29
Pseudomonadota; Psychrobacter	692	0	0	992	0	InRegAd	42
Pseudomonadota; Escherichia-Shigella	690	0	153	0	7	CtrlJuv	10
Pseudomonadota; Providencia alcalifaciens DSM 30120	665	3	0	0	410	MaleFF	74

* Abbreviation for the ANCOM W-value, which is an estimate of the number of sub-hypotheses that have passed for a given species.

4. Discussion

This study documents taxonomic community and biodiversity differences in microbial communities in a wild pinniped gut in relation to presence at/near marine salmon aquaculture activity and a shared habitat with human development. Overall, alpha diversity differences seen between male seals sampled near or at fish farms and all other seals, suggest that microbiome biodiversity shifts seen in the males foraging at salmon farms and seals in the region may play a larger role in shaping fur seal gut microbiomes than age class does. However, beta diversity differences suggest that community makeup, rather than overall biodiversity, was implicated in the community diversity difference seen between groups of seals. The differences between pups and all adults were minor compared to the regionality or their foraging at or near farms, the differences witnessed were expected based on previous research indicating the community shifting as an animal ages [41]. The analysis of composition revealed several differentially abundant ASVs. Of the ones most associated with males captured at the farms or from the seal scats collected in the region around farms, one was an apparently a novel genus and a few others were potential pathogens, which could be used as microbes for biocontamination source tracking efforts.

One phylum where the differences between the males captured at the farms and the rest of seal groups was most obvious was *Bacteroidota*. At the family level within the *Bacteroidota* phylum, the differences between all adult groups sampled, indicating that the lower taxonomic differences in the present phyla was very different between seal groups, with males captured on the farms having higher diversity. This suggests that pups have a more basal microbial community which then shifts as a seal ages. Previously published work on fur seal pup gut microbiota in the region supports this and shows a change in microbial community between pup and adult fur seals [41,55]. We also noticed a similar proportion of AVSs in *Campylobacterota* between males captured at the farms and control juveniles; in all other aspects, the community diversity of pups overlapped with all other groups of seals. This may be an artifact of the representative bacteria in this phylum possibly degrading outside of the host faster than others and, therefore, the swabs from males at the farms and the live control pups were able to pick up this phylum up more readily. This limitation of comparing scat to swabs was unavoidable, but from our results, this difference appears minimal compared to the impacts of variables related to the seal groups tested.

Foraging on/near farmed salmon appears to have some effect on the biodiversity and taxonomic community of the seal microbiome, as seen by the differences in ASV richness, Shannon diversity index, Bray–Curtis dissimilarity distances, and the ANCOM results. Even when accounting for confounding factors, this trend appeared throughout the results. Due to the nature of sampling wild seals, we attempted to correct for the major confounding effects of sample type in all analysis, including the ANCOM model. It appears that taxonomic and bio-diversity differences transcend the effect of swabs vs. scat collection, and we found that the Bray–Curtis distances were partially explained by including animal status/sample type in the final model along with region and month (Figure 3). This suggests that sampling scat from haul-outs, feces from necropsied seals, and swabs from live handled seals does have some effect on the microbial community found via microbiome analysis. However, as the majority of samples cluster together, overlapping through their demographic groups regardless of sample type, with divergence in some samples from males captured at the farms and from scats collected in the region around farms, we believe that the driving factor is related to biological differences, not sample type/animal status. The majority of samples clustering together is expected as these groups annually intermix at the rookery islands, and even males captured at the farms spend different amounts of time in the region. When observing the Bray–Curtis distances via the NMDS ordination plot, there was a group of males caught at the farms that were more divergent from the rest of seal groups, regardless of sample type, and a group of scat samples of both control adults and adults in the regions around farms diverged from the main cluster. The diversity of these particular samples could be due to the life history of those seals, sample type, or a biological factor, such as age, sex, or other factor that we were not able to measure and account for due to the nature of sampling wild animals in a wild system. In the future, maintaining the same sample type from all individuals would be preferred, when possible, and sampling across the entire year and evenly across the sexes or limiting samples to males only would strengthen future results.

While this study included samples across the Tasmanian habitat, age range, and variable distances from salmon farms, it was limited to seals from the end of 2017 through the beginning of 2019. Therefore, changes over time and sex-specific differences were not captured, and as it is possible that seasonality could play a role, we included month in the PERMANOVA model selection, and it was significantly related to diversity differences; however, its inclusion had a smaller effect than seal group or animal status/sample type. Due to the life histories of Australian fur seals, it is impossible to separate seasonality or month from sampling groups other than haul-out scats, as pups can only be sampled during summer during pupping season when females leave them on the rookeries while foraging [56]. It is likely that male seals are over-represented compared to females in the haul-out samples in regions around farms, due to their natural behavior to forage in more southerly regions for most of the year [57]. However, as they are the only sex that forage at the farms, this limitation may have led to capturing of more fine-scale changes based on male diet and regionality rather than overall differences between sexes. Not knowing the sex from scat samples could lead to confounding results, but sex-biased results were not analyzed in this study. Further work should also include multiple years, representative samples from the rest of the species range, the targeted sampling of adult females and males in known proportions, and/or the pair metabarcoding of microbiomes with sex identification molecular methods [58].

The salmon microbiota samples included in this study were a representative sampling of all farmed salmon in Tasmania in net pens, as known from previous research on the entire salmon population; therefore, the differences seen between the seals and salmon are likely representative of overall differences between these two populations in the region, but likely would be improved if salmon gut microbiota could be sampled from the same seals that were captured over the same time period from the farms [59,60]. Such work would improve the ability to determine fine-scale taxonomic diversity similarities between seal and salmon microbial communities, which we were not able to do during this study. More broadly, the taxonomic analysis we performed revealed one phyla which was over-represented and more diverse in seals foraging at fish farms, compared to all other groups of seals. The increase in *Bacteroidota* indicates that these bacteria may be more prevalent in the area around farmed salmon leases and/or urban areas either from exposure through predation or merely through environmental exposure.

Community differences were analyzed further by ANCOM and these results were compared to ASVs in farmed salmon microbiomes. This revealed that male seals from fish farms appear to have bacterial taxa that are not typical in the rest of the fur seal population sampled during this study. An unclassified genus of bacteria designated as the unclassified *Weeksellaceae* genus was also successfully cultured and was found to be considerably more prevalent in the males sampled from fish farms. This taxon belongs to the family *Weeksellaceae* within *Bacteroidota* and clearly represents a novel group that potentially could have biological significance in seals but requires further study. Currently there are no known sequences that have a greater than 90% similarity in the SILVA database for the 16S rRNA V1-V3 region; therefore, the taxonomy is completely novel. The most similar species, *O. rhinotracheale*, is interestingly an aerobic respiratory pathogen of birds and not at all plausibly found in seal guts, nor was the percent identity high enough for us to consider it to be the same genus. The relatedness to genus *Faecalibacter* (*F. macacae*—Chen et al. IJSEM 2020), which includes strains from the Assam macaque, suggests that the general group the novel seal sequences falls into includes gut microbiome taxa, which remain so far unstudied in any detail. Through ANCOM, it was revealed the ASVs assigned to the cultured isolate sequences of the novel unclassified *Weeksellaceae* genus were also found more prevalently in males from salmon farms than any other group of seal and were only found in limited numbers in both adults in and outside of the regions around salmon farms. While it is interesting that the unclassified *Weeksellaceae* genus sequences appear over-represented in the males sampled at fish farms, it should be noted that none of the ASVs in salmon gut samples assigned to this family were in the same novel genus. This suggests that while this novel genus is more likely to be present in male seals foraging at the farms, it is highly unlikely to be derived from the salmon gut microbiota. Therefore, it is unclear where this novel genus identified in males at the farms is derived from. It may be from the environment, another form of prey, from the seal microbiomes themselves, or potentially predating on farmed salmon, but if the latter is the case, more evidence of it being associated with skin or gill microbiota would be required.

There were 13 ASVs that were conserved specifically between seals foraging near salmon farms and salmon microbiomes identified through the ANCOM compared to all ASVs found in the salmon gut microbiomes. None of these were of the unclassified *Weeksellaceae*, but several other species of *Pseudomonadota* were found as being shared between salmon and seals differentially. Specifically, ASVs representing *Neisseria* spp., *Providencia alcalifaciens*, *Psychrobacter* spp., and members of the genus *Clostridium sensu stricto* (cluster 1) were significantly more frequently found in seals sampled at the farms or in the regions around the farms than in the control seals. *Providencia alcalifaciens*, a Gram-negative bacterium, sometimes associated with diarrheal infections in animals and humans, was found to be more abundant in seals at or near farms and also in the gut microbiomes of salmon [61]. As this bacterium was isolated from soil, its presence both in aquaculture salmon and wild fur seals may be due to multi-species fecal run-off from terrestrial sources nearby, but more research, for instance, involving using environmental DNA to identify it in water and terrestrial sources, would be necessary [61]. Another differentially abundant taxa worth noting was the Gram-negative genus *Psychrobacter*, which includes species commonly found in saline environments and usually grows at low temperatures [62,63]. Since certain species of *Psychrobacter* have been identified as opportunistic, and in some cases virulent, pathogens in humans and animals, including opportunistic infections in farmed salmonids, the further identification of these species in seals could lead to a better understanding pathogens and transmission risk at this interface [63–65]. None of the ASVs that were assigned to the genus *Psychrobacter* had a species assignment, which may suggest the existence of a novel species that have yet to be cultured and speciated, occurring in fur seal microbiomes. The V1–V3 region of the 16S rRNA gene, does not allow for species resolution for all members of this genus.

Among these ASVs, the fish pathogen *Photobacterium damselae* subsp. *piscicida* was identified, along with members of *Clostridium*-assigned ASVs, as differentially abundant and present in both seals and salmon [66]. These species may be of concern for the health of wild and aquaculture fish populations, and potentially the marine mammals and the humans that interact with them, as all are at risk of the incidental infection of virulent strains in other regions where these bacteria are found; however, little evidence was obtained during this study to ascertain the level of risk [67–72].

The potential for pathogen transmission to and from aquaculture salmon and human development should be studied further. This study was limited to only one wild population of fur seals, which are known to forage regularly on farmed salmon stocks, while other seabird, marine mammal, shark, and wild fish species also interact with farmed salmon. Other limitations arose from the nature of sampling wild fur seals, as we could only swab captured seals and were permitted only to sample pups during pup surveys and the males that were already being captured at the farms for transport; therefore, we relied on scats for the rest of the microbiome analysis. While the majority of the diversity differences appeared to be related to proximity to farms, there certainly were differences which arose from this different type of sampling. It does appear in our study that those differences were likely minimal, which is useful for researchers in the field.

The full extent of risk from urban runoff and/or aquaculture activities to wildlife is not known from this study, and the risk to salmon exposed to regular wild seal predation and runoff from human development is largely unknown, beyond what potential pathogens may be carried and shed by fur seals in the region. This study suggests that foraging on farmed salmon and in the region around fish farms and urban areas affects the diversity and taxonomic make-up of microbial communities in seals. The full extent of bacterial sharing between fur seals and farmed salmon is not currently known and could not be elucidated by this study. Further monitoring of potential pathogens in this system should be undertaken at this interface in Tasmania and at similar wild predator and aquaculture interfaces globally.

Supplementary Materials: The following supporting information can be downloaded at: https://www.mdpi.com/article/10.3390/oceans4020014/s1, Figure S1: Shannon diversity of seal microbiomes by animal status at collection.; Figure S2. Shannon diversity of seal microbiomes by sample type upon collection.; Figure S3. Shannon diversity of seal microbiomes by the region in which they were collected.; Figure S4. Heatmap of 72 ASVs aggregated at the taxon level (39 taxa) found in significantly higher abundance in the adults sampled in the region around farms (InRegAd) than any of the other three groups of seals tested.; Figure S5. Richness (left panel) and Shannon diversity (right panel) differences between seal groups and salmon gut microbiomes. Figure S6. PCoA plot of Bray–Curtis distances between all seal and salmon samples with colors representing seal groups and salmon.

Author Contributions: Conceptualization, E.D., M.-A.L. and J.P.B.; methodology, E.D., J.P.B. and R.J.M.; software, R.J.M.; formal analysis, E.D. and R.J.M.; investigation, E.D.; resources, W.A.S., J.P.B. and M.-A.L.; data curation, E.D. and R.J.M.; writing—original draft preparation, E.D.; writing—review and editing, W.A.S., M.-A.L. and E.S.; visualization, E.D.; supervision, W.A.S. and J.P.B.;

funding acquisition, W.A.S., J.P.B., M.-A.L. and E.D. All authors have read and agreed to the published version of the manuscript.

Funding: Funding for this project was provided Tassal Operations Pty Ltd.; the University of Tasmania; Tasmania Institute of Agriculture; University of California, Davis; Laboratory of Woutrina Smith; the Australian Department of Education for Endeavour fellowship funding; and the University of Tasmania, College of Science and Engineering grant funding.

Institutional Review Board Statement: The seal and salmon related study protocols used were approved by the Ethics Committee of the University of Tasmania and by the Department of Primary Industries, Parks, Water (DPIPWE)and the Environment now called the Department of Natural Resources and Environment Tasmania (NRET). Collection of seal fecal samples was approved under DPIPWE/NRET permit number FA17115. The handling and sampling of adult male seals at fish farms was approved under DPIPWE/NRET-granted scientific permit FA17158 and under University of Tasmania Animal Ethics permit A0016754. Fecal sampling of Atlantic salmon was approved under University of Tasmania Animal Ethics permit A0015452.

Data Availability Statement: The metadata and analytical scripts are accessible on github https://github.com/erdagnese/AFS-salmon-microbiome (accessed on 31 May 2023). The fastq files for seal data generated in this study are deposited under NCBI Project accession number PRJNA979964 (SAMN 35653418-35653644). The fastq files for the salmon DNA sequence data used are stored under NCBI BioProject Accession number PRJNA434528 and sequence read archive study number SRP133157.

Acknowledgments: We would like to acknowledge the Palawa/Pakana people as the traditional owners and ongoing custodians of Lutruwita (Tasmania) and pay our respects to their elder's past, present, and future. We would like to thank The Marine Conservation Program and Princess Melikoff Trust Marine Mammal Conservation Program within the Department of Primary Industries, Parks, Water and Environment, for sampling support for the fur seals around Tasmania, specifically Sam Thalmann, Annie Phillips, and Cloe Cummings. We thank staff at Tassal Pty Ltd. for their assistance in sampling the adult male seals. farms was varied out with.

Conflicts of Interest: The authors declare no conflict of interest.

References

1. Bowen, W.; Lidgard, D. Vertebrate Predator Control: Effects on Prey Populations in Terrestrial and Aquatic Ecosystems. 2011. Available online: https://publications.gc.ca/collections/collection_2013/mpo-dfo/Fs70-5-2011-028-eng.pdf (accessed on 28 March 2023).
2. Gulland, J.A. Seals and fisheries: A case for predator control? *Trends Ecol. Evol.* **1987**, *2*, 102–104. [CrossRef] [PubMed]
3. Kemper, C.; Pemberton, D.; Cawthorn, M.; Heinrich, S.; Mann, J.; Würsig, B.; Shaughnessy, P. Aquaculture and Marine Mammals: Co-Existence or Conflict? In *Marine Mammals: Fisheries, Tourism and Management Issues*; CSIRO: Canberra, Australia, 2003.
4. Australian Bureau of Agricultural Resource Economics and Sciences (ABARES). *Australian Fisheries and Aquaculture Statistics 2017*; ABARES: Canberra, Australia, 2018. [CrossRef]
5. Patterson, H.; Parsa, M.; Larcombe, J.; Noriega, R.; Emery, T.; Georgeson, L.; Marton, N.; Williams, A.; Woodhams, J.; Helidoniotis, F. *Fishery Status Reports 2019*; ABARES: Canberra, Australia, 2019.
6. Kirkwood, R.; Pemberton, D.; Gales, R.; Hoskins, A.J.; Mitchell, T.; Shaughnessy, P.D.; Arnould, J.P.Y.; Kirkwood, R.; Pemberton, D.; Gales, R.; et al. Continued population recovery by Australian fur seals. *Mar. Freshw. Res.* **2010**, *61*, 695–701. [CrossRef]
7. Deagle, B.E.; Kirkwood, R.; Jarman, S.N. Analysis of Australian fur seal diet by pyrosequencing prey DNA in faeces. *Mol. Ecol.* **2009**, *18*, 2022–2038. [CrossRef]
8. Gales, R.; Pemberton, D. Diet of the Australian fur seal in Tasmania. *Mar. Freshw. Res.* **1994**, *45*, 653–664. [CrossRef]
9. Nash, C.E.; Iwamoto, R.N.; Mahnken, C.V.W. Aquaculture risk management and marine mammal interactions in the Pacific Northwest. *Aquaculture* **2000**, *183*, 307–323. [CrossRef]
10. Würsig, B.; Gailey, G.A. Marine mammals and aquaculture: Conflicts and potential resolutions. In *Responsible Marine Aquaculture*; CABI: Wallingford, UK, 2002; pp. 45–59. [CrossRef]
11. Pemberton, D.; Shaughnessy, P.D. Interaction between seals and marine fish-farms in Tasmania, and management of the problem. *Aquat. Conserv. Mar. Freshw. Ecosyst.* **1993**, *3*, 149–158. [CrossRef]
12. Robinson, S.; Terauds, A.; Gales, R.; Greenwood, M. Mitigating fur seal interactions: Relocation from Tasmanian aquaculture farms. *Aquat. Conserv.* **2008**, *18*, 1180–1188. [CrossRef]
13. Scott, T.M.; Rose, J.B.; Jenkins, T.M.; Farrah, S.R.; Lukasik, J. Microbial source tracking: Current methodology and future directions. *Appl. Environ. Microbiol.* **2002**, *68*, 5796–5803. [CrossRef]

14. González-Fernández, A.; Symonds, E.M.; Gallard-Gongora, J.F.; Mull, B.; Lukasik, J.O.; Rivera Navarro, P.; Badilla Aguilar, A.; Peraud, J.; Brown, M.L.; Mora Alvarado, D.; et al. Relationships among microbial indicators of fecal pollution, microbial source tracking markers, and pathogens in Costa Rican coastal waters. *Water Res.* **2021**, *188*, 116507. [CrossRef]
15. Geldreich, E.E.; Litsky, W. Fecal coliform and fecal streptococcus density relationships in waste discharges and receiving waters. *C R C Crit. Rev. Environ. Control* **1976**, *6*, 349–369. [CrossRef]
16. Murillo, T.; Schneider, D.; Heistermann, M.; Daniel, R.; Fichtel, C. Assessing the drivers of gut microbiome composition in wild redfronted lemurs via longitudinal metacommunity analysis. *Sci. Rep.* **2022**, *12*, 21462. [CrossRef]
17. Mathai, P.P.; Staley, C.; Sadowsky, M.J. Sequence-enabled community-based microbial source tracking in surface waters using machine learning classification: A review. *J. Microbiol. Methods* **2020**, *177*, 106050. [CrossRef]
18. Paruch, L.; Paruch, A.M. An overview of microbial source tracking using host-specific genetic markers to identify origins of fecal contamination in different water environments. *Water* **2022**, *14*, 1809. [CrossRef]
19. Ahmed, W. Limitations of library-dependent microbial source tracking methods. *Water (AWA)* **2007**, *34*, 96–101.
20. Bourne, D.G.; Dennis, P.G.; Uthicke, S.; Soo, R.M.; Tyson, G.W.; Webster, N. Coral reef invertebrate microbiomes correlate with the presence of photosymbionts. *ISME J.* **2013**, *7*, 1452–1458. [CrossRef]
21. Kembel, S.W.; Jones, E.; Kline, J.; Northcutt, D.; Stenson, J.; Womack, A.M.; Bohannan, B.J.; Brown, G.Z.; Green, J.L. Architectural design influences the diversity and structure of the built environment microbiome. *ISME J.* **2012**, *6*, 1469–1479. [CrossRef]
22. Knights, D.; Kuczynski, J.; Charlson, E.S.; Zaneveld, J.; Mozer, M.C.; Collman, R.G.; Bushman, F.D.; Knight, R.; Kelley, S.T. Bayesian community-wide culture-independent microbial source tracking. *Nat. Methods* **2011**, *8*, 761–763. [CrossRef] [PubMed]
23. Boukerb, A.M.; Noël, C.; Quenot, E.; Cadiou, B.; Chevé, J.; Quintric, L.; Cormier, A.; Dantan, L.; Gourmelon, M. Comparative analysis of fecal microbiomes from wild waterbirds to poultry, cattle, pigs, and wastewater treatment plants for a microbial source tracking approach. *Front. Microbiol.* **2021**, *12*, 697553. [CrossRef] [PubMed]
24. Holyoak, M.; Leibold, M.A.; Holt, R.D. *Metacommunities: Spatial Dynamics and Ecological Communities;* University of Chicago Press: Chicago, IL, USA, 2005.
25. Miller, E.T.; Svanbäck, R.; Bohannan, B.J.M. Microbiomes as metacommunities: Understanding host-associated microbes through metacommunity ecology. *Trends Ecol. Evol.* **2018**, *33*, 926–935. [CrossRef] [PubMed]
26. Unno, T.; Jang, J.; Han, D.; Kim, J.H.; Sadowsky, M.J.; Kim, O.-S.; Chun, J.; Hur, H.-G. Use of barcoded pyrosequencing and shared OTUs to determine sources of fecal bacteria in watersheds. *Environ. Sci. Technol.* **2010**, *44*, 7777–7782. [CrossRef] [PubMed]
27. Unno, T.; Di, D.Y.W.; Jang, J.; Suh, Y.S.; Sadowsky, M.J.; Hur, H.-G. Integrated online system for a pyrosequencing-based microbial source tracking method that targets Bacteroidetes 16S rDNA. *Environ. Sci. Technol.* **2012**, *46*, 93–98. [CrossRef] [PubMed]
28. Couch, C.; Sanders, J.; Sweitzer, D.; Deignan, K.; Cohen, L.; Broughton, H.; Steingass, S.; Beechler, B. The relationship between dietary trophic level, parasites and the microbiome of Pacific walrus (*Odobenus rosmarus* divergens). *Proc. Biol. Sci.* **2022**, *289*, 20220079. [CrossRef] [PubMed]
29. Raza, S.; Kim, J.; Sadowsky, M.J.; Unno, T. Microbial source tracking using metagenomics and other new technologies. *J. Microbiol.* **2021**, *59*, 259–269. [CrossRef]
30. Barko, P.C.; McMichael, M.A.; Swanson, K.S.; Williams, D.A. The gastrointestinal microbiome: A review. *J. Vet. Intern. Med.* **2018**, *32*, 9–25. [CrossRef]
31. Grice, E.A.; Segre, J.A. The skin microbiome. *Nat. Rev. Microbiol.* **2011**, *9*, 244–253. [CrossRef]
32. Christian, N.; Whitaker, B.K.; Clay, K. Microbiomes: Unifying animal and plant systems through the lens of community ecology theory. *Front. Microbiol.* **2015**, *6*, 869. [CrossRef]
33. Xu, Z.; Knight, R. Dietary effects on human gut microbiome diversity. *Br. J. Nutr.* **2015**, *113*, S1–S5. [CrossRef]
34. Bornbusch, S.L.; Greene, L.K.; Rahobilalaina, S.; Calkins, S.; Rothman, R.S.; Clarke, T.A.; LaFleur, M.; Drea, C.M. Gut microbiota of ring-tailed lemurs (*Lemur catta*) vary across natural and captive populations and correlate with environmental microbiota. *Anim. Microbiome* **2022**, *4*, 29. [CrossRef]
35. Martinez-Guryn, K.; Leone, V.; Chang, E.B. Regional Diversity of the Gastrointestinal Microbiome. *Cell Host Microbe* **2019**, *26*, 314–324. [CrossRef] [PubMed]
36. Campbell, T.P.; Sun, X.; Patel, V.H.; Sanz, C.; Morgan, D.; Dantas, G. The microbiome and resistome of chimpanzees, gorillas, and humans across host lifestyle and geography. *ISME J.* **2020**, *14*, 1584–1599. [CrossRef]
37. Benson, A.K.; Kelly, S.A.; Legge, R.; Ma, F.; Low, S.J.; Kim, J.; Zhang, M.; Oh, P.L.; Nehrenberg, D.; Hua, K.; et al. Individuality in gut microbiota composition is a complex polygenic trait shaped by multiple environmental and host genetic factors. *Proc. Natl. Acad. Sci. USA* **2010**, *107*, 18933–18938. [CrossRef]
38. Ley, R.E.; Lozupone, C.A.; Hamady, M.; Knight, R.; Gordon, J.I. Worlds within worlds: Evolution of the vertebrate gut microbiota. *Nat. Rev. Microbiol.* **2008**, *6*, 776–788. [CrossRef] [PubMed]
39. Hills, R.D.; Pontefract, B.A.; Mishcon, H.R.; Black, C.A.; Sutton, S.C.; Theberge, C.R. Gut microbiome: Profound implications for diet and disease. *Nutrients* **2019**, *11*, 1613. [CrossRef] [PubMed]
40. Pacheco-Sandoval, A.; Schramm, Y.; Heckel, G.; Brassea-Pérez, E.; Martínez-Porchas, M.; Lago-Lestón, A. The Pacific harbor seal gut microbiota in Mexico: Its relationship with diet and functional inferences. *PLoS ONE* **2019**, *14*, e0221770. [CrossRef] [PubMed]
41. Smith, S.C.; Chalker, A.; Dewar, M.L.; Arnould, J.P.Y. Age-related differences revealed in Australian fur seal *Arctocephalus pusillus doriferus* gut microbiota. *FEMS Microbiol. Ecol.* **2013**, *86*, 246–255. [CrossRef]

42. Faith, J.J.; Guruge, J.L.; Charbonneau, M.; Subramanian, S.; Seedorf, H.; Goodman, A.L.; Clemente, J.C.; Knight, R.; Heath, A.C.; Leibel, R.L.; et al. The long-term stability of the human gut microbiota. *Science* **2013**, *341*, 1237439. [CrossRef]
43. De Leaniz, C.G.; Gajardo, G.; Consuegra, S. From Best to Pest: Changing perspectives on the impact of exotic salmonids in the southern hemisphere. *Syst. Biodivers.* **2010**, *8*, 447–459. [CrossRef]
44. Buschmann, A.H.; Cabello, F.; Young, K.; Carvajal, J.; Varela, D.A.; Henríquez, L. Salmon aquaculture and coastal ecosystem health in Chile: Analysis of regulations, environmental impacts and bioremediation systems. *Ocean Coast. Manag.* **2009**, *52*, 243–249. [CrossRef]
45. Allen, H.K.; Bayles, D.O.; Looft, T.; Trachsel, J.; Bass, B.E.; Alt, D.P.; Bearson, S.M.D.; Nicholson, T.; Casey, T.A. Pipeline for amplifying and analyzing amplicons of the V1–V3 region of the 16S rRNA gene. *BMC Res. Notes* **2016**, *9*, 380. [CrossRef]
46. Huse, S.M.; Ye, Y.; Zhou, Y.; Fodor, A.A. A core human microbiome as viewed through 16S rRNA sequence clusters. *PLoS ONE* **2012**, *7*, e34242. [CrossRef]
47. Zarkasi, K.Z.; Taylor, R.S.; Abell, G.C.J.; Tamplin, M.L.; Glencross, B.D.; Bowman, J.P. Atlantic salmon (*Salmo salar* L.) gastrointestinal microbial community dynamics in relation to digesta properties and diet. *Microb. Ecol.* **2016**, *71*, 589–603. [CrossRef]
48. Bolyen, E.; Rideout, J.R.; Dillon, M.R.; Bokulich, N.A.; Abnet, C.C.; Al-Ghalith, G.A.; Alexander, H.; Alm, E.J.; Arumugam, M.; Asnicar, F.; et al. Reproducible, interactive, scalable and extensible microbiome data science using QIIME 2. *Nat. Biotechnol.* **2019**, *37*, 852–857. [CrossRef]
49. Hall, M.; Beiko, R.G. 16S rRNA Gene Analysis with QIIME2. In *Microbiome Analysis*; Methods in Molecular Biology; Humana Press: Clifton, NJ, USA, 2018; Volume 1849, pp. 113–129. [CrossRef]
50. Callahan, B.J.; McMurdie, P.J.; Rosen, M.J.; Han, A.W.; Johnson, A.J.A.; Holmes, S.P. DADA2: High-resolution sample inference from Illumina amplicon data. *Nat. Methods* **2016**, *13*, 581–583. [CrossRef] [PubMed]
51. McMurdie, P.J.; Holmes, S. phyloseq: An R package for reproducible interactive analysis and graphics of microbiome census data. *PLoS ONE* **2013**, *8*, e61217. [CrossRef] [PubMed]
52. Oksanen, J.; Blanchet, F.G.; Kindt, R.; Legendre, P.; Minchin, P.; O'Hara, R.; Simpson, G.; Solymos, P.; Stevenes, M.; Wagner, H. *Vegan: Community Ecology Package*, R Package Version 2.0-2; 2014. Available online: https://cran.r-project.org/web/packages/vegan/vegan.pdf (accessed on 28 March 2023).
53. Gloor, G.B.; Macklaim, J.M.; Pawlowsky-Glahn, V.; Egozcue, J.J. Microbiome Datasets Are Compositional: And This Is Not Optional. *Front. Microbiol.* **2017**, *8*, 2224. [CrossRef]
54. Mandal, S.; Van Treuren, W.; White, R.A.; Eggesbø, M.; Knight, R.; Peddada, S.D. Analysis of composition of microbiomes: A novel method for studying microbial composition. *Microb. Ecol. Health Dis.* **2015**, *26*, 27663. [CrossRef]
55. Grosser, S.; Sauer, J.; Paijmans, A.J.; Caspers, B.A.; Forcada, J.; Wolf, J.B.W.; Hoffman, J.I. Fur seal microbiota are shaped by the social and physical environment, show mother-offspring similarities and are associated with host genetic quality. *Mol. Ecol.* **2019**, *28*, 2406–2422. [CrossRef]
56. Pemberton, D.; Kirkwood, R.J. Pup production and distribution of the Australian fur seal, *Arctocephalus pusillus doriferus*, in Tasmania. *Wildl. Res.* **1994**, *21*, 341–351. [CrossRef]
57. Kirkwood, R.; Lynch, M.; Gales, N.; Dann, P.; Sumner, M. At-sea movements and habitat use of adult male Australian fur seals (*Arctocephalus pusillus* doriferus). *Can. J. Zool.* **2006**, *84*, 1781–1788. [CrossRef]
58. Schwarz, D.; Spitzer, S.M.; Thomas, A.C.; Kohnert, C.M.; Keates, T.R.; Acevedo-Gutiérrez, A. Large-scale molecular diet analysis in a generalist marine mammal reveals male preference for prey of conservation concern. *Ecol. Evol.* **2018**, *8*, 9889–9905. [CrossRef]
59. Zarkasi, K.Z.; Abell, G.C.J.; Taylor, R.S.; Neuman, C.; Hatje, E.; Tamplin, M.L.; Katouli, M.; Bowman, J.P. Pyrosequencing-based characterization of gastrointestinal bacteria of Atlantic salmon (*Salmo salar* L.) within a commercial mariculture system. *J. Appl. Microbiol.* **2014**, *117*, 18–27. [CrossRef] [PubMed]
60. Neuman, C.; Hatje, E.; Zarkasi, K.Z.; Smullen, R.; Bowman, J.P.; Katouli, M. The effect of diet and environmental temperature on the faecal microbiota of farmed Tasmanian Atlantic Salmon (*Salmo salar* L.). *Aquac. Res.* **2016**, *47*, 660–672. [CrossRef]
61. Shah, M.M.; Odoyo, E.; Ichinose, Y. Epidemiology and pathogenesis of *Providencia alcalifaciens* infections. *Am. J. Trop. Med. Hyg.* **2019**, *101*, 290–293. [CrossRef] [PubMed]
62. Bozal, N.; Montes, M.J.; Tudela, E.; Guinea, J. Characterization of several *Psychrobacter* strains isolated from Antarctic environments and description of Psychrobacter luti sp. nov. and *Psychrobacter fozii* sp. nov. *Int. J. Syst. Evol. Microbiol.* **2003**, *53*, 1093–1100. [CrossRef] [PubMed]
63. Bowman, J.P. The Genus Psychrobacter. In *The Prokaryotes: A Handbook on the Biology of Bacteria Volume 6: Pseudomonadota: Gamma Subclass*; Dworkin, M., Falkow, S., Rosenberg, E., Schleifer, K.-H., Stackebrandt, E., Eds.; Springer: New York, NY, USA, 2006; pp. 920–930.
64. Hisar, O.; Yanik, T.; Hisar, S.A. Clinical and pathological investigation of *Psychrobacter immobilis* infection in rainbow trout (*Oncorhynchus mykiss*, Walbaum). *Isr. J. Aquac.* **2002**, *54*, 189–196.
65. Romanenko, L.A.; Lysenko, A.M.; Rohde, M.; Mikhailov, V.V.; Stackebrandt, E. *Psychrobacter maritimus* sp. nov. and *Psychrobacter arenosus* sp. nov., isolated from coastal sea ice and sediments of the Sea of Japan. *Int. J. Syst. Evol. Microbiol.* **2004**, *54*, 1741–1745. [CrossRef]
66. Romalde, J.L. *Photobacterium damselae* subsp. *piscicida: An integrated view of a bacterial fish pathogen. Int.* Microbiol. **2002**, 5, 3–9. [CrossRef]
67. Austin, H.K.; Austin, D.A. *Bacterial Fish Pathogens*; Springer: Dordrecht, The Netherlands, 2012.

68. Di Francesco, G.; Cammà, C.; Curini, V.; Mazzariol, S.; Proietto, U.; Di Francesco, C.E.; Ferri, N.; Di Provvido, A.; Di Guardo, G. Coinfection by *Ureaplasma* spp., *Photobacterium damselae* and an *Actinomyces*-like microorganism in a bottlenose dolphin (*Tursiops truncatus*) with pleuropneumonia stranded along the Adriatic coast of Italy. *Res. Vet. Sci.* **2016**, *105*, 111–114. [CrossRef]
69. Dryden, M.; Legarde, M.; Gottlieb, T.; Brady, L.; Ghosh, H.K. *Vibrio damsela* wound infections in Australia. *Med. J. Aust.* **1989**, *151*, 540–541. [CrossRef]
70. Fouz, B.; Larsen, J.; Nielsen, B.; Barja, J.; Toranzo, A. Characterization of *Vibrio damsela* strains isolated from turbot Scophthalmus maximus in Spain. *Dis. Aquat. Organ.* **1992**, *12*, 155–166. [CrossRef]
71. Rivas, A.J.; Lemos, M.L.; Osorio, C.R. *Photobacterium damselae* subsp. damselae, a bacterium pathogenic for marine animals and humans. *Front. Microbiol.* **2013**, *4*, 283. [CrossRef] [PubMed]
72. Tang, W.M.; Wong, J.W. Necrotizing fasciitis caused by *Vibrio damsela*. *Orthopedics* **1999**, *22*, 443–444. [CrossRef] [PubMed]

MDPI AG
Grosspeteranlage 5
4052 Basel
Switzerland
Tel.: +41 61 683 77 34

Oceans Editorial Office
E-mail: oceans@mdpi.com
www.mdpi.com/journal/oceans

www.ingramcontent.com/pod-product-compliance
Lightning Source LLC
Chambersburg PA
CBHW061500210926
51850CB00008B/1776